Systemic chemotherapy for prostate cancer was first studied in the 1970s. The initial regimen of cyclophosphamide and 5-fluorouracil was quickly joined by multiple other regimens of systemic chemotherapy drugs.

Unlike many other forms of cancer, prostate cancer does not respond well to chemotherapy alone. Chemotherapy is, however, used in conjunction with hormone therapy and other medications. In 1996, the FDA approved mitoxantrone on the basis of its palliative effect becoming the first chemotherapeutic drug used in combination with steroids to fight hormone refractory prostate cancer. In 2004, the FDA approved docetaxel (Taxotere) along with prednisone steroids for prostate cancer that did not respond to hormone therapy anymore.

Recently new hormonal and chemotherapeutic agents have been approved for the management of castration-resistant prostate cancer (CRPC) including abiraterone and cabazitaxel. It has also been shown that vaccine therapy for prostate cancer is feasible, leading to the introduction of sipuleucel-T.

In the surgical field, radical retropubic prostatectomy (RRP) was developed for the first time in 1983 by Patrick Walsh. This surgical approach allowed removal of the prostate with preservation of both continence and erectile function, thanks to the preservation of neurovascular bundles.

Over the last decade, open surgery for the prostate is gradually being substituted with laparoscopic and robotic surgery.

New technologies are continually developing for the treatment of prostate cancer, with the promise to offer patients minimally invasive therapies in which cancer is eradicated (e.g. focal therapy), while normal physiological functions are maintained after treatment.

2.2 Epidemiology of Prostate Cancer

2.2.1 Incidence and Mortality

Cancer of the prostate (PCa) is now recognized as one of the most important medical problems facing the male population, being the most frequently diagnosed cancer in men. Furthermore, PCa is currently the second most common cause of cancer death in men (Jemal et al. 2011).

In Europe, PCa is the most common solid neoplasm, with an incidence of 370,733 cases, 21.8 % of the total (12 % of all male cancers), according to data of the *International Agency for Research on Cancer* (Ferlay et al. 2010; Bray et al. 2010) (Figs. 2.1 and 2.2).

In the United States (US), an estimated 240,890 new cases of prostate cancer will occur during 2011 (Fig. 2.2). For reasons that remain unclear, incidence rates are significantly higher in African-Americans than in whites (Fig. 2.3).

Incidence rates for prostate cancer changed substantially between the mid-1980s and mid-1990s, in large part reflecting changes in prostate cancer screening with the prostate-specific antigen (PSA) blood test. Since 1998, incidence rates have remained relatively stable (Jemal et al. 2011).

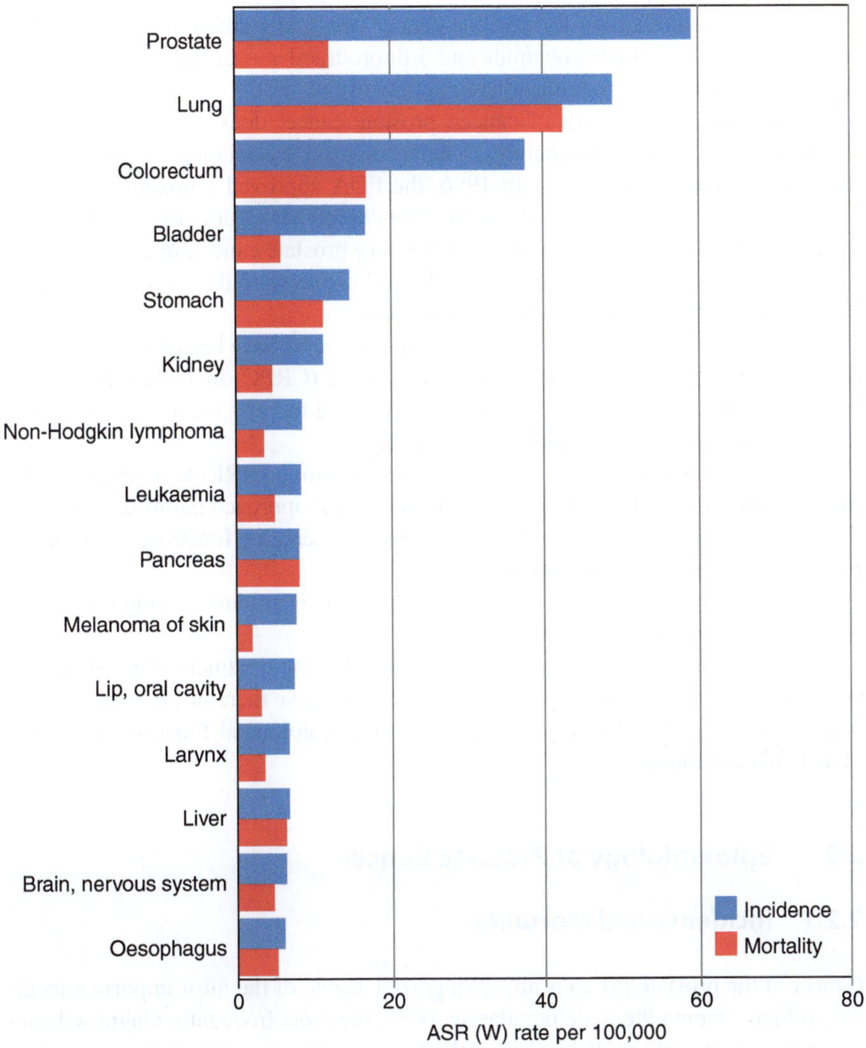

Fig. 2.1 Incidence and mortality for major cancer types

Prostate cancer affects elderly men more often than young men. It is therefore a bigger health concern in developed countries with their greater proportion of elderly men. Thus, in developed countries about 15 % of male cancers are PCa compared to 4 % of in underdeveloped countries (Quinn and Babb 2002).

There are large regional differences in incidence rates of PCa, which vary by more than 25-fold worldwide; the highest rates are in Australia/New Zealand (104.2 per 100,000), Western and Northern Europe, and Northern America, partly due to the practice of prostate-specific antigen (PSA) testing and subsequent biopsy, having become widespread in those regions. Incidence rates are relatively high in

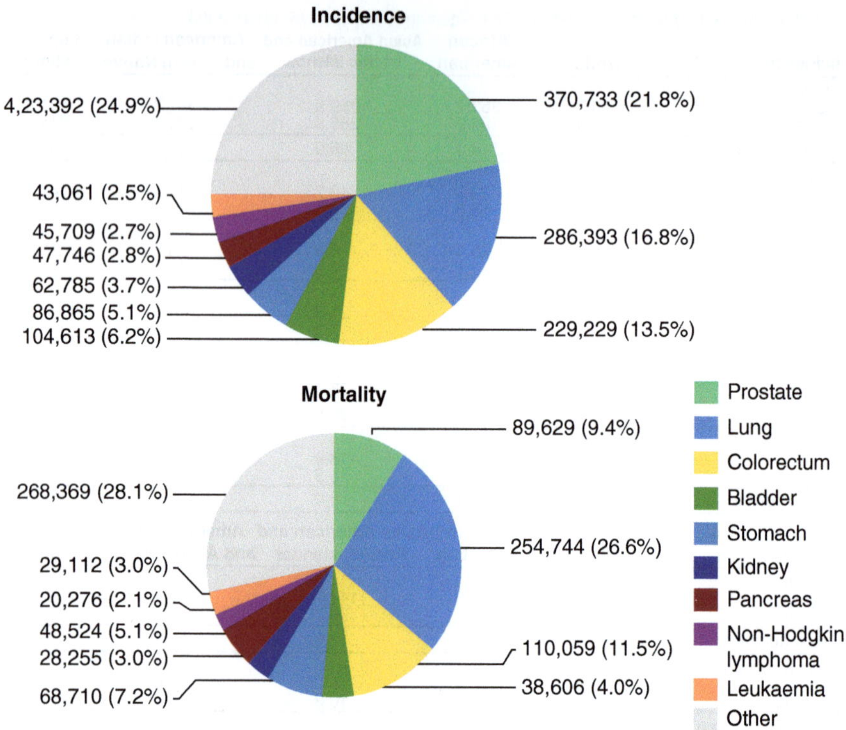

Fig. 2.2 Incidence and mortality for different types of cancer

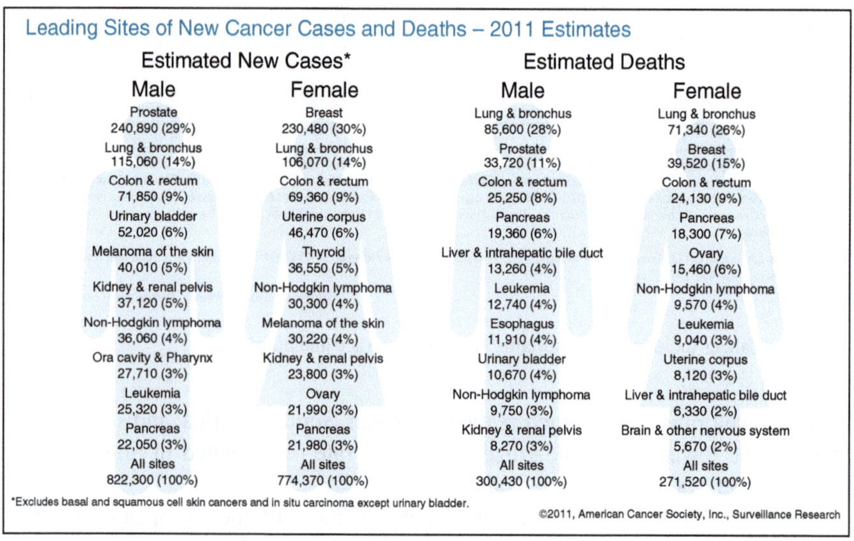

Leading Sites of New Cancer Cases and Deaths – 2011 Estimates

Estimated New Cases*		Estimated Deaths	
Male	Female	Male	Female
Prostate 240,890 (29%)	Breast 230,480 (30%)	Lung & bronchus 85,600 (28%)	Lung & bronchus 71,340 (26%)
Lung & bronchus 115,060 (14%)	Lung & bronchus 106,070 (14%)	Prostate 33,720 (11%)	Breast 39,520 (15%)
Colon & rectum 71,850 (9%)	Colon & rectum 69,360 (9%)	Colon & rectum 25,250 (8%)	Colon & rectum 24,130 (9%)
Urinary bladder 52,020 (6%)	Uterine corpus 46,470 (6%)	Pancreas 19,360 (6%)	Pancreas 18,300 (7%)
Melanoma of the skin 40,010 (5%)	Thyroid 36,550 (5%)	Liver & intrahepatic bile duct 13,260 (4%)	Ovary 15,460 (6%)
Kidney & renal pelvis 37,120 (5%)	Non-Hodgkin lymphoma 30,300 (4%)	Leukemia 12,740 (4%)	Non-Hodgkin lymphoma 9,570 (4%)
Non-Hodgkin lymphoma 36,060 (4%)	Melanoma of the skin 30,220 (4%)	Esophagus 11,910 (4%)	Leukemia 9,040 (3%)
Ora cavity & Pharynx 27,710 (3%)	Kidney & renal pelvis 23,800 (3%)	Urinary bladder 10,670 (4%)	Uterine corpus 8,120 (3%)
Leukemia 25,320 (3%)	Ovary 21,990 (3%)	Non-Hodgkin lymphoma 9,750 (3%)	Liver & intrahepatic bile duct 6,330 (2%)
Pancreas 22,050 (3%)	Pancreas 21,980 (3%)	Kidney & renal pelvis 8,270 (3%)	Brain & other nervous system 5,670 (2%)
All sites 822,300 (100%)	All sites 774,370 (100%)	All sites 300,430 (100%)	All sites 271,520 (100%)

*Excludes basal and squamous cell skin cancers and in situ carcinoma except urinary bladder.

©2011, American Cancer Society, Inc., Surveillance Research

Fig. 2.3 Leading sites of new cancer cases and deaths per gender

Cancer Incidence and Mortality Rates* by Site, Race and Ethnicity, US, 2000–2004

Incidence	White	African American	Asian American and Pacific Islander	American Indian and Alaska Native†	Hispanic/ Latino‡§
All sites					
Males	556.7	663.7	359.9	321.2	421.3
Females	423.9	396.9	285.8	282.4	314.2
Breast (female)	132.5	118.3	89.0	69.8	89.3
Colon & rectum					
Males	60.4	72.6	49.7	42.1	47.5
Females	44.0	55.0	35.3	39.6	32.9
Kidney & renal pelvis					
Males	18.3	20.4	8.9	18.5	16.5
Females	9.1	9.7	4.3	11.5	9.1
Liver & bile duct					
Males	7.9	12.7	21.3	14.8	14.4
Females	2.9	3.8	7.9	5.5	5.7
Lung & bronchus					
Males	81.0	110.6	55.1	53.7	44.7
Females	54.6	53.7	27.1	36.7	25.2
Prostate	161.4	255.5	96.5	68.2	140.8
Stomach					
Males	10.2	17.5	18.9	16.3	16.0
Females	4.7	9.1	10.8	7.9	9.6
Uterine cervix	8.5	11.4	8.0	6.6	13.8
Mortality	**White**	**African American**	**Asian American and Pacific Islander**	**American Indian and Alaska Native†**	**Hispanic/ Latino‡§**
All sites					
Males	234.7	321.8	141.7	187.9	162.2
Females	161.4	189.3	96.7	141.2	106.7
Breast (female)	25.0	33.8	12.6	16.1	16.1
Colon & rectum					
Males	22.9	32.7	15.0	20.6	17.0
Females	15.9	22.9	10.3	14.3	11.1
Kidney & renal pelvis					
Males	6.2	6.1	2.4	9.3	5.4
Females	2.8	2.8	1.1	4.3	2.3
Liver & bile duct					
Males	6.5	10.0	15.5	10.7	10.8
Females	2.8	3.9	6.7	6.4	5.0
Lung & bronchus					
Males	72.6	95.8	38.3	49.6	36.0
Females	42.1	39.8	18.5	32.7	14.6
Prostate	25.6	62.3	11.3	21.5	21.2
Stomach					
Males	5.2	11.9	10.5	9.6	9.1
Females	2.6	5.8	6.2	5.5	5.1
Uterine cervix	2.3	4.9	2.4	4.0	3.3

*Per 100,000, age adjusted to the 2000 US standard population. †Data based on Contract Health Service Delivery Areas (CHSDA), 624 counties Comprising 54% of the US American Indian/Alaska Native population: for more information, please see: Espey DK, Wu XC, Swan J, et al. Annual report to the nation on the status of cancer, 1975–2004, featuring cancer in American Indians and Alaska Natives. ‡Persons of Hispanic/Latino origin may be of any race. §Data unavailable from the Alaska Native Registry and Kentucky. ¶Data unavailable form Minnesota, New Hampshire, and North Dakota.

Source: Ries LAG, Melbert D. Krapcho M. et al (eds.). *SEER Cancer Statistics Review, 1975–2004*, National Cancer Institude, Bethesda, MD, www.seer.cancer.gow/csr/1975_2004/, 2007.

American Cancer Society, Surveillance Research, 2008

Fig. 2.4 Cancer incidence and mortality rates by site, race and ethnicity

certain developing regions such as the Caribbean, South America, and sub-Saharan Africa. The lowest age-standardized incidence rate is estimated in South Central Asia (4.1 per 100,000) (Figs. 2.4 and 2.5).

Incidence has been uniformly increasing in most of the European countries, although in a few of them (Sweden, Finland, and the Netherlands), incidence has begun to fall during the last 3–4 years. Incidence rates were highest in Ireland,

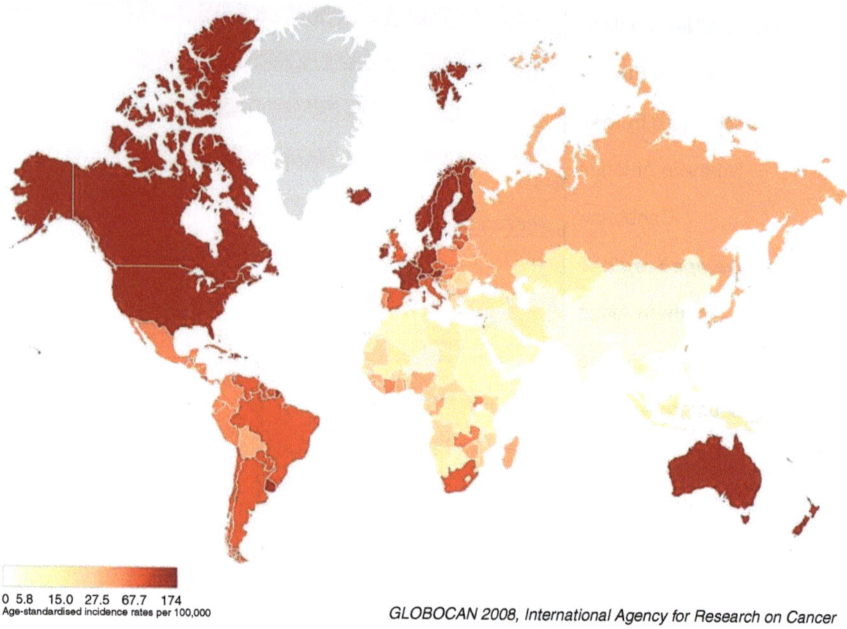

Fig. 2.5 Global differences in the incidence of prostate cancer

France, Belgium, and Northern European countries (Norway, Sweden, Iceland, and Finland). Rates were lower in a range of Central, Eastern, and Southern European countries, including Turkey, Greece, Romania, and Bulgaria. At least part of the fivefold difference between countries with the highest and lowest incidence rates is due to under-registration of prostate cancer in some countries as well as the use of sensitive diagnostic tests for early detection in others (Ferlay et al.2010) (Fig. 2.6).

Prostate cancer is currently the second most common cause of cancer death in men (Jemal et al. 2011). Because PSA testing has a much greater effect on incidence than on mortality, there is less variation in mortality rates worldwide (tenfold) than is observed for incidence (25-fold), and the number of deaths from prostate cancer is almost the same in developed and developing regions. Mortality rates are generally high in predominantly black populations (Caribbean, 26.3 per 100,000 and sub-Saharan Africa, 18–19 per 100,000), very low in Asia (e.g. 2.5 per 100,000 in Eastern Asia), and intermediate in Europe and Oceania (Jemal et al. 2011) (Fig. 2.7).

Mortality rates increased slowly for most countries between 1985 and 1995 (Quinn and Babb 2002). The CONCORD study (Coleman et al. 2008), a worldwide population-based analysis of cancer survival in five continents, analysed international differences in survival for breast, colorectal, and prostate cancer. Using data from cancer registries, age-standardized 5-year survival rates were found to vary greatly, ranging from 80 % or higher in the United States (92 %), Australia, and Canada to less than 40 % in Denmark, Poland, and Algeria (Coleman et al. 2008).

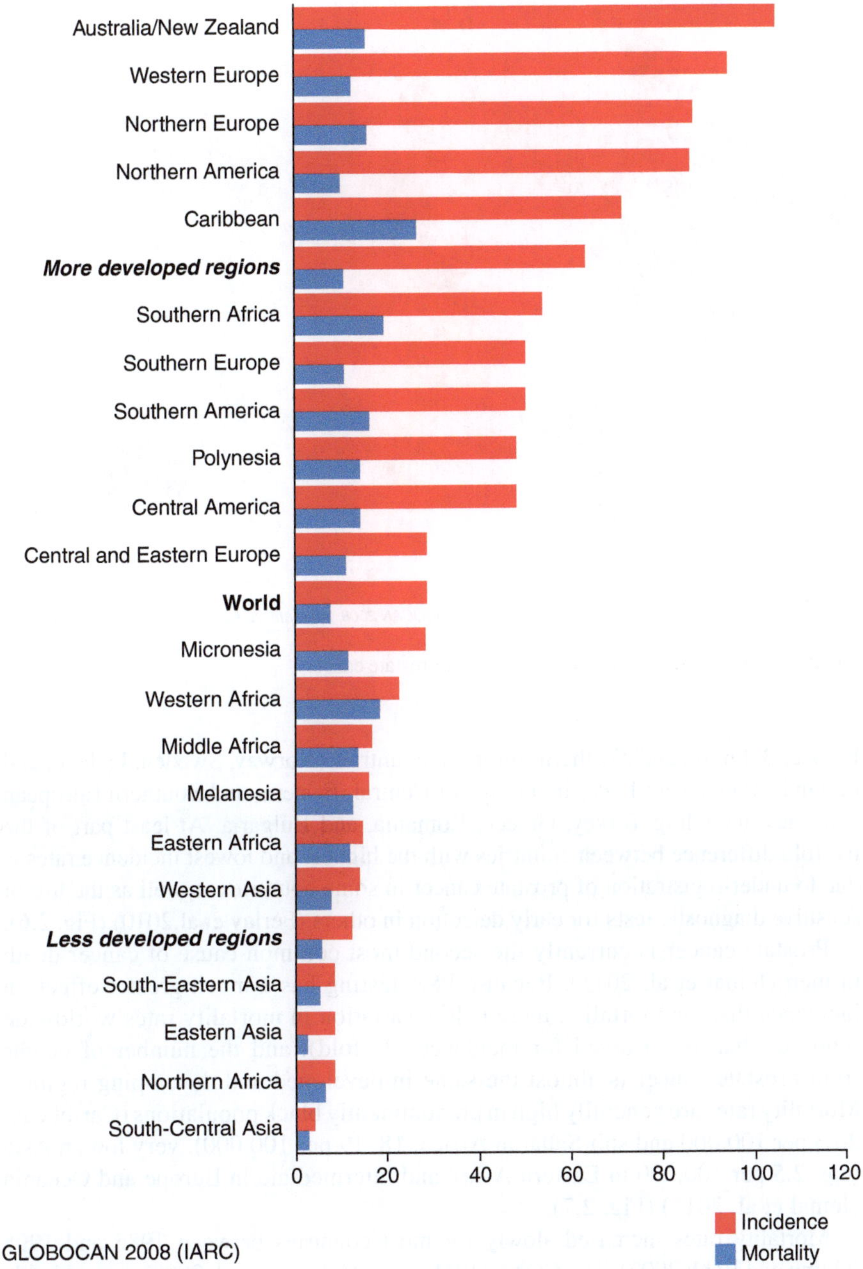

Fig. 2.6 Regional differences in the incidence and mortality of prostate cancer

Possible explanations for the worldwide and ethnic variations in prostate cancer incidence and mortality could be due to the access and quality of health care, the accuracy of cancer registries, and the PSA screening performance.

Interstitial Prostate Brachytherapy

György Kovács • Peter Hoskin
Editors

Interstitial Prostate Brachytherapy

LDR-PDR-HDR

Springer

Editors
György Kovács
Interdisciplinary Brachytherapy Unit
University of Lübeck
Lübeck
Germany

Peter Hoskin
Mount Vernon Cancer Centre
University College London
London
United Kingdom

ISBN 978-3-642-36498-3 ISBN 978-3-642-36499-0 (eBook)
DOI 10.1007/978-3-642-36499-0
Springer Heidelberg New York Dordrecht London

Library of Congress Control Number: 2013941242

Printed on acid-free paper

Springer is part of Springer Science+Business Media (www.springer.com)

Contents

Introduction

1

György Kovács and Peter Hoskin

Prostate cancer continues to increase in incidence in both the developed and the developing world. As population awareness increases in parallel and ready access to diagnostic tests such as serum PSA increases in prevalence, the majority of cases are now diagnosed as localized disease in the low- to intermediate-risk groups. These patients are presented with a bewildering array of possible treatment options, all valid approaches seeking to control or eradicate the cancer whilst inflicting few or no side effects. Amongst these, brachytherapy has emerged as a popular choice with now proven efficacy and a very favourable toxicity profile.

Brachytherapy for localized prostate cancer is not new. There are numerous well-written books and review publications on the market. Why this publication now? The story starts in the early 1990s when there was a hard fight between urologists performing radical prostatectomy and brachytherapy experts performing already well-established low-dose-rate (seed) treatments as well the relatively new high-dose-rate (HDR remote afterloader) implants. The need of a less emotional and more scientific argument in the discussion between "prostate treatment experts" was growing, and the idea was born to start an interdisciplinary teaching course supported by the European Association of Urology (EAU) as well by the GEC-ESTRO (the Brachytherapy Committee of ESTRO). Foreseen were regular teaching courses at three places: Leeds/UK, Utrecht/NL and Kiel/D. The interdisciplinary teaching staff represented the urology, diagnostic radiology, brachytherapy and brachytherapy medical physics communities interested in prostate treatment. The courses proved to be highly successful with many participants who gained a firm

G. Kovács (✉)
Interdisciplinary Brachytherapy Unit,
University of Lübeck, Lübeck, Germany
e-mail: kovacsluebeck@gmail.com

P. Hoskin
Mount Vernon Cancer Centre,
University College London, London, UK
e-mail: peterhoskin@nhs.net

G. Kovács, P. Hoskin (eds.), *Interstitial Prostate Brachytherapy*,
DOI 10.1007/978-3-642-36499-0_1, © Springer-Verlag Berlin Heidelberg 2013

understanding of the issues from interaction with the experts and first-hand observation of live implants.

Over the 13 years that the course has been running, it has evolved considerably based on continued interaction between the teachers old and new and detailed evaluation of the course assessments completed by students after each session. This experience from many years of interdisciplinary teaching is now summarized in this book with the hope that it will serve a useful support for both beginners and those with experience in the brachytherapy of localized prostate cancer.

The structure of the content is based around the lectures given on the current and immediate past courses. Thus the early chapters describe the history of brachytherapy and the epidemiology of prostate cancer. This is followed by important aspects of diagnosis with particular emphasis on imaging which is fundamental to modern brachytherapy which is based on real-time image-guided implantation. The role of modern functional imaging as we seek to gain ever more accurate information on the precise disposition of malignancy in the prostate gland is also covered. Subsequent chapters cover the selection of patients and technical aspects of brachytherapy using both low-dose-rate and high-dose-rate sources including aspects of radiation protection. The wealth of data now published defining the efficacy and toxicity of prostate brachytherapy is then presented, and the final chapters cover the difficult and often neglected area of the management of brachytherapy related toxicity.

We hope the reader will find this as a comprehensive overview of the subject; however, it cannot replace the discussion and practical demonstrations experienced on a course such as that held by GEC-ESTRO or replace the need for close mentorship in those embarking upon prostate brachytherapy for the first time.

Epidemiology and Prevention of Prostate Cancer

2

Ernesto R. Cordeiro, Bertrand Tombal, and Theo M. de Reijke

2.1 Historical Facts on Prostate Cancer (Marx and Karenberg 2010; Josef and Karenberg 2009; Androutsos 2005)

The first knowledge of the prostate as an anatomic entity dates back to the third century BC, when *Herophilus of Chalcedon* (*c. 330–260 BC*) first described the anatomical features of the prostatic gland as we know it today.

In 1817, *George Langstaff* (*1780–1846*), a London physician, provided the first genuine description of prostate cancer to appear in the medical literature. Later that same century, the French surgeon *Stanislas Tanchou* observed that only 5 of 9,118 cancer deaths in Paris from 1830 to 1840 were due to prostate cancer. *John Adams* (*1806–1877*), a surgeon at the London Hospital, described in 1853 the first case of prostate cancer established by histological examination. Adams noted in his report that this condition was "a very rare disease" and also stated, "…of the treatment, unfortunately, little of a satisfactory nature can be said".

The works of the English urologist *Sir Henry Thompson* (*1820–1904*), collected in his book "Bladder Tumors", were for over 40 years the most important reference in terms of urological pathology. He was the first to identify that cancerous cells in

Ernesto R. Cordeiro, MD.
Department of Urology, Academic Medical Center,
University of Amsterdam, Meibergdreef 9,
1105 AZ, Amsterdam, The Netherlands
e-mail: e.r.cordeiro@amc.uva.nl, ercordeiro@hotmail.com

B. Tombal
Department of Urology, Saint-Luc University Hospital,
Hippocrates 10, B-1200 Brussels
e-mail: bertrand.tombal@uclouvain.be

T.M. de Reijke, MD, PhD, FEBU (✉)
Department of Urology, Academic Medical Center,
Meibergdreef 9, 1105 AZ, Amsterdam, The Netherlands
e-mail: t.m.dereyke@amc.uva.nl

G. Kovács, P. Hoskin (eds.), *Interstitial Prostate Brachytherapy*,
DOI 10.1007/978-3-642-36499-0_2, © Springer-Verlag Berlin Heidelberg 2013

urine were associated with a positive diagnosis of cancer. In 1859 he was awarded the *Jackson Prize* by the Royal College of Surgeons of England for his contribution to the understanding of the prostate's anatomy, physiology, pathology, and cancer.

In 1867, a German-Austrian surgeon, *Christian Albert Theodor Billroth* (*1829–1894*), performed the first perineal prostatectomy for prostate cancer, and from then on, numerous surgical techniques for the prostatic surgery appeared, drastically reducing mortality rates for these types of interventions. *Billroth* also described several surgical procedures not only for urological diseases but for digestive and gynaecological disorders.

A British surgeon, Arthur *Ferguson McGill* (*1850–1890*), performed in 1867 the first suprapubic prostatectomy, reaching in a few years 37 cases with promising results. A few years later, the North American surgeon *William Belfield* (*1856–1929*) provided some interesting new surgical approaches for prostate cancer, having reached in 1890 up to 80 cases with low complication rate.

The North American surgeon *Hugh Hampton Young* (*1870–1945*) was the last of the nineteenth century's celebrities to contribute to the urological field. He is considered the pioneer of modern urology, having introduced numerous surgical procedures and devices, like "the punch", a type of urethroscope with an inner cutting steel tube, which gained wide acceptance among the urological community. In 1904 he performed the first radical prostatectomy for prostate cancer in the John Hopkins Hospital. He was recognized not only as an extraordinary and innovative surgeon but also as a world authority on the treatment of prostate cancer.

At the turn of the twentieth century, one of the treatments developed for not only prostate cancer but cancer in general was that of radiation therapy. Radium implants were used in the early twentieth century to treat prostate cancer, and this is the first form of radiation treatment used in prostate cancer history.

External beam radiation became more popular as stronger radiation sources became available in the middle of the twentieth century. Brachytherapy with implanted seeds was first described in 1965.

Another important development in the world of prostate cancer history came from the Canadian physician and physiologist *Charles Brenton Huggins* (*1901–1997*). In 1941, he published studies in which he used oestrogen to oppose testosterone production in men with metastatic prostate cancer. Huggins was awarded the Nobel Prize in physiology and medicine for this discovery of the effects of "chemical castration" in 1966. This led to the development of other hormone treatments for prostate cancer and other hormone-treatable cancers. Two of the most common hormone treatments used today are leuprolide and goserelin; these developments began in the late 1970s with the discovery of the neurohormone GnRH by both a Polish-born American endocrinologist *Andrzej Viktor Schally* (*1926–*) and a French-born American biologist *Roger Guillemin* (*1924–*), who won the Nobel Prize in medicine for their work in 1977. Today hormone therapy remains the mainstay for systemic treatment of advanced prostate cancer. New discoveries on the resistance mechanisms to castration have led to the discovery of newer drugs such as abiraterone, a steroid synthesis inhibitor, and MDV3100, a new antiandrogen.

At the same time, the various methods of radiation therapy and chemotherapy also continued to develop.

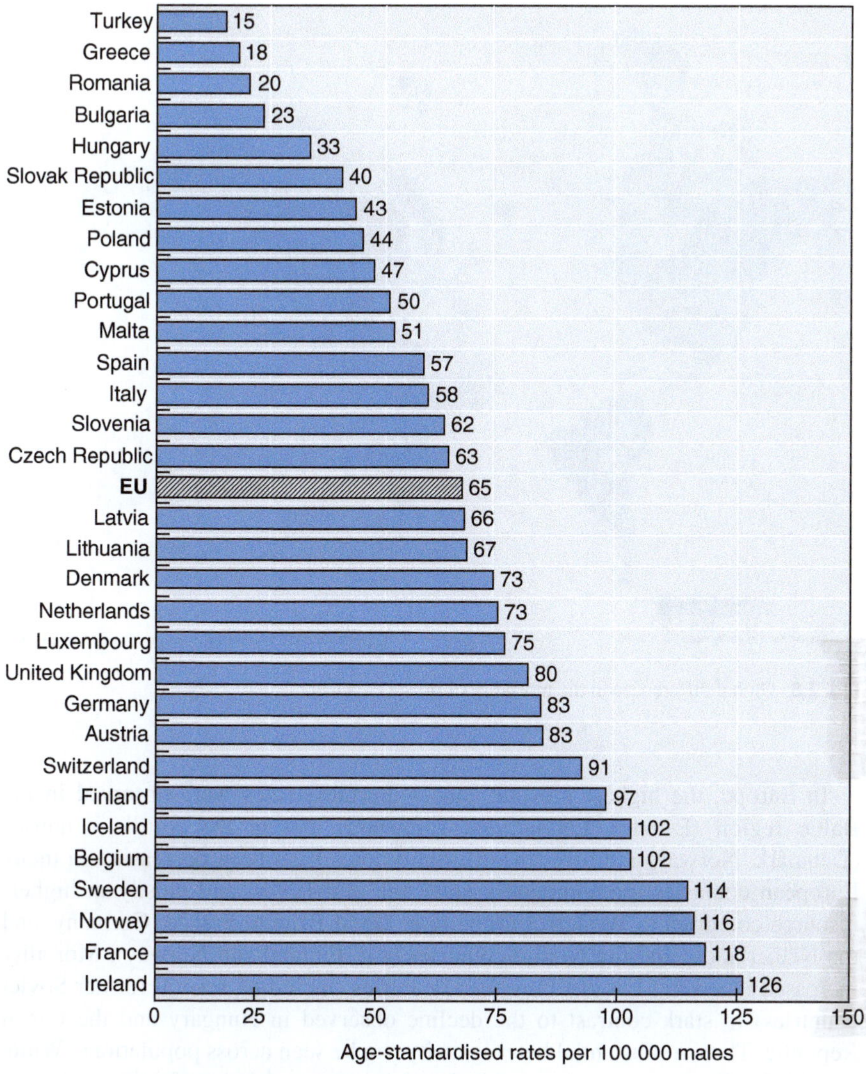

Source: OECD Health Data 2010; Ferlay et al. (2010).
StateLink http://dx.doi.org/10.1787/888932336103

Fig. 2.7 Differences in the incidence of prostate cancer among European countries

In the USA, with an estimated 33,720 deaths in 2011, prostate cancer is the second leading cause of cancer death in men (Fig. 2.8). Prostate cancer death rates have been decreasing since the mid-1990s in both African-Americans and whites. Although death rates have decreased more rapidly among African-American than white men, rates in African-Americans remain more than twice as high as those in whites (Jemal et al. 2011) (Fig. 2.3).

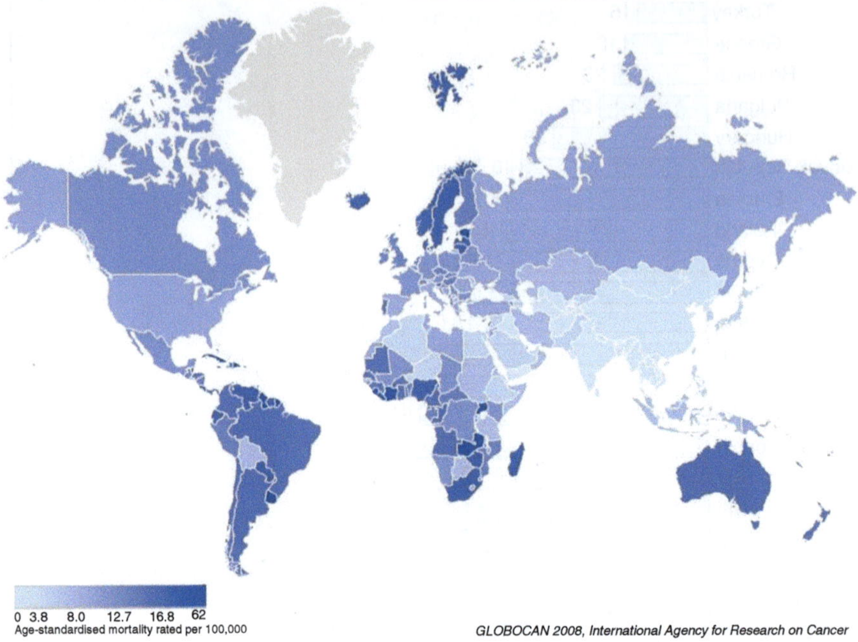

Fig. 2.8 Global differences in the mortality of prostate cancer

In Europe, the highest prostate cancer mortality rates were observed in the Baltic region (Estonia, Latvia, and Lithuania) and in the Nordic countries (Denmark, Norway, and Sweden). Mortality rates have been decreasing in many European countries, predominantly since the mid-1990s, and mostly in higher-resource countries in Western Europe (e.g. Great Britain, France, Germany, and the Netherlands) and the Nordic countries (e.g. Finland and Norway). Mortality increased in several Eastern European countries (including several former Soviet countries) in stark contrast to the decline observed in Hungary and the Czech Republic. Thus, a substantial heterogeneity can be seen across populations. While it is likely that there are geographical and temporal variations in the quality of reporting of the underlying cause of death in Europe, regional patterns emerge, with downward turns in several Western, Northern, and, lately, Eastern European countries.

There appears however little relation between the increasing incidence and decreasing mortality in the recent past, consistent with an effect of over-diagnosis or detection of indolent tumours via PSA testing. In contrast, uniformly increasing mortality trends persist in a number of Central and Eastern European countries, other than the Czech Republic and Hungary. It remains unclear as to what extent such trends reflect true changes in risk or are a result of increasing detection of latent disease (Bray et al. 2010) (Fig. 2.9).

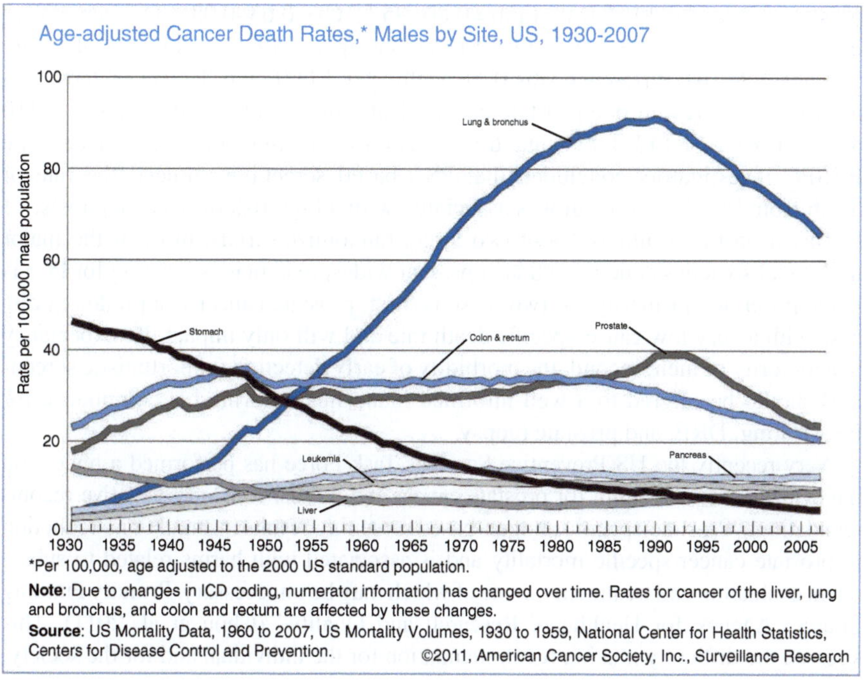

Fig. 2.9 Age-adjusted cancer death rates by site

2.2.2 Effect of PSA-Based Screening on Mortality

The benefit of PSA-based screening on overall mortality remains one of the hottest and most intensively debated topics in modern urology.

Recently, the results of two large randomized trials assessing the effect of PSA screening on prostate cancer mortality were published (Djavan 2011).

The first examined results from the Prostate, Lung, Colorectal, and Ovarian (PLCO) Cancer Screening Trial and reported on 76,693 men aged 55–74 years at 10 US centres receiving either annual screening or usual care (Strope and Andriole 2010). After 7 years of follow-up, no difference in prostate cancer mortality was detected between the groups, with an incidence of 2.0 deaths per 10,000 person-years (50 deaths) in the screening group and 1.7 deaths per 10,000 person-years (44 deaths) in the control group (rate ratio = 1.13; 95 % CI = 1.16–1.29). The data at 10 years were 67 % complete and consistent with these overall findings. The PLCO project team concluded that PCa-related mortality was very low and not significantly different between the two study groups.

The second trial, the European Randomized Study of Screening for Prostate Cancer (ERSPC), included 162,243 men between the ages of 55 and 69 years randomized to either PSA screening every 4 years or no screening (Schroder 2008). After a median follow-up of 9 years, screening reduced the rate of death from prostate cancer by 20 % (rate ratio = 0.80; 95 % CI = 0.65–0.98). The cumulative incidence of PCa was 8.2 % in the screened group and 4.8 % in the control group. The absolute risk difference was 0.71 deaths per 1,000 men. However, they estimated that to prevent one prostate cancer death, one would need to screen 1,410 men (95 % CI = 1,142–1,721) and treat 48 additional cases of prostate cancer. The ERSPC investigators concluded that PSA-based screening reduced the rate of death from PCa by 20 % but was associated with a high risk of over-diagnosis.

Based on the results of these two large, randomized trials, most of the major urological societies concluded that at present widespread mass screening for PCa is not appropriate, primarily for two reasons: first, prostate cancer is an indolent disease with a very low cause-specific death rate and will only impact life expectancy in a minority of men; second, the morbidity of early detection (opportunistic screening) should be offered to a well-informed man, thus ensuring his compliance for PSA testing, DRE, and prostate biopsy.

Very recently the US Preventive Services Task Force has performed a review on the evidence for screening for prostate cancer and published a very negative recommendation: "Prostate-specific antigen-based screening results in small or no reduction in prostate cancer-specific mortality and is associated with harms related to subsequent evaluation and treatments, some of which may be unnecessary. Primary Funding Source: Agency for Healthcare Research and Quality" (Chou et al. 2011). This stressed the difficulty making recommendation for the individual and for the society.

2.3 Risk Factors

The factors that determine the risk of developing clinical PCa are not well known, although a few have been identified. There are three well-established risk factors for PCa: increasing age, ethnic origin, and heredity (Heidenreich et al. 2011). Nevertheless, several exogenous risk factors have been described that may be involved in the development of prostate cancer and include environmental factors, dietary intake, obesity and physical activity, smoking and alcohol consumption, vasectomy and sexual activity, and steroid hormones.

2.3.1 Age

Age is the strongest risk factor for prostate cancer. Prostate cancer is very rare before the age of 40, but the chance of having prostate cancer rises rapidly after age 50. The median age at diagnosis is 68 years, with 63 % diagnosed after the age of 65. At 85 years of age, the cumulative risk of clinically diagnosed prostate cancer ranges from 0.5 to 20 % worldwide, despite autopsy evidence of microscopic lesions in approximately 30 % of men in the fourth decade, 50 % of men in the sixth decade, and more than 75 % of men older than 85 years (Sakr 1999).

2.3.2 Race/Ethnicity

It has been accepted that there might be subtle biologic differences among populations, but disease-related differences observed between groups may be more likely due to environmental exposure, diet, lifestyle, and attitudes toward health care than of differences in genetic structure or function.

It is noteworthy that African-American men have the highest reported incidence of prostate cancer in the world, with a relative incidence of 1.6 compared with white men in the United States (Crawford 2003). Although African-Americans have experienced a greater decline in mortality than white men since the early 1990s, their death rates remain more than 2.4 times higher than whites.

On the other hand, prostate cancer occurs less often in Asian-American and Hispanic/Latino men than in non-Hispanic whites. The reasons for these racial and ethnic differences remain unclear (Fig. 2.3).

2.3.3 Heredity, Familial Aggregation, and Genetic Background

Prostate cancer seems to occur in some families, which suggests that in some cases there may be an inherited or genetic factor. The risk of developing prostate cancer is at least doubled when a first-line relative has PCa. If two or more first-line relatives are affected, the risk increases 5- to 11-fold (Kalish et al. 2000). A small subpopulation of individuals with PCa (about 9 %) has true hereditary PCa. This is defined as three or more affected relatives or at least two relatives who have developed early onset disease (55 years of age or younger), usually 6–7 years prior to spontaneous cases (Carter et al. 1992).

Scientists have found several inherited genes that seem to increase prostate cancer risk, but they probably account for only a small number of cases overall. Genetic testing for most of these genes is not yet available. Recently, some common gene variations have been linked to the risk of prostate cancer. Studies to confirm these results are needed to see if testing for the gene variants will be useful in predicting prostate cancer risk.

Some inherited genes raise the risk for more than one type of cancer. For example, inherited mutations of the *BRCA1* (17q21) and *BRCA2* (13q12) genes are the reason that breast and ovarian cancers are much more common in some families. Mutations in these genes may also increase prostate cancer risk in some men, with a cumulative risk of 30 % by 80 years of age, but they account for a very small percentage of prostate cancer cases (Gallagher et al. 2010).

Linkage studies have identified a number of candidate prostate cancer susceptibility genes, including *RNaseL* (hereditary prostate cancer-1 [*HPC1*] region, 1q23-25), *ELAC2* (*HPC2* region, 17p), and *MSR1* (8p22–23). Many other genetic loci and prostate cancer susceptibility genes have been more recently identified, and the list is continually growing (Simard et al. 2002).

To date, only limited information about the prostate cancer genetic background has been discovered. With the continued improvement in genetic technology, it is likely that the number of known susceptibility genes will increase.

2.4 Environmental Factors

It has been accepted that the environment also plays an important role in modulating prostate cancer risk around the world. As it was described before, the incidence of clinical PCa differs widely between different geographical areas, being high in the USA and Northern Europe and low in Southeast Asia (Quinn and Babb 2002). However, if Japanese men move from Japan to Hawaii, their risk of PCa increases; if they move to California, their risk increases even more, approaching that of American men. Japanese and Chinese men in the United States have a higher risk of developing and dying from prostate cancer than do their relatives in Japan and China (Hsing et al. 2000). It is important to note, however, that Asian-Americans have a lower prostate cancer incidence than white or African-American men, indicating that genetics still plays a role in determining prostate cancer predisposition.

2.4.1 Dietary Intake

It has been shown that prostate cancer incidence and mortality rates around the world correlate highly with the average level of fat consumption, especially for polyunsaturated fats (Bostwick et al. 2004). The Western diet, together with other lifestyle factors such as physical activity levels, may be a significant risk factor in the development of prostate cancer. The Western diet tends to be high in animal products and processed, refined foods, resulting in a high intake of saturated fats, processed polyunsaturated fats (such as the *trans* fats), and refined carbohydrates. In addition, the Western diet, usually high in meats, is often low in fresh vegetables, fruit, pulses, and whole grains, resulting in a low intake of fibre and phytonutrients that may protect against prostate cancer. Overall, the Western diet is often calorie dense but lacking in certain essential nutrients. Furthermore, meats and dairy products contain other constituents such as zinc (Zn) and calcium (Ca) that may affect prostate cancer risk. Some studies have suggested that men who consume a lot of calcium (through food or supplements) may have a higher risk of developing advanced prostate cancer. Most studies have not found such a link with the levels of calcium found in the average diet, and it is important to note that calcium is known to have other important health benefits (Butler et al. 2010). By contrast, in Far Eastern countries, e.g. Japan and China, where the incidence of prostate cancer is lower, the traditional diet is mainly plant-based and minimally processed or refined. Relatively small amounts of animal products accompany vegetables, fruit, and other plant foods, and overall the diet is lower in calories than the Western diet but is likely to contain greater amounts of certain essential nutrients. Particular foods that feature more heavily in a traditional Far Eastern diet may have an impact on prostate cancer risk including green tea, soy, and cruciferous vegetables (Bostwick et al. 2004).

2.4.2 Obesity and Physical Activity

Obesity has been suggested to be a risk factor for prostate cancer because of its common occurrence in middle-aged men and clear links to colon and breast cancer risk (Hsing et al. 2007). White fat in mammals serves not only as an important energy reservoir but also as an endocrine organ, with secretion of cytokines and agents with cytokine-like activity (tumour necrosis factor [TNF]-α, interleukin [IL]-1β, IL-6, IL-8, IL-10, transforming growth factor [TGF]-β), as well as their soluble receptors (Baillargeon and Rose 2006). Treatment of obesity through reduction in fat intake and increased exercise has been shown to reduce oxidative stress, suggesting that lifestyle modification could be important in reducing the risk of prostate cancer (Wright et al. 2007). Some studies have found that high levels of physical activity, particularly in older men, may lower the risk of advanced prostate cancer. More research in this area is needed (Nilsen et al. 2006).

Recently, large prospective studies, examining the association between obesity and prostate cancer risk by stage and/or grade at diagnosis, suggested that obesity was associated with a lower risk of low-grade disease but a greater risk of high-grade disease (Wright et al. 2007; Gong et al. 2006; Rodriguez et al. 2007). This could be related to detection bias given that the lower serum PSA usually associated to obese patients could lead to fewer prostate biopsies. An association has also been observed in obesity with higher serum estradiol, insulin, free IGF-1, and leptin levels and lower free testosterone and adiponectin levels, which have also been associated with more aggressive prostate cancer (Hsing et al. 2007).

2.4.3 Smoking

It has been suggested that smoking may increase the risk of death from prostate cancer (Daniell 2010). Nevertheless, a clear dose–response relationship has not been demonstrated and will need to be confirmed by further investigations.

2.4.4 Alcohol Consumption

Although initial studies suggested an association between alcohol intake and prostate cancer risk, to date, it has not been possible to determine an association between total alcohol intake and the incidence of prostate cancer, suggesting that alcohol does not contribute appreciably to the aetiology of this disease. The only noticeable exception goes to red wine: a population-based case–control study conducted in King County, WA (USA), pointed out that each additional glass of red wine consumed per week showed a statistically significant 6 % decrease in relative risk (OR = 0.94; 95 % CI = 0.90–0.98) and there was evidence for a decline in risk estimates across increasing categories of red wine intake (trend $P = 0.02$). No clear associations were seen for consumption of beer or liquor (Schoonen et al. 2005).

2.4.5 Vasectomy and Sexual Activity

Some earlier studies had suggested that men who underwent vasectomy, especially at an early age, had a higher risk for prostate cancer (Giovannucci et al. 1993). But most recent studies have not been able to find a strong association between vasectomy and prostate cancer (Dennis et al. 2002a; Cox et al. 2002). The biologic mechanism by which vasectomy might predispose to cancer is unknown, although presence of antisperm antibodies, decreased seminal androgen concentrations, or secretory activity have been proposed.

Studies have also suggested a protective association between prostate cancer and frequency of ejaculation; the existence of a protective effect existed for men in their 20s and 40s, when a reported frequency greater than or equal to 21 ejaculations per month was found (Giles et al. 2004; Leitzmann et al. 2004). The biologic basis for this effect is not known.

2.4.6 Infection and Prostate Cancer

It has been suggested that the association between sexual activity and the exposure to sexually transmitted infections (HPV, gonorrhoea, or Chlamydia), possibly by leading to prostatic inflammation, may increase the risk of developing prostate cancer (Nelson et al. 2004).

Two meta-analyses examining 34 case–control studies reported statistically significant associations of prostate cancer with a history of sexually transmitted infection (RR = 1.4) or prostatitis (OR = 1.57) (Dennis et al. 2002b). Supportive evidence is provided by studies demonstrating positive associations of antibodies against syphilis, human papillomavirus (HPV), and human herpesvirus-8 (HHV-8) with prostate cancer (De Marzo et al. 2007). However, recent studies assessing the association between infection and prostate cancer have shown mixed results (Sutcliffe et al. 2006; Sarma et al. 2006). To date, no firm conclusions have been reached and further investigations are needed.

2.4.7 Steroid Hormones

Androgens play an important role in prostate carcinogenesis, as supported by the historical observation that the majority of prostate cancers initially respond to androgen-deprivation therapy and more recently by results of the Prostate Cancer Prevention Trial, which indicated that inhibition of the conversion of testosterone to the more potent dihydrotestosterone by finasteride reduces the incidence of prostate cancer by approximately 25 % (Thompson et al. 2007). High serum androgen levels have long been hypothesized to be a risk factor for prostate cancer. However, studies examining this association have been inconsistent, with some studies finding an association between specific hormones and prostate cancer risk.

Estrogens have been postulated to play a role in prostate cancer initiation and progression. Historically, estrogens have been considered protective against prostate cancer and have been used as a treatment for advanced disease. However, there is increasing evidence that estrogens may act as procarcinogens in the prostate (Dorgan et al. 1998). However, the association between serum oestrogen levels and prostate cancer risk is still inconsistent.

Leptins, peptide hormones produced by adipocytes which contribute to body weight and fat deposits, may be associated with development of prostate cancer by stimulation of androgen-independent prostate cancer cell lines (Ribeiro et al. 2004).

An association has been suggested between vitamin D and the risk of the development of prostate cancer. Lower serum vitamin D levels could be related to higher risk of prostate cancer (Giovannucci 1998). On the other hand, prostate cancer cells express the vitamin D receptor, and several studies have demonstrated an antiproliferative effect of vitamin D on prostate cancer cell lines (Chen and Holick 2003). All these findings need further investigations.

2.5 Prevention of Prostate Cancer

It has been accepted that carcinogenesis occurs slowly during a prolonged interval, from precursor lesions to the development of malignant cells. Theoretically, this provides the opportunity to intervene before malignancy is established, through lifestyle changes (dietary alterations, smoking cessation, exercise) or by chemoprevention, defined as the use of natural or synthetic agents that reverse, inhibit, or prevent the development of cancer (Boyle and Severi 1999). Effective chemoprevention requires the use of non-toxic agents that inhibit specific molecular steps in the carcinogenic pathway (Boyle and Severi 1999). As PCa is extremely common and generally slow to progress, it is regarded as an ideal candidate for chemoprevention. At present, the 5-alpha-reductase inhibitors finasteride and dutasteride have been identified as preventive agents. Today, chemopreventive agents may be appropriate for high-risk patients like those with high-grade prostatic intraepithelial neoplasia (PIN) and other high-risk groups such as patients with elevated prostate-specific antigen (PSA) and negative biopsy, rapid PSA velocity, and a family history of PCa. Although larger randomized controlled studies are needed and epidemiologic evidence should be placed in a clinical context, physicians must be aware of these preventive opportunities in PCa care (Van and Tombal 2011).

Numerous observations in the epidemiologic literature during the last years suggest associations between various dietary, lifestyle, genetic, and nontraditional factors and the risk for developing prostate cancer. Several randomized studies for prevention of PCa with pharmaceutical agents, dietary modifications, and supplements have been published. Some of the most promising dietary nutrients and supplements as well as the most important clinical studies on their effect on PCa are summarized in this review.

2.5.1 Prostate Cancer Prevention Trial

The most significant event in chemoprevention of prostate cancer occurred with the publication of the results of the Prostate Cancer Prevention Trial (PCPT) (Thompson et al. 2003a). This study, carried out in 1993, was the first large-scale population-based trial to test a chemopreventive strategy in men at risk for prostate cancer. The PCPT study was designed on the theoretical basis that androgens are required for the development of prostate cancer and men with a congenital deficiency of type 2 5α-reductase are unaffected by benign prostatic hyperplasia (BPH) and prostate cancer. Hypothetically, treatment with finasteride would lower intraprostatic DHT levels and thereby prevent prostate cancer. In this study, 18,882 men older than or equal to 55 years of age with a normal digital rectal examination (DRE) and a prostate-specific antigen (PSA) level of less than or equal to 3.0 ng/mL were randomly assigned to treatment with finasteride (5 mg/day) or a placebo for 7 years. Prostate biopsy was recommended if the annual PSA level, adjusted for the effect of finasteride, exceeded 4.0 ng/mL or if the DRE was abnormal. The primary endpoint was the prevalence of prostate cancer during the 7 years of the study, as diagnosed by either for-cause biopsies (abnormal DRE or PSA) or end-of-study biopsy. The results of the study were satisfactory, and the trial was stopped approximately 15 months early because the primary endpoint of a 25 % risk reduction on the finasteride arm was reached, and sensitivity analyses suggested that additional follow-up would not change that outcome.

The prevalence of prostate cancer was reduced by 24.8 % (hazard ratio [HR]=0.75; 95 % CI=18.6–30.6) from 24.4 to 18.4 % in those participants randomized to finasteride compared with placebo. The prevalence of Gleason grade 7–10 tumours was higher in the finasteride group than the placebo group (6.4 % vs. 5.1 %; HR=1.27; 95 % CI=1.07–1.50). The risk reduction associated with finasteride among risk groups was of the same general magnitude, but sexual side effects were more common with finasteride, whereas urinary symptoms were more common within the placebo group.

Deciding whether or not the advantages of taking finasteride outweigh the potential disadvantages is not a simple task. Several studies analysed the cost-effectiveness relation for chemoprevention with finasteride, but none suggested that it may be cost-effective in high-risk populations (Klein 2005; Unger et al. 2004; Zeliadt et al. 2005).

In addition to the prevention of prostate cancer, 5α-reductase inhibitors (5ARIs) have other benefits that need to be considered. Finasteride improves the sensitivity of PSA and DRE in prostate cancer screening (Thompson et al. 2006), decreases the risk of high-grade prostatic intraepithelial neoplasia (HR=0.79; 95 % CI=0.70–0.89), and may be effective in the treatment and possible prevention of chronic nonbacterial prostatitis (Thompson et al. 2007). Furthermore, benign prostatic hyperplasia (BPH) treatment trials in patients with moderate to severe lower urinary tract symptoms demonstrated reduction in symptom scores, reduction in the risk of acute urinary retention, and reduction in the risk of surgical intervention due to BPH progression (Wilt et al. 2008).

2.5.2 Other 5α-Reductase Inhibitors

A second large-scale trial of another 5ARI, dutasteride, was carried out in 2005. This agent inhibits both type 1 and type 2 forms of 5α-reductase, is anti-androgenic, promotes death of prostate cancer cell lines, and has been shown to reduce the risk of prostate cancer in men treated for lower urinary tract symptoms related to benign prostatic enlargement when compared with placebo (Andriole et al. 2004). The eligibility for the Reduction by Dutasteride of Prostate Cancer Events (REDUCE) trial included men aged older than or equal to 50 and younger than or equal to 75 years who had PSA scores from 2.5 to 10 ng/mL, prostate volumes less than or equal to 80 cc, and one prior negative prostate biopsy within 6 months of enrolment, thus representing a group at high risk for cancer on subsequent biopsy (Andriole et al. 2004). The primary endpoint of REDUCE was the prevalence of cancer on study-mandated prostate biopsies performed at 2 and 4 years after entry. The trial recruited 8,231 men, of whom 6,726 (82.6 %) underwent at least one biopsy and 1,516 (22.5 %) were diagnosed with prostate cancer. Dutasteride reduced the risk of prostate cancer over 4 years by 23 % (857 in the placebo arm vs. 659 in the dutasteride arm, $P<0.0001$). Interestingly, no significant increase in Gleason sum 8–10 tumours was observed in the study (19 in the placebo arm vs. 29 in dutasteride arm, $P=0.15$). Preliminary analyses also suggested that dutasteride enhanced the utility of PSA as a diagnostic test for prostate cancer, demonstrated beneficial effects on BPH outcomes (relative risk reductions of 77 % for acute urinary retention and 73 % for BPH-related surgery), and was generally well tolerated (15 % drug-related adverse events in the placebo arm vs. 22 % in dutasteride arm). The fact that the results of REDUCE were congruent with those of the PCPT with respect to the magnitude of risk reduction, benefits for BPH endpoints, minimal toxicity, and no issues related to tumour grade suggests a class effect for 5 ARIs and that these agents should be used more liberally for prevention of prostate cancer (Musquera et al. 2008). In December 2010, the FDA's Oncologic Drugs Advisory Committee (ODAC) voted against recommending finasteride and dutasteride for the indication to reduce prostate cancer risk because in the view of the ODAC members, the risk for more aggressive tumours outweighed the potential for chemoprevention. Panelists and FDA reviewers shared three main concerns about the drugs: the risk of exposing currently healthy people to an increased risk for high-grade tumours, the fact that risk reduction was only in low-grade tumours, and the doubt that the supporting clinical studies are generalizable to clinical practice in the American population.

2.5.3 Selenium and Vitamin E

Selenium is an essential trace element found in vegetables, grains, red meat, fish, poultry, and eggs. The concentration of selenium in the vegetables depends on how much of the mineral was in the soil where the plants grew. Selenium is distributed in body tissues and helps to make special proteins, called antioxidant enzymes, which play a role in preventing cell damage. Epidemiologic evidence provides support for a cancer prevention effect.

The strongest evidence for a protective effect of selenium came from the Nutritional Prevention of Cancer Trial, a randomized study of oral selenized yeast in patients with nonmelanoma skin cancer, which showed a 65 % reduction in the prostate cancer incidence when compared with the placebo (Clark et al. 1998). In that trial, 1,312 participants took the equivalent of 200 mcg yeast per day versus the placebo, and with a mean follow-up of 4.5 years, the incidence of prostate cancer was reduced in the selenium arm by 65 % compared with the placebo. Of note, the effect was strongest for those with a PSA value less than 4 ng/mL and those with the lowest serum selenium levels at study entry (Duffield-Lillico et al. 2003).

Vitamin E is an essential lipid-soluble antioxidant found in plant oils such as soy, corn, and olive oil. Other sources include nuts, seeds, and green leafy vegetables. It protects cells from free radicals. Several forms of vitamin E have been identified. The most active form with the highest bioavailability in human tissues is alpha-tocopherol. The body is not capable of producing this substance, and it must be consumed in the diet or supplements for proper health. Alpha-tocopherol may influence the development of cancer through several mechanisms, including induction of cell cycle arrest, and through direct antiandrogen activity (Thompson et al. 2003b). The Alpha-Tocopherol, Beta-Carotene Cancer Prevention Trial (ATBC), a randomized, placebo-controlled trial of α-tocopherol (50 mg/day) and beta-carotene (20 mg/day) alone or in combination in male smokers, with a primary endpoint of lung cancer incidence and mortality, on secondary analysis found a statistically significant 32 % reduction in prostate cancer incidence and a 41 % lower mortality in those receiving α-tocopherol (Albanes 2000).

2.5.4 Selenium and Vitamin E Cancer Prevention Trial (SELECT)

The accumulated epidemiologic and biologic evidence that selenium and vitamin E might prevent prostate cancer led to the design of SELECT, the Selenium and Vitamin E Cancer Prevention Trial (Klein et al. 2001; Lippman et al. 2009). SELECT was a phase III, randomized, double-blind, placebo-controlled, population-based trial designed to test the efficacy of selenium and vitamin E alone and in combination in the prevention of prostate cancer. It was the largest cancer prevention trial ever performed. Eligibility criteria include age greater than or equal to 50 years for African-Americans, age older than or equal to 55 years for whites, a DRE not suspicious for cancer, serum PSA less than or equal to 4 ng/mL, and normal blood pressure. Randomization was equally distributed among the four study arms (selenium + placebo, vitamin E + placebo, selenium + vitamin E, and placebo + placebo). Although the study duration was planned for 12 years, as a result of a planned interim analysis in August 2008, an independent data and safety monitoring committee recommended discontinuation of the study because the data demonstrated no effect on the risk of prostate cancer by either agent alone or in combination and no chance of a beneficial effect of the hypothesized magnitude with continued supplementation (Lippman et al. 2009). There were statistically nonsignificant increased risks of prostate cancer in the vitamin E group (HR = 1.13; 99 % CI = 0.95–1.35; P = 0.06) and type 2 diabetes mellitus in the selenium group (RR = 1.07; 95 % CI = 0.94–1.22; P = 0.16). However, neither of these findings was observed in the combination group.

More mature results were published recently, reflecting the final data collected by the study sites on their participants on July 5, 2011. The final report includes 54,464 additional person-years of follow-up and 521 additional cases of prostate cancer since the primary report. Compared with the placebo (referent group) in which 529 men developed prostate cancer, 620 men in the vitamin E group developed prostate cancer (hazard ratio [HR] = 1.17; 99 % CI = 1.004–1.36; P = 0.008), as did 575 in the selenium group (HR = 1.09; 99 % CI = 0.93–1.27; P = 0.18) and 555 in the selenium plus vitamin E group (HR = 1.05; 99 % CI = 0.89–1.22; P = 0.46). Compared with placebo, the absolute increase in risk of prostate cancer per 1,000 person-years was 1.6 for vitamin E, 0.8 for selenium, and 0.4 for the combination (Klein et al. 2011).

This study therefore indicates that supplementation with vitamin E significantly increased the risk of prostate cancer among healthy men.

There are many attempts to explain the negative results of SELECT. The high dose of vitamin E (400 IU/D of the alpha-tocopherol form) in SELECT may have been less effective than a lower dose such as the eightfold lower 50 IU/D of the ATBC study (Lippman et al. 2009).

In SELECT, 200 μg of L-selenomethionine was chosen, whereas in the NPC trial, 200 μg of high-Se yeast contained only 20 % of L-selenomethionine (Duffield-Lillico et al. 2003). Another drawback of SELECT is the absence of selection of patients since it is likely that personal predispositions may enhance or hinder the benefit of supplementation. For example, several studies have suggested that vitamin E is more protective against PCa in smokers, and in SELECT less than 60 % of men were current or former smokers, whereas in the ATBC study all men were smokers. As for Se, genetic susceptibilities exist that may confer different benefit to Se supplementation. Chan et al. have assessed manganese superoxide dismutase (SOD2) gene variants and plasma Se in 489 patients with localized/locally advanced PCa (Chan et al. 2009). SOD2 is an endogenous mitochondrial enzyme that metabolizes reactive oxygen species and superoxide anions to oxygen and hydrogen peroxide. Several polymorphisms of SOD2 have been identified, including a single nucleotide permutation that encodes either an alanine (A) or a valine (V). SOD2 genotype alone was not associated with disease aggressiveness, whereas higher versus lower Se levels were associated with a slightly increased likelihood of presenting with aggressive disease (RR = 1.35; 95 % CI = 0.99–1.84). There was evidence of an interaction between SOD2 and Se levels such that among men with the AA genotype, higher Se levels were associated with a reduced risk of presenting with aggressive disease (RR = 0.60; 95 % CI = 0.32–1.12), whereas among men with a V allele, higher Se levels were associated with an increased risk of aggressive disease (for VV or VA men, RR = 1.82; 95 % CI = 1.27–2.61; P for interaction <0.007).

But clearly one of the more consistent hypotheses is that the positive effects of Se in the NPC study and of vitamin E in the ATBC trial could have been due to chance in secondary analyses. Recent results from the Prostate Cancer Prevention Trial found no significant association between vitamin E and Se and the incidence of PCa (Kristal et al. 2010).

2.6 Other Agents

2.6.1 Isoflavones

Isoflavones, a subclass of the flavonoids, are plant-derived compounds with weak estrogenic activity and therefore classified as phyto-oestrogens. Phyto-oestrogens have been suggested to have a preventive effect against various cancers. Soy foods are a rich source of isoflavones. Isoflavone intake in Asian countries is approximately 50 mg daily, which is about ten times higher than intake in Western countries. Migration studies and lower prostate cancer rates in Asian men with higher soy intake also support the role of soy as an anticancer agent.

The main isoflavones found in most soy products are genistein, daidzein, and glycitein. It has been suggested that these may inhibit benign and malignant prostatic epithelial cell growth, downregulate androgen-regulated genes, and reduce tumour growth in animal models (Castle and Thrasher 2002; Kurahashi et al. 2007; Lee et al. 2003). Also a protective effect of isoflavones against PCa development has been demonstrated. Among these effects, isoflavones possess weak oestrogen activity, inhibit tyrosine protein kinases, block angiogenesis, and reduce serum testosterone levels. They also inhibit 5-alpha-reductase, an enzyme that metabolizes testosterone to dihydrotestosterone. Isoflavones are further metabolized in humans to many different intermediates, such as equol, perhaps the best-studied metabolite of daidzein. Equol is ten times more potent than daidzein in retarding PCa growth (Lampe 2010).

Although there exists overwhelming data from epidemiologic studies, case–control studies, and in vitro/vivo data that soy isoflavone may be a promising chemopreventive agent against PCa, there have been no published prospective randomized clinical studies with sufficient statistical power to assess whether isoflavone supplementation can reduce PCa development or delay PCa progression. Shortcomings of many studies published to date are small patient numbers, lack of randomization, short-term isoflavone administration, and possibly insufficient doses. Current ongoing clinical trials may help us understand the role of soy in the prevention of PCa.

2.6.2 Lycopene

Lycopene is a carotenoid without vitamin A activity that gives the red colour to tomatoes and tomato-derived products. It is also available in other red fruits and vegetables such as red carrots, watermelons, pink grapefruit, and papayas. Lycopene is a highly unsaturated acyclic isomer of β-carotene, is the predominant carotenoid in human plasma, and possesses potent antioxidant activity. Lycopene has been shown to inhibit the growth of benign and malignant prostatic epithelial cells in vitro, having shown anticancer properties (Etminan et al. 2004). There is epidemiologic evidence that consumption of tomato products or lycopene is associated with a lower risk of prostate cancer (Giovannucci et al. 2002). A meta-analysis of observational studies reported a 23 % reduction in prostate cancer risk with high tomato and lycopene intake (Etminan et al. 2004). However, nested case–control studies prospectively examined the intake of several tomato-containing foods in men and found no correlation with the incidence of prostate cancer (Giovannucci 2002). To date, results of initial clinical trials have demonstrated potential beneficial activity of lycopene in prostate cancer; however, further investigations and phase III trials examining the role of lycopene in prostate cancer prevention are required.

2.6.3 Polyphenols

Polyphenols are the largest group of constituents found in tea. Green tea contains catechins, a category of water-soluble polyphenolic substances. The four principal catechins are (−)-epicatechin (EC), (−)-epicatechin-3-gallate (ECG), (−)-epigallocatechin (EGC), and (−)-epigallocatechin-3-gallate (EGCG) (Boehm et al. 2009). EGCG, found in the highest concentration in green tea, is the most studied and most active of all green tea catechins

(GTC) for the inhibition of oncogenesis and reduction of oxidative stress. Its mechanism of action has not yet been fully determined. Several epidemiologic studies have focused on the lower incidence of PCa in Asian populations where green tea is consumed regularly as compared with Western populations, suggesting that green tea is protective against PCa. To date, various small randomized trials have been conducted with mixed results (Boehm et al. 2009; Adhami et al. 2009; Jian et al. 2004; Kikuchi et al. 2006; Kurahashi et al. 2008). Confirmatory trials are needed to better assess the role of green tea consumption in prostate cancer prevention.

2.6.4 Resveratrol

Because tumours develop resistance to chemotherapeutic agents, the cancer research community continues to search for effective chemosensitizers. One promising possibility is to use dietary agents that sensitize tumours to the chemotherapeutics. Resveratrol (3,5,4′-trihydroxystilbene), a natural stilbenoid present in red wine, grapes, berries, peanuts, and dietary supplements as well as polyhydroxy analogues of resveratrol have potential cancer chemopreventive properties. It has sensitization/enhancing activities against tumour cells when used in combination with standard cancer chemotherapeutics (Hsieh et al. 2011). It has been suggested that resveratrol can sensitize tumour cells to chemotherapeutic agents. The tumours shown to be sensitized by resveratrol include lung carcinoma, acute myeloid leukaemia, promyelocytic leukaemia, multiple myeloma, prostate cancer, oral epidermoid carcinoma, and pancreatic cancer. The chemotherapeutic agents include vincristine, adriamycin, paclitaxel, doxorubicin, cisplatin, gefitinib, 5-fluorouracil, velcade, and gemcitabine. The chemosensitization of tumour cells by resveratrol appears to be mediated through its ability to modulate multiple cell-signalling molecules, including drug transporters, cell survival proteins, cell proliferative proteins, and members of the NF-kappa B and STAT3 signalling pathways. Overall, studies suggest that resveratrol can be used to sensitize tumours to standard cancer chemotherapeutics.

Recent findings strongly suggest that a suppressor of cytokine signalling (SOCS-3), an antiapoptotic molecule that is upregulated in PCa, is one of the proteins that influence the ability of resveratrol and tumour necrosis factor (TNF)-related apoptosis-inducing ligand (TRAIL) to cause programmed cell death in PCa (Hsieh et al. 2011; Horndasch and Culig 2011). Future studies should concentrate on the determination of molecular mechanisms of chemosensitization and of resveratrol combinations by clinically relevant in vivo studies and demonstration of safety and effectiveness of combinations in humans.

2.6.5 Statins

Statins are widely used cholesterol-lowering drugs given for the treatment and prevention of atherosclerotic cardiovascular disease. They inhibit 3-hydroxy-3-methylglutaryl coenzyme A (HMG-CoA), the rate-limiting enzyme in cholesterol biosynthesis. Statins are hypothesized to play a role in the prevention of cancer by inhibiting inflammation, angiogenesis, cell proliferation, migration/adhesion, invasion, and/or preferentially promoting apoptosis in tumour cells (Moyad 2005). Observational studies of statin use demonstrate high heterogeneity with mixed results for prostate cancer risk.

Hypothetically, some of the mechanisms that increase the risk of cardiovascular disease (CVD) may also increase the risk or progression of prostate cancer. Numerous recent lifestyle interventions that reduce cholesterol have also been found to have a potential impact on reducing the risk of prostate cancer. Recent studies of statins and other heart healthy agents have found a secondary potential for exhibiting a reduced risk or progression of prostate cancer. Statin users have been shown to have lower serum PSA levels than non-users (Papadopoulos et al. 2011), suggesting an anticancer effect but potentially leading to fewer prostate biopsies and thus biasing results of epidemiologic and clinical studies. Statin users tend to be healthier and more medically compliant than non-users, making them more likely to undergo PSA screening and possibly resulting in earlier cancer detection. Further research is needed to help determine the role, if any, of statins in the prevention of prostate cancer.

Conclusion

The prostate carcinogenesis is a complex process. All constitutional, behavioural, molecular, and environmental factors continuously interact in different proportions in the organism during the years, and their effects become manifest through the diagnosis of prostate cancer. Several clinical trials and epidemiological studies have been carried out around the world in order to achieve a better understanding of the prostate carcinogenesis and its circumstances.

Prostate cancer is an attractive target for chemoprevention because of its ubiquity, treatment-related morbidity, long latency between premalignant lesions and clinically evident cancer, and defined molecular pathogenesis. High-risk patients should be the target of chemoprevention, as the risk of PCa may justify the cost and potential side effects of these agents. It also seems reasonable to believe that chemoprevention strategies are more effective in high-risk groups. Nevertheless, the identification of high-risk groups is at this moment not easy. Patients with isolated HGPIN on prostate biopsies constitute a unique and well-demarcated risk group for PCa. Prognostic, randomized data on chemopreventive strategies in HGPIN are scarce but seem promising. Other high-risk groups include those above 40 years of age, elevated PSA levels, rapid PSA velocity, sub-Saharan African ethnicity, a family history of PCa or with specific genes, or obese men with insulin resistance and those who would benefit from early diagnosis and treatment with at least 10–15 years of life expectancy.

Future research should focus on determining the target population for PCa chemoprevention. Large prospective randomized studies are required to define the benefits of chemopreventive agents for PCa. Some of these studies are ongoing and results are eagerly awaited. Combinations of chemopreventive agents for PCa should be carefully investigated.

References

Adhami VM, Siddiqui IA, Sarfaraz S, Khwaja SI, Hafeez BB, Ahmad N et al (2009) Effective prostate cancer chemopreventive intervention with green tea polyphenols in the TRAMP model depends on the stage of the disease. Clin Cancer Res 15(6):1947–1953

Albanes D (2000) Prostate cancer: epidemiology and prevention. Nestle Nutr Workshop Ser Clin Perform Programme 4:55–62

Andriole GL, Roehrborn C, Schulman C, Slawin KM, Somerville M, Rittmaster RS (2004) Effect of dutasteride on the detection of prostate cancer in men with benign prostatic hyperplasia. Urology 64(3):537–541

Androutsos G (2005) Carcinoma of the prostate. A historical account. J BUON 10(1):135–144

Baillargeon J, Rose DP (2006) Obesity, adipokines, and prostate cancer (review). Int J Oncol 28(3):737–745

Boehm K, Borrelli F, Ernst E, Habacher G, Hung SK, Milazzo S et al (2009) Green tea (Camellia sinensis) for the prevention of cancer. Cochrane Database Syst Rev 3:CD005004

Bostwick DG, Burke HB, Djakiew D, Euling S, Ho SM, Landolph J et al (2004) Human prostate cancer risk factors. Cancer 101(10 Suppl):2371–2490

Boyle P, Severi G (1999) Epidemiology of prostate cancer chemoprevention. Eur Urol 35(5–6): 370–376
Bray F, Lortet-Tieulent J, Ferlay J, Forman D, Auvinen A (2010) Prostate cancer incidence and mortality trends in 37 European countries: an overview. Eur J Cancer 46(17):3040–3052
Butler LM, Wong AS, Koh WP, Wang R, Yuan JM, Yu MC (2010) Calcium intake increases risk of prostate cancer among Singapore Chinese. Cancer Res 70(12):4941–4948
Carter BS, Beaty TH, Steinberg GD, Childs B, Walsh PC (1992) Mendelian inheritance of familial prostate cancer. Proc Natl Acad Sci USA 89(8):3367–3371
Castle EP, Thrasher JB (2002) The role of soy phytoestrogens in prostate cancer. Urol Clin North Am 29(1):71–81, viii–ix
Chan JM, Oh WK, Xie W, Regan MM, Stampfer MJ, King IB et al (2009) Plasma selenium, manganese superoxide dismutase, and intermediate- or high-risk prostate cancer. J Clin Oncol 27(22):3577–3583
Chen TC, Holick MF (2003) Vitamin D and prostate cancer prevention and treatment. Trends Endocrinol Metab 14(9):423–430
Chou R, Croswell JM, Dana T, Bougatsos C, Blazina I, Fu R et al (2011) Screening for prostate cancer: a review of the evidence for the U.S. Preventive Services Task Force. Ann Intern Med 155(11):762–771
Clark LC, Dalkin B, Krongrad A, Combs GF Jr, Turnbull BW, Slate EH et al (1998) Decreased incidence of prostate cancer with selenium supplementation: results of a double-blind cancer prevention trial. Br J Urol 81(5):730–734
Coleman MP, Quaresma M, Berrino F, Lutz JM, De AR, Capocaccia R et al (2008) Cancer survival in five continents: a worldwide population-based study (CONCORD). Lancet Oncol 9(8):730–756
Cox B, Sneyd MJ, Paul C, Delahunt B, Skegg DC (2002) Vasectomy and risk of prostate cancer. JAMA 287(23):3110–3115
Crawford ED (2003) Epidemiology of prostate cancer. Urology 62(6 Suppl 1):3–12
Daniell HW (2010) Smoking, obesity, and statin therapy in the prognosis of prostate cancer. J Clin Oncol 28(31):e643–e646
De Marzo AM, Platz EA, Sutcliffe S, Xu J, Gronberg H, Drake CG et al (2007) Inflammation in prostate carcinogenesis. Nat Rev Cancer 7(4):256–269
Dennis LK, Dawson DV, Resnick MI (2002a) Vasectomy and the risk of prostate cancer: a meta-analysis examining vasectomy status, age at vasectomy, and time since vasectomy. Prostate Cancer Prostatic Dis 5(3):193–203
Dennis LK, Lynch CF, Torner JC (2002b) Epidemiologic association between prostatitis and prostate cancer. Urology 60(1):78–83
Djavan B (2011) Screening for prostate cancer: practical analysis of the ERSPC [corrected] and PLCO trials. Eur Urol 59(3):365–369
Dorgan JF, Albanes D, Virtamo J, Heinonen OP, Chandler DW, Galmarini M et al (1998) Relationships of serum androgens and estrogens to prostate cancer risk: results from a prospective study in Finland. Cancer Epidemiol Biomarkers Prev 7(12):1069–1074
Duffield-Lillico AJ, Dalkin BL, Reid ME, Turnbull BW, Slate EH, Jacobs ET et al (2003) Selenium supplementation, baseline plasma selenium status and incidence of prostate cancer: an analysis of the complete treatment period of the Nutritional Prevention of Cancer Trial. BJU Int 91(7):608–612
Etminan M, Takkouche B, Caamano-Isorna F (2004) The role of tomato products and lycopene in the prevention of prostate cancer: a meta-analysis of observational studies. Cancer Epidemiol Biomarkers Prev 13(3):340–345
Ferlay J, Parkin DM, Steliarova-Foucher E (2010) Estimates of cancer incidence and mortality in Europe in 2008. Eur J Cancer 46(4):765–781
Gallagher DJ, Gaudet MM, Pal P, Kirchhoff T, Balistreri L, Vora K et al (2010) Germline BRCA mutations denote a clinicopathologic subset of prostate cancer. Clin Cancer Res 16(7):2115–2121
Giles GG, Severi G, English DR, Hopper JL (2004) Frequency of ejaculation and risk of prostate cancer. JAMA 292(3):329

Giovannucci E (1998) Dietary influences of 1,25(OH)2 vitamin D in relation to prostate cancer: a hypothesis. Cancer Causes Control 9(6):567–582

Giovannucci E (2002) A review of epidemiologic studies of tomatoes, lycopene, and prostate cancer. Exp Biol Med (Maywood) 227(10):852–859

Giovannucci E, Ascherio A, Rimm EB, Colditz GA, Stampfer MJ, Willett WC (1993) A prospective cohort study of vasectomy and prostate cancer in US men. JAMA 269(7):873–877

Giovannucci E, Rimm EB, Liu Y, Stampfer MJ, Willett WC (2002) A prospective study of tomato products, lycopene, and prostate cancer risk. J Natl Cancer Inst 94(5):391–398

Gong Z, Neuhouser ML, Goodman PJ, Albanes D, Chi C, Hsing AW et al (2006) Obesity, diabetes, and risk of prostate cancer: results from the prostate cancer prevention trial. Cancer Epidemiol Biomarkers Prev 15(10):1977–1983

Heidenreich A, Bellmunt J, Bolla M, Joniau S, Mason M, Matveev V et al (2011) EAU guidelines on prostate cancer. Part 1: screening, diagnosis, and treatment of clinically localised disease. Eur Urol 59(1):61–71

Horndasch M, Culig Z (2011) SOCS-3 antagonizes pro-apoptotic effects of TRAIL and resveratrol in prostate cancer cells. Prostate 71(12):1357–1366

Hsieh TC, Huang YC, Wu JM (2011) Control of prostate cell growth, DNA damage and repair and gene expression by resveratrol analogues, in vitro. Carcinogenesis 32(1):93–101

Hsing AW, Tsao L, Devesa SS (2000) International trends and patterns of prostate cancer incidence and mortality. Int J Cancer 85(1):60–67

Hsing AW, Sakoda LC, Chua S Jr (2007) Obesity, metabolic syndrome, and prostate cancer. Am J Clin Nutr 86(3):s843–s857

Jemal A, Bray F, Center MM, Ferlay J, Ward E, Forman D (2011) Global cancer statistics. CA Cancer J Clin 61(2):69–90

Jian L, Xie LP, Lee AH, Binns CW (2004) Protective effect of green tea against prostate cancer: a case–control study in southeast China. Int J Cancer 108(1):130–135

Josef MF, Karenberg A (2009) History of the term prostate. Prostate 69(2):208–213

Kalish LA, McDougal WS, McKinlay JB (2000) Family history and the risk of prostate cancer. Urology 56(5):803–806

Kikuchi N, Ohmori K, Shimazu T, Nakaya N, Kuriyama S, Nishino Y et al (2006) No association between green tea and prostate cancer risk in Japanese men: the Ohsaki Cohort Study. Br J Cancer 95(3):371–373

Klein EA (2005) Can prostate cancer be prevented? Nat Clin Pract Urol 2(1):24–31

Klein EA, Thompson IM, Lippman SM, Goodman PJ, Albanes D, Taylor PR et al (2001) SELECT: the next prostate cancer prevention trial. Selenium and Vitamin E Cancer Prevention Trial. J Urol 166(4):1311–1315

Klein EA, Thompson IM Jr, Tangen CM, Crowley JJ, Lucia MS, Goodman PJ et al (2011) Vitamin E and the risk of prostate cancer: the Selenium and Vitamin E Cancer Prevention Trial (SELECT). JAMA 306(14):1549–1556

Kristal AR, Arnold KB, Neuhouser ML, Goodman P, Platz EA, Albanes D et al (2010) Diet, supplement use, and prostate cancer risk: results from the prostate cancer prevention trial. Am J Epidemiol 172(5):566–577

Kurahashi N, Iwasaki M, Sasazuki S, Otani T, Inoue M, Tsugane S (2007) Soy product and isoflavone consumption in relation to prostate cancer in Japanese men. Cancer Epidemiol Biomarkers Prev 16(3):538–545

Kurahashi N, Sasazuki S, Iwasaki M, Inoue M, Tsugane S (2008) Green tea consumption and prostate cancer risk in Japanese men: a prospective study. Am J Epidemiol 167(1):71–77

Lampe JW (2010) Emerging research on equol and cancer. J Nutr 140(7):1369S–1372S

Lee MM, Gomez SL, Chang JS, Wey M, Wang RT, Hsing AW (2003) Soy and isoflavone consumption in relation to prostate cancer risk in China. Cancer Epidemiol Biomarkers Prev 12(7):665–668

Leitzmann MF, Platz EA, Stampfer MJ, Willett WC, Giovannucci E (2004) Ejaculation frequency and subsequent risk of prostate cancer. JAMA 291(13):1578–1586

Lippman SM, Klein EA, Goodman PJ, Lucia MS, Thompson IM, Ford LG et al (2009) Effect of selenium and vitamin E on risk of prostate cancer and other cancers: the Selenium and Vitamin E Cancer Prevention Trial (SELECT). JAMA 301(1):39–51

Marx FJ, Karenberg A (2010) Uro-words making history: ureter and urethra. Prostate 70(9): 952–958

Moyad MA (2005) Heart healthy equals prostate healthy equals statins: the next cancer chemoprevention trial. Part I. Curr Opin Urol 15(1):1–6

Musquera M, Fleshner NE, Finelli A, Zlotta AR (2008) The REDUCE trial: chemoprevention in prostate cancer using a dual 5alpha-reductase inhibitor, dutasteride. Expert Rev Anticancer Ther 8(7):1073–1079

Nelson WG, De Marzo AM, DeWeese TL, Isaacs WB (2004) The role of inflammation in the pathogenesis of prostate cancer. J Urol 172(5 Pt 2):S6–S11

Nilsen TI, Romundstad PR, Vatten LJ (2006) Recreational physical activity and risk of prostate cancer: a prospective population-based study in Norway (the HUNT study). Int J Cancer 119(12):2943–2947

Papadopoulos G, Delakas D, Nakopoulou L, Kassimatis T (2011) Statins and prostate cancer: molecular and clinical aspects. Eur J Cancer 47(6):819–830

Quinn M, Babb P (2002) Patterns and trends in prostate cancer incidence, survival, prevalence and mortality. Part I: international comparisons. BJU Int 90(2):162–173

Ribeiro R, Lopes C, Medeiros R (2004) Leptin and prostate: implications for cancer prevention – overview of genetics and molecular interactions. Eur J Cancer Prev 13(5):359–368

Rodriguez C, Freedland SJ, Deka A, Jacobs EJ, McCullough ML, Patel AV et al (2007) Body mass index, weight change, and risk of prostate cancer in the Cancer Prevention Study II Nutrition Cohort. Cancer Epidemiol Biomarkers Prev 16(1):63–69

Sakr WA (1999) Prostatic intraepithelial neoplasia: a marker for high-risk groups and a potential target for chemoprevention. Eur Urol 35(5–6):474–478

Sarma AV, McLaughlin JC, Wallner LP, Dunn RL, Cooney KA, Schottenfeld D et al (2006) Sexual behavior, sexually transmitted diseases and prostatitis: the risk of prostate cancer in black men. J Urol 176(3):1108–1113

Schoonen WM, Salinas CA, Kiemeney LA, Stanford JL (2005) Alcohol consumption and risk of prostate cancer in middle-aged men. Int J Cancer 113(1):133–140

Schroder FH (2008) Screening for prostate cancer (PC) – an update on recent findings of the European Randomized Study of Screening for Prostate Cancer (ERSPC). Urol Oncol 26(5):533–541

Simard J, Dumont M, Soucy P, Labrie F (2002) Perspective: prostate cancer susceptibility genes. Endocrinology 143(6):2029–2040

Strope SA, Andriole GL (2010) Prostate cancer screening: current status and future perspectives. Nat Rev Urol 7(9):487–493

Sutcliffe S, Giovannucci E, De Marzo AM, Leitzmann MF, Willett WC, Platz EA (2006) Gonorrhea, syphilis, clinical prostatitis, and the risk of prostate cancer. Cancer Epidemiol Biomarkers Prev 15(11):2160–2166

Thompson IM, Goodman PJ, Tangen CM, Lucia MS, Miller GJ, Ford LG et al (2003a) The influence of finasteride on the development of prostate cancer. N Engl J Med 349(3):215–224

Thompson IM, Basler JA, Leach R, Troyer D, Klein E, Brawley O (2003b) Challenges and opportunities to the design and implementation of chemoprevention trials for prostate cancer. Urol Oncol 21(1):73–78

Thompson IM, Chi C, Ankerst DP, Goodman PJ, Tangen CM, Lippman SM et al (2006) Effect of finasteride on the sensitivity of PSA for detecting prostate cancer. J Natl Cancer Inst 98(16): 1128–1133

Thompson IM, Pauler AD, Chi C, Goodman PJ, Tangen CM, Lippman SM et al (2007) Prediction of prostate cancer for patients receiving finasteride: results from the Prostate Cancer Prevention Trial. J Clin Oncol 25(21):3076–3081

Unger JM, LeBlanc M, Thompson IM, Coltman CA Jr (2004) The person-years saved model and other methodologies for assessing the population impact of cancer-prevention strategies. Urol Oncol 22(4):362–368

Van PH, Tombal B (2011) Chemoprevention of prostate cancer with nutrients and supplements. Cancer Manag Res 3:91–100

Wilt TJ, MacDonald R, Hagerty K, Schellhammer P, Kramer BS (2008) Five-alpha-reductase inhibitors for prostate cancer prevention. Cochrane Database Syst Rev 2:CD007091

Wright ME, Chang SC, Schatzkin A, Albanes D, Kipnis V, Mouw T et al (2007) Prospective study of adiposity and weight change in relation to prostate cancer incidence and mortality. Cancer 109(4):675–684

Zeliadt SB, Etzioni RD, Penson DF, Thompson IM, Ramsey SD (2005) Lifetime implications and cost-effectiveness of using finasteride to prevent prostate cancer. Am J Med 118(8):850–857

3 Imaging Localised Prostate Carcinoma

Brendan M. Carey

Historically, imaging has played a relatively small role in the management of clinically localised prostate cancer. In more recent years, medical imaging techniques, such as transrectal ultrasound (TRUS), computed tomography (CT), magnetic resonance imaging (MRI), and radioisotopes scanning, have facilitated and improved the clinical staging of patients with prostate cancer and help inform the decision regarding the most appropriate treatment for an individual patient. Newer functional imaging techniques including positron emission tomography (PET) together with molecular imaging developments will further improve our ability to diagnose, stage, and appropriately treat men with prostate cancer.

The clinical presentation, diagnosis, and management of prostate cancer are changing. The widespread use of prostate-specific antigen (PSA) screening has led to a dramatic downstaging of prostate cancer at diagnosis. Today, prostate cancers are generally smaller and of lower stage at diagnosis than in the past. Newer treatment techniques, including brachytherapy, offer the possibility of better local treatment with less morbidity and improved outcomes; more accurate imaging has contributed to better staging and subsequent stratification of patients into different treatments. The optimum management of prostate cancer today is as patient specific as current tumour characterisation techniques permit. Our ability to predict biologic aggressiveness of prostate cancers is still based on the histological appearance of prostate cancer (Gleason grading system) which has significant limitations. However, new biomarkers for prostate cancer are becoming clinically available and will hopefully improve our ability to predict the biological behaviour of the tumour and therefore impact on management in the future. Because younger men are being treated, the long-term consequences of all active therapies need consideration, since these younger patients may have to live with any treatment-associated morbidity including brachytherapy and complications for many years.

B.M. Carey
Radiologist, Institute of Oncology, St. James Hospital, Leeds, UK
e-mail: brendan.carey@btinternet.com

G. Kovács, P. Hoskin (eds.), *Interstitial Prostate Brachytherapy*,
DOI 10.1007/978-3-642-36499-0_3, © Springer-Verlag Berlin Heidelberg 2013

In the early years of prostate brachytherapy, clinical nomograms (Partin et al. 2001) based on the combination of the presenting serum PSA level, digital rectal examination (DRE) findings, and the results of systematic TRUS-guided prostate biopsy and biopsy-based Gleason score were a valuable clinical aid in patient selection for brachytherapy. The sensitivities and specificities, however, of these nomograms are problematic and still lack comprehensive external validation and extrapolation to different patient groups. DRE has a low overall sensitivity (37 %) and low positive predictive value in men with PSA < 3 ng/mL (Schröder et al. 1998). PSA measurement has yielded higher detection rates than has DRE, but its specificity remains low at 36 % (Schröder et al. 2008; Catalona et al. 1994). When DRE results are positive or when the PSA level is elevated, systematic sextant TRUS biopsy with a minimum of four additional cores from lateral peripheral zones or from a suspicious area is generally recommended to be performed initially (Donovan et al. 2003). Systematic random prostate biopsy is prone to undersampling (35 % cancers missed on first biopsy, Djavan et al. 2001) and underestimation of Gleason grade in 46 % of cases (Noguchi et al. 2001). Clinical reliance on nomograms alone may lead to inaccurate risk assessments and suboptimal treatment choices. They should be considered together with the imaging findings in the appropriate selection of patients for brachytherapy.

All forms of active treatment for prostate cancer benefit from accurate tumour characterisation and staging in order to select the optimum therapeutic approach from a range of different options, both surgical and nonsurgical. Many different imaging techniques can be used at various stages during the management of prostate cancer, providing morphological, structural, metabolic, and functional information about the cancer, its treatment, and treatment response. The earlier detection of prostate cancer in today's PSA era has brought new challenges to clinical assessment and treatment selection—challenges compounded by variability in the natural history of prostate cancer itself. Although the cancers of today may be smaller, a wide range of aggressiveness remains and our ability to predict the clinical progress of an individual patient and his prostate cancer is still limited. The natural history of prostate cancer is remarkably heterogeneous and still not completely understood. Autopsy and early follow-up observational studies have suggested that approximately one in three men ≥ 50 years old will show histological evidence of prostate cancer; a significant portion of these tumours are small and possibly clinically insignificant, although others are aggressive and potentially lethal . The controversy regarding the potential overtreatment of 'clinically insignificant' prostate tumours is balanced by the infrequent but worrying finding of higher-volume, higher-grade cancers recognised on final surgical pathology after radical prostatectomy. The challenge is to distinguish between the two and manage accordingly. We have made progress, based on the mature clinical outcome data of various treatment regimes, better imaging, and developments in molecular biology. Prostate cancer imaging has evolved into a process where the selection of an imaging technique for an individual patient with prostate cancer should be based on the questions that need to be answered for that particular patient. Ultimately, the imaging workup should identify those patients who are most likely to benefit from their treatment (e.g. brachytherapy)—who are likely to have a good outcome in terms of disease control and associated treatment-related morbidity.

Low-dose rate (LDR) and high-dose rate (HDR) brachytherapies have accepted roles in the modern management of localised prostate cancer in the appropriate clinical setting. Although brachytherapy remains a therapy dominated by the skill of the operator in terms of source placement, imaging guidance is crucial to optimum therapy and the integration of prostate imaging with prostate brachytherapy is evident at all stages of the treatment process. Substantial progress has been made in the imaging of prostate cancer, particularly in MRI. These advances are beginning to translate into better treatment selection and better image-guided therapies; brachytherapy can and will increasingly benefit from these improvements now and in the future. The integration of molecular imaging into current techniques will further enhance our ability to stage, appropriately select, and follow up patients who are treated with brachytherapy. The planning, delivery, and verification of radiation therapy are based on contemporary and evolving imaging techniques. Advances in brachytherapy techniques have benefited considerably from better TRUS, and the development of complex radiation dosimetry demands optimum tumour identification and staging. Brachytherapy in the future may require identification of different anatomic subregions within the prostate and/or tumour to define targets for differential dose delivery in order to selectively increase the dose to specific tumour-bearing regions. Brachytherapy may involve simultaneous delivery of different dose prescriptions to multiple cancer foci within the prostate. This will require detailed knowledge about tumour location, volume, and full extent. Furthermore, information regarding tumour biology (tumour aggressiveness, angiogenesis, and hypoxia) is becoming available (Hricak et al. 2007). Functional imaging can already identify foci of hypoxic tumour clones that might benefit from higher local radiation doses. On the basis of such anatomic and metabolic information, patient-specific parametric images have been used to dose paint during brachytherapy planning.

Prostate brachytherapy is an operator-dependent treatment that requires training and the necessary skills to ensure safety and good outcomes. It also requires familiarity with the normal anatomy of the prostate gland shown in Fig. 3.1. Understanding prostate and tumour anatomy is an essential part of image interpretation, and it is important that the operator is familiar with basic prostate anatomy and tumour morphology as demonstrated on imaging with TRUS and MRI.

The prostate has four glandular zones, each with its own ductal system (McNeal 1981). Histologically, the prostate gland consists of glandular (acinar) and nonglandular elements. The major nonglandular elements are the prostatic urethra and the anterior fibromuscular stroma. The glandular prostate consists of outer and inner components, which are differentiated by location, duct anatomy, and histological characteristics. The inner prostate consists of the periurethral glandular tissue and the transition zone, whereas the outer prostate consists of the central and peripheral zones. The periurethral glands compose less than 1 % of the glandular prostate. The transition zone constitutes only about 5 % of the glandular prostate in young men. The central zone forms most of the glandular tissue at the prostate base and makes up about 25 % of the total volume of glandular prostate tissue. The peripheral zone

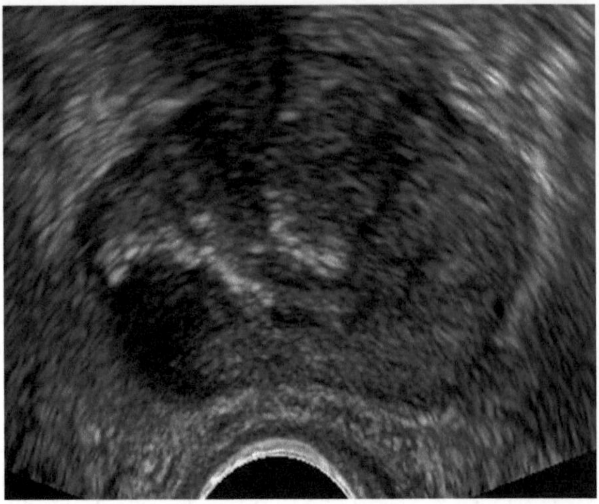

Fig. 3.1 TRUS showing right-sided hypoechoic tumour in peripheral gland

is the major glandular component of the prostate, composing 70 % of the prostate in healthy young men. The junction of the transition and peripheral zones is marked by a visible linear boundary, which is often referred to as the prostate pseudocapsule or surgical capsule. The anterior part of the prostate is composed mainly of a nonglandular fibromuscular stroma, which is continuous with detrusor fibres. Towards the apex of the gland, this fibromuscular tissue blends with striated muscle from the levator. The neurovascular bundles run craniocaudally along the posterolateral aspects of the prostate. In the radiology literature, terms *central gland* (which refers collectively to the periurethral, central, and transition zones) and *peripheral gland* (which includes only the peripheral zone) are used as well, especially when describing the sonographic appearance of prostate zonal anatomy.

3.1 Transrectal Ultrasound

The development of TRUS (Watanabe et al. 1968) marked the beginning of prostate imaging, and it remains a standard technique for the assessment of prostate cancer, primarily used for biopsy guidance and image-guided treatments including brachytherapy source placement (Aigner et al. 2010). Despite several decades of improvements in TRUS technology, TRUS is still unreliable in differentiating normal prostate gland from cancer tissue, and TRUS biopsy can miss 20–30 % of clinically significant cancers (Rifkin et al. 1990; Norberg et al. 1997; Beerlage et al. 2001). A normal TRUS scan cannot be relied upon therefore to exclude cancer. In addition, Gleason grades derived from biopsy specimens are frequently upgraded in after prostatectomy, and there is a substantial risk of false-negative biopsy results, despite multiple attempts in some cases. Even with systematic sampling, under-diagnosis of the extent of prostate cancer can still occur with

TRUS biopsy (Novara et al. 2010; Ashley et al. 2008). The anterior gland is often under-sampled and although cancers are less frequent here, the consequences of not recognising the true location of all cancers in the gland are obvious in terms of potential under-dosage.

Cancer, depending on its size, grade, and location, usually appears hypoechoic relative to the normal peripheral zone of the prostate on TRUS scanning (Fig. 3.1); only about 1 % of prostate cancers appear hyperechoic. Some palpable cancers are not visible at TRUS, and some visible cancers are not palpable on DRE. Moreover, there are other causes of hypoechoic foci such as prostatitis, benign prostatic hyperplasia (BPH), and atrophy, all of which lower the specificity of TRUS. Many early-stage prostate tumours cannot be identified on TRUS at the time of brachytherapy, and this only highlights the importance of accurate whole-gland dosimetry, particularly the anterior gland. The advent of PSA screening and better awareness of prostate cancer as a health issue for men have generated a shift towards smaller, early-stage cancers that are just not visible even with state-of-the-art TRUS. Many hypoechoic areas do not prove to be malignant on biopsy; therefore, TRUS alone, without biopsy, has limited value in the detection of cancer. Earlier studies performed in the 1980s, when cancers tended to be larger (stage T3) and more easily palpated, reported sensitivity in the range of 80 % for detecting T3 disease on TRUS. Accuracy improved when TRUS findings were interpreted in conjunction with DRE findings and PSA levels to estimate the likelihood of extraprostatic T3 extension. Today, however, prostate cancers present when they are smaller and local extension is uncommon, particularly in patients who are candidates for brachytherapy. Clinical nomograms are useful but provide no information about the presence or location of the cancer or the site of any T3 spread (Figs. 3.2 and 3.3) which may be very relevant to the choice of brachytherapy (LDR or HDR) and the potential for boost treatment.

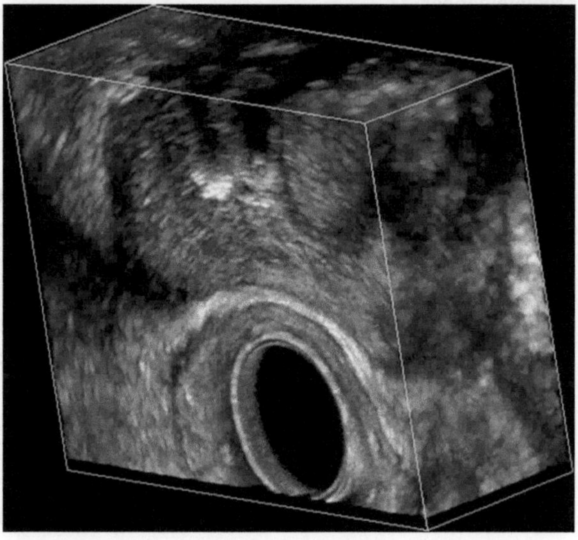

Fig. 3.2 3D TRUS with T3a tumour on left side

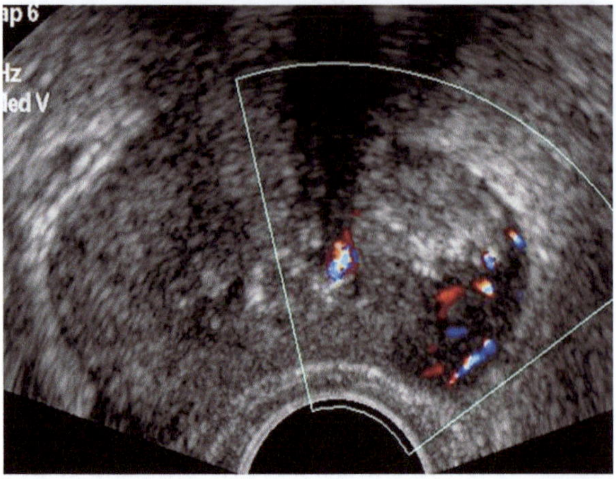

Fig. 3.3 Hypervascular left peripheral gland tumour on colour Doppler TRUS

TRUS has many advantages, including portability, ease of use, lack of ionising radiation, and low cost, and it has been almost exclusively utilised as the imaging technique for brachytherapy planning and treatment delivery. TRUS is also rarely useful in demonstrating extracapsular extension of prostate cancer and seminal vesicle invasion, unless gross extension is present. These are important issues for HDR brachytherapy and correlation with other imaging such as MRI may be the only way to ensure the correct choice of brachytherapy technique.

Various TRUS enhancements, such as colour Doppler, colour power Doppler, and contrast-enhancement (microbubble), have been developed and newer techniques such as TRUS elastography and HistoScanning™ are becoming more available. These TRUS imaging techniques have had varying degrees of clinical uptake, reflecting lack of clarity as to their exact value and contribution to the imaging of prostate cancer.

Colour Doppler TRUS (Fig. 3.3) exploits changes in the microvascular environment of prostate cancer although with only small improvements in sensitivity and specificity. The addition of colour Doppler and/or power Doppler can increase the rate of tumour visualisation by detecting regions of hypervascularity, but many small tumour foci do not have sufficient angiogenesis to result in noticeable changes on colour or power Doppler modes (Ashley et al. 2008; Cornud et al. 2000; Halpern et al. 2002). Contrast-enhanced TRUS with microbubbles has been found to provide higher sensitivity for the detection of cancer foci than standard TRUS and has increased the detection rate of clinically significant prostate cancer in several studies (Kuligowska et al. 2001). Furthermore, contrast-enhanced TRUS with microbubbles has been shown to detect prostate cancer in patients with previous negative TRUS biopsies and persistently rising serum PSA values. The imaging appearances on contrast-enhanced TRUS can, however, be subtle, short-lived, and overlap with those seen in patients with prostatitis. Additionally, contrast-enhanced TRUS with microbubbles (Halpern et al. 2005) for prostate cancer imaging has not been extensively tested or validated, and because of the large (approx. 5–10 μm)

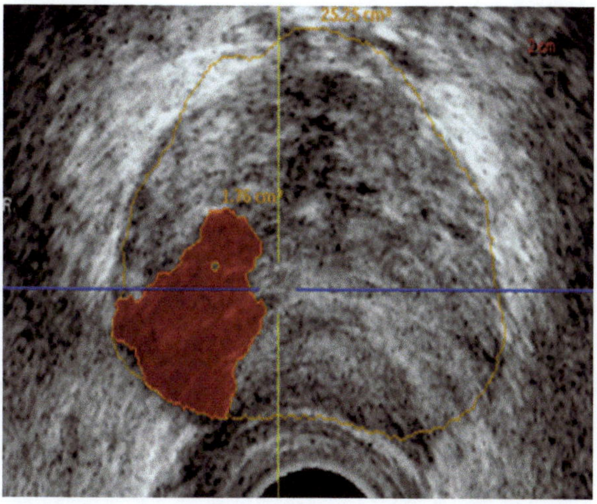

Fig. 3.4 HistoScan™ showing abnormal area of tumour in right lobe of gland (Courtesy of Advanced Medical Diagnostics. HistoScanning™ and its name derivations are registered trademarks of Advanced Medical Diagnostics SA/NV)

size of microbubbles, often only the vessels themselves are seen and, unlike contrast-enhanced MRI, the leakage into the tumour tissue cannot be well visualised. Routine use of these TRUS techniques cannot be advocated for prostate brachytherapy patient selection or treatment guidance at the present time.

Malignant prostate tissue can be firm to palpation, with limited elasticity and compressibility. The clinical application of sonoelastography in the diagnosis of prostate carcinoma is based on the fact that prostate tumour tissue has a greater stiffness than surrounding normal prostate tissue. TRUS elastography of the prostate using TRUS is based on this lack of compressibility in malignant tissue as a targeting strategy. Elastography is among a number of new technologies under development for improvement in prostate cancer detection. Tissue elasticity (Tomoaki et al. 2009; Eggert et al. 2008, 2010; Salomon et al. 2009; Trabulsi et al. 2010) has been developed as a qualitative biomarker for prostate cancer, and sonoelastography is an emerging imaging technique for providing qualitative as well as quantitative measurements of the stiffness of prostate tissue. Recent studies show significant improvements using the latest generation of TRUS elastography scanners. Although elastography is a promising new development in TRUS imaging, prospective studies are needed to define its applications. It is however not a perfect technique and is particularly sensitive to operator-induced vibration and other artefacts.

HistoScanning™ is a new ultrasound application that utilises advanced tissue characterisation algorithms to visualise the position and extent of abnormal prostate tissue, suspected of being malignant (Fig. 3.4). HistoScanning technology uses radiofrequency (RF) ultrasound data directly from the transducer that is capable of acquiring volume RF (native) data in a standard way. This contains much more information than the grey-level data displayed on the ultrasound monitor and is not affected by any of the scanner user settings (Simmons et al. 2012; Braeckmann et al. 2008a, b; Spethmann et al. 2010). There is potential to use these newer TRUS techniques to aid radiotherapy treatment planning and assist delineation of future brachytherapy intraprostatic boost treatments. They may also hold potential for focal brachytherapy.

In addition to whatever contribution is possible in terms of detecting, biopsying, and staging the prostate cancer, there are some specific details available from the TRUS study that are applicable to prostate brachytherapy. Prostate volume, generally not of major importance to the surgeon prior to radical surgery or to the clinical oncologist planning external beam radiotherapy, is an essential information in assessing patients for possible brachytherapy treatment. Patients with very large glands, generally greater than 60 cc, may be technically impossible to implant adequately because of pubic arch interference. Large median lobes may also be an issue for brachytherapy and can be assessed with TRUS. The shape of the pubic arch can be assessed as part of the TRUS evaluation of potential brachytherapy patients. Patients who have had a transurethral prostate resection (TURP) may have a higher risk of incontinence after brachytherapy because of microvascular damage to the urethral blood supply and increased radiation dose to the central part of the prostate. TRUS can be used to measure the residual central gland volume following TURP in order to help decide if brachytherapy is technically feasible or even advisable. Whilst peripheral source placement can reduce the dose to the urethra itself, a large cavity may preclude adequate source placement in the prostate, and it might not be possible to achieve the requisite dose to the whole gland with brachytherapy alone. The urethra is generally assumed to be a midline structure for the purpose of radiotherapy planning: an asymmetric urethra, as may occur with benign prostatic hyperplasia or regrowth of central gland tissue into a TURP cavity, can distort the urethra and may even preclude a safe prostate implant. The course and shape of the urethra are important details that the TRUS operator needs to record and communicate to the radiation physicists who are planning the brachytherapy treatment radiotherapy in order to avoid inadvertent trauma to this structure.

3.2 Magnetic Resonance Imaging

In the mid-1980s, the first prostate magnetic resonance imaging (MRI) examinations were performed. Since then, MRI has evolved from a promising new technique into a mature prostate imaging technique. MRI can now provide functional as well as anatomical information about the prostate gland and prostate cancer tissue. Anatomical T2-weighted MRI scans should ideally be complemented by functional MRI techniques such as dynamic contrast-enhanced MRI (DCE-MRI), diffusion-weighted imaging (DW-MRI), and proton magnetic spectroscopic imaging (MRS). All of these MRI sequences are increasingly being integrated in a single multiparametric MRI (mpMRI) examination which is becoming the standard imaging approach in prostate cancer.

T2-weighted sequences are the basis for prostate MRI. These have high spatial resolution and, thus, can differentiate the normal intermediate- to high-signal-intensity peripheral zone from the low-signal-intensity central and transition zones especially in younger men (Hricak et al. 1987). With ageing, BPH expands the transition

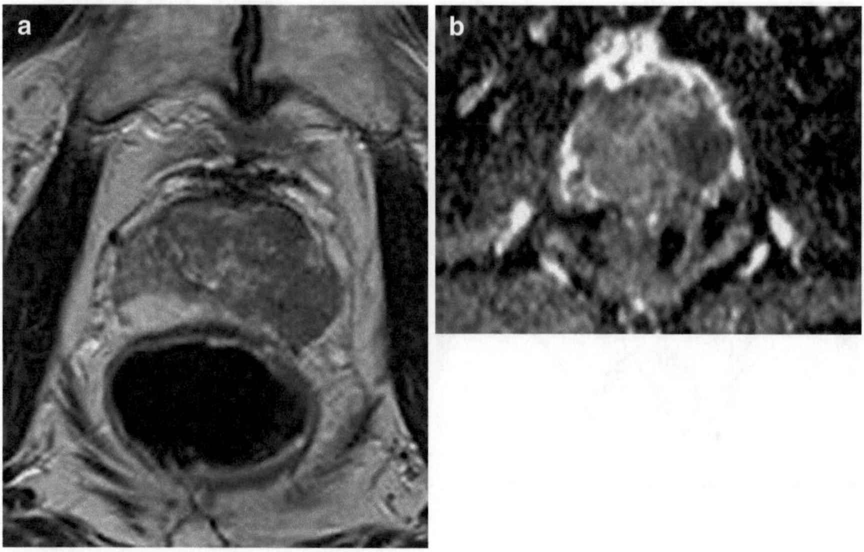

Fig. 3.5 Left sided carcinoma on T2 MRI (**a**) and on DWI-MRI (**b**)

zone and alters the size and signal intensity of this prostate zone on MRI imaging. Because of this transition zone expansion, the remainder of the compressed central zone is often difficult to evaluate on MRI, and small cancers in this area of the gland may be very difficult to indentify. On T2-MRI, prostate cancer can appear as an area of low signal intensity within the high signal intensity of the normal peripheral zone (Fig. 3.5). The degree of signal intensity reduction may vary with the Gleason score of the tumour: higher Gleason score components (4/5) have shown lower signal intensities than does lower Gleason score components 2/3 (Wang et al. 2008), but this is unreliable in all cases. The density and the growth pattern of the cancer are additional factors that may also influence T2-weighted signal intensity (Langer et al. 2008). A limitation of T2-weighted imaging is that focal areas of low signal intensity in the peripheral zone do not always represent cancer. Benign abnormalities such as hyperplasia, chronic prostatitis, focal prostate atrophy, scars, postirradiation fibrosis, hormonal treatment effects, and postbiopsy haemorrhage may mimic tumour tissue. Low-signal-intensity lesions in the peripheral zone that are wedge shaped and a diffuse area of altered signal without mass suggest a benign aetiology (Cruz et al. 2002). Haemorrhage following TRUS biopsy also interferes with signal interpretation on MRI. Because of the anticoagulant effect of abundant citrate in normal tissue in the peripheral zone, blood products may persist 4–6 weeks or longer after prostate biopsy, leading to low signal intensity on T2-weighted images and high signal on T1-weighted images. Ideally, MRI should be avoided for 6–8 weeks after prostate biopsy to minimise artefacts due to postbiopsy haemorrhage (Qayyum et al. 2004).

Due to the presence of BPH, cancers in the central and transition zones are more difficult to recognise on MRI scanning (Fig. 3.6). BPH may have signal intensity

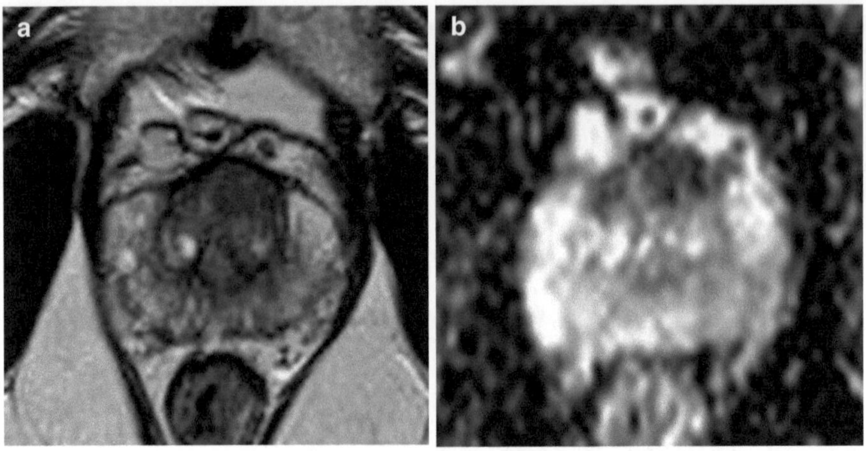

Fig. 3.6 Left Transition Zone tumour on (**a**) T2-MRI and (**b**) DWI-MRI

Fig. 3.7 Left-sided T3a tumour extension on T2-MRI

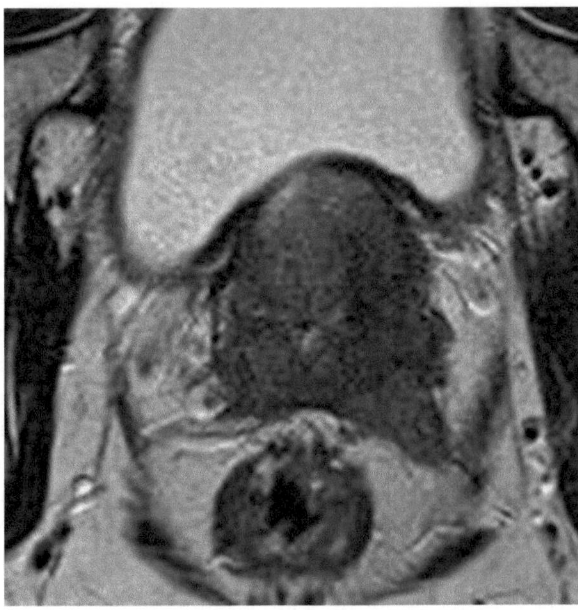

similar to that of prostate cancer on T2-weighted images. However, it has been reported that features such as homogeneously low T2-weighted signal intensity, ill-defined irregular edges, invasion into the urethra, or the anterior fibromuscular stroma and lenticular shape are useful signs for discriminating tumours in the transition zone (Akin et al. 2006). Furthermore, high inter- and intraobserver variability may lead to under- or overestimation of cancer stage and potentially confound the interpretation of studies based on T2-weighted MRI.

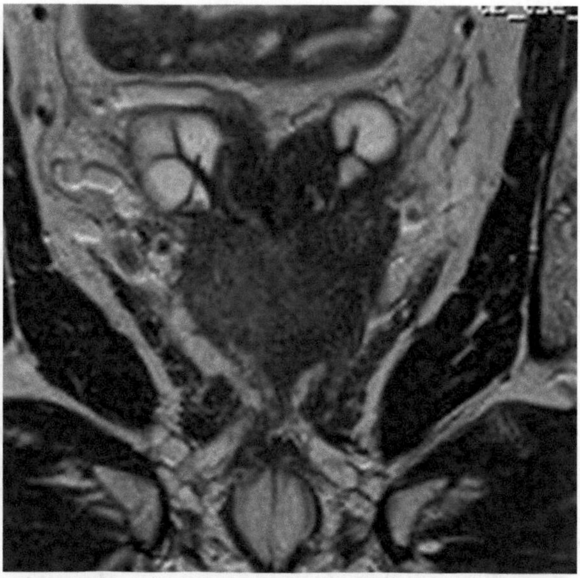

Fig. 3.8 Invasion of seminal vesicles (T3b) on T2-MRI

The main application of T2-weighted MRI is in the local staging of prostate cancer. The most widely used criteria for extracapsular spread are asymmetry of the neurovascular bundles, obliteration of the rectoprostatic angle, irregular bulging of the prostatic margin, and low-signal tumour in the periprostatic fat (Fig. 3.7). Low-signal tumour in the normally high-signal seminal vesicles is indicative of (T3b) on T2-MRI (Fig. 3.8). The issue of whether or not to use an endorectal coil for the MRI examination continues to be debated, and there is no clear consensus. Most centres, however, do not routinely use an endorectal coil and reserve this for dedicated MRS or further identification of the neurovascular structures prior to surgery. In recent years, technologic developments with higher field strengths improved pelvic phased-array coils, and multiparametric MRI techniques have improved staging accuracy considerably. However, accuracy results vary between different studies (Lee et al. 2010).

3.3 Dynamic Contrast-Enhanced MRI

Dynamic contrast-enhanced MRI provides information about tumour angiogenesis. Tumours are associated with the production and release of angiogenic factors such as vascular permeability factor or vascular endothelial growth factor in response to local areas of hypoxia or lack of nutrients (Bonekamp and Macura 2008). This increases the number of small vessels in the tumour tissue, and these vessels also have greater permeability than do normal vessels. Furthermore, because the volume of interstitial space is greater than normal in tumour tissue, there is a larger contrast gradient between the plasma and the interstitial tissue. This changes the overall enhancement pattern of the tumour tissue (Alonzi

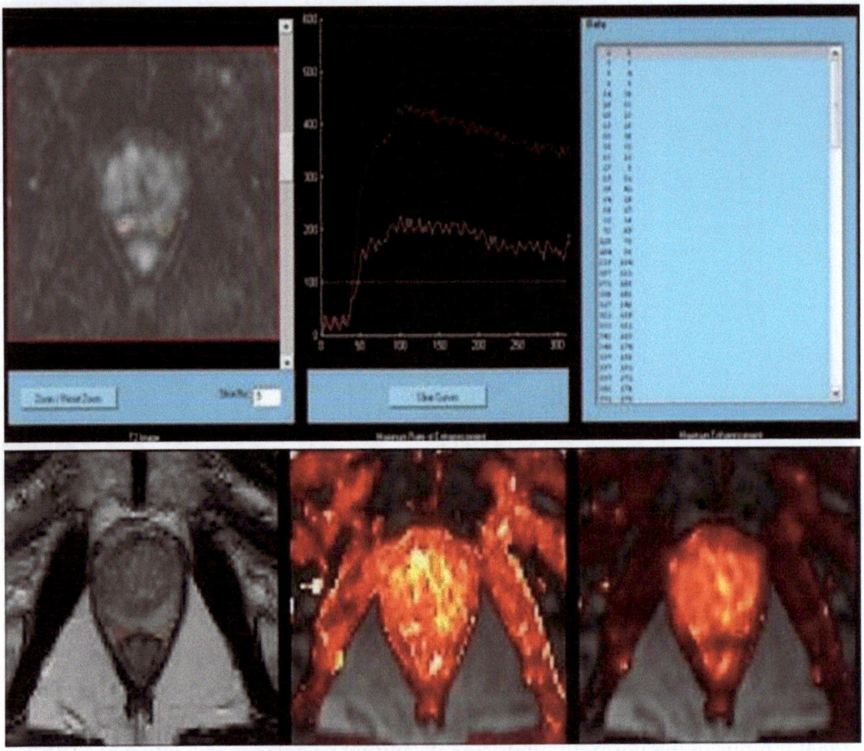

Fig. 3.9 DCE-MRI perfusion map

et al. 2007). It has been shown that the various contrast-enhancement parameters such as mean transit time, blood flow, permeability surface area, and interstitial volume are significantly greater in tumour tissue than in normal tissue (Padhani et al. 2005). DCE-MRI for prostate cancer is based on the measurable differences in these parameters, and it exploits the dynamic uptake and rapid washout of a gadolinium chelate contrast agent (Fig. 3.9). The prostate is a highly vascularised organ and therefore simple comparison of pre- and post-gadolinium images is usually insufficient to discriminate areas of tumour within the gland. Dynamic contrast-enhanced MR imaging consists of a series of fast T1-weighted sequences acquired before and after rapid injection of low-molecular-weight gadolinium chelates. It is an MRI technique that yields high temporal resolution with low spatial resolution, thereby enabling depiction of the early enhancement phase of a prostate tumour.

Interpretation of signal intensity changes on T1-weighted DCE-MRI to assess contrast agent uptake can be performed qualitatively, semiquantitatively, or quantitatively (Fig. 3.9). With the application of multicompartmental modelling, time-intensity curves of contrast enhancement can be plotted, pharmacokinetic parameters

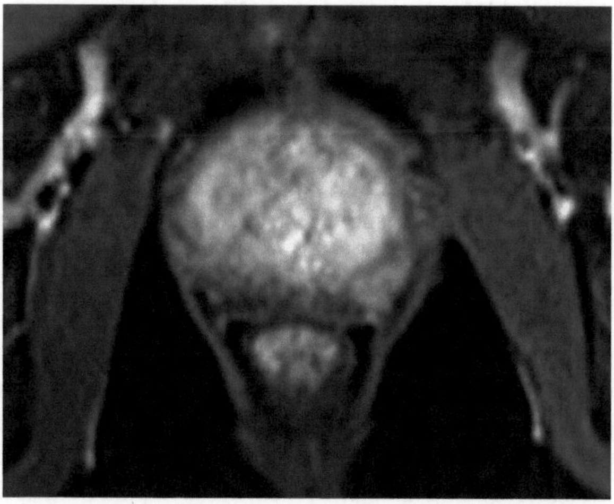

Fig. 3.10 DCE-MRI showing area of abnormal tumoural enhancement in left peripheral gland

can be calculated, and parametric images can be coregistered with T2-weighted images. However, this technique requires specialist software for quantitative measurement of the various functional parameters. DCE-MRI does have some limitations: suboptimal depiction of transitional zone tumours in patients with hypervascular benign prostatic hyperplasia. In addition, there is as yet no consensus with regard to the best acquisition protocol and the optimal perfusion parameter for differentiating cancer from normal prostate tissue. Contrast washout is a semiquantitative parameter that analyses the curve pattern after the first peak of enhancement. Semiquantitative parameters have the advantages of being fast, relatively simple to calculate, and of being available on most modern MRI scanners. They may, however, be influenced by individual MR scanner set-up. Prostate tumours tend to enhance earlier, faster, to a greater magnitude and show more rapid and earlier contrast washout compared with healthy prostate tissue. This characteristic makes DCE-MRI a sensitive technique for prostate cancer localisation. Calculated quantitative parameters can be displayed as colour-overlay maps on anatomic T2-weighted MR images in order to correlate DCE-MRI images with prostate anatomy. One of the limitations of DCE-MRI is the discrimination of cancer from prostatitis in the peripheral gland and from highly vascularised BPH nodules in the transition zone. Other problems are a limited use of standardised software for calibration and analysis, the shortage of uniform commercially available tools for pharmacokinetic analysis, and the lack of consensus in acquisition protocols. DCE-MRI is an accurate functional MRI technique that has a role in the assessment of early prostate cancer. In a multiparametric MR imaging examination, the high sensitivity of DCE-MRI may be used for initial evaluation of potential tumour locations (Fig. 3.10). Other functional MRI techniques may subsequently be added to increase specificity for prostate cancer localisation and provide information for potential brachytherapy targeted therapy.

3.4 Diffusion-Weighted Imaging

DWI is a fast, simple, and readily available MRI technique for imaging prostate cancer. Diffusion is the process of thermally induced random molecular displacement, also known as Brownian motion. The diffusion properties of a tissue are related to the amount of interstitial free water and tissue permeability. In general, malignant tissue tends to have more restricted diffusion than does normal tissue because of higher cell densities and profusion of intra- and intercellular membranes in cancer (Jacobs et al. 2008; Franiel et al. 2008). In prostate cancer, normal glandular architecture is disrupted and replaced by a disordered mesh of cancer cells and fibrotic stroma. These tissue changes inhibit the movement of water macromolecules, causing restriction of diffusion and reduction of the measured apparent diffusion coefficients (ADC) values in the cancer tissue. DWI uses proton diffusion properties in water to produce image contrast. Image sequences are acquired by applying motion-encoding gradients, which cause phase shifts in mobile protons, depending on the direction and quantity of their movement. The *b* value and the apparent diffusion coefficient (ADC) are components in this equation. Whilst the *b* value expresses the amount of diffusion weighting, the ADC value reflects the movement of the water molecules within the interpulse time. Because ADC quantifies the flow as well as the distance a water molecule has moved, it represents both capillary perfusion and diffusion characteristics. In particular, within the transition zone, high *b* values may help improve differentiation of BPH from prostate cancer.

Healthy prostate tissue in the peripheral zone, which is rich in tubular structures, allows extensive diffusion of water molecules within these gland tubules. Consequently, ADC values in healthy peripheral zone tissues are usually high. Prostate cancer tissue destroys the normal glandular structure of the prostate and replaces ducts. It also has a higher cellular density than does healthy prostate peripheral zone tissue. On ADC maps, therefore, prostate cancer often shows lower ADC values in comparison with the surrounding healthy peripheral zone prostate tissue (Fig. 3.11). Because the acquired ADC value depends on the specific pulse sequence parameters (especially the *b* values), the specific MRI system used, and the magnetic field strength, the ADCs of healthy and cancerous tissue have varied among reported studies. Furthermore, there is an overlap in the ADCs of healthy tissue and those of prostate cancer, within and between subjects, which limits the determination of any single threshold ADC for malignancy (Zelhof et al. 2009; Katahira et al. 2011).

The reported differences in detection accuracies with DWI might be explained by the different tissue composition in different anatomic zones of the prostate. It may be that a higher degree of proton movement in the extracellular water compartment occurs as opposed to that in intracellular water, where movement is restricted by cell membranes or other intracellular structures. As a result of variation in glandular tissue within a healthy peripheral zone to muscular or fibrous tissue within the transition zone, the ratio of intracellular to extracellular water also differs. This variation might also explain the variability of ADC values in healthy prostate tissue that have been reported. DWI-MRI of the prostate has the limitation of low spatial resolution even at 3 T but does reflect cellular density, which makes the technique potentially useful to determine tumour aggressiveness. DWI, being a technique for measuring proton

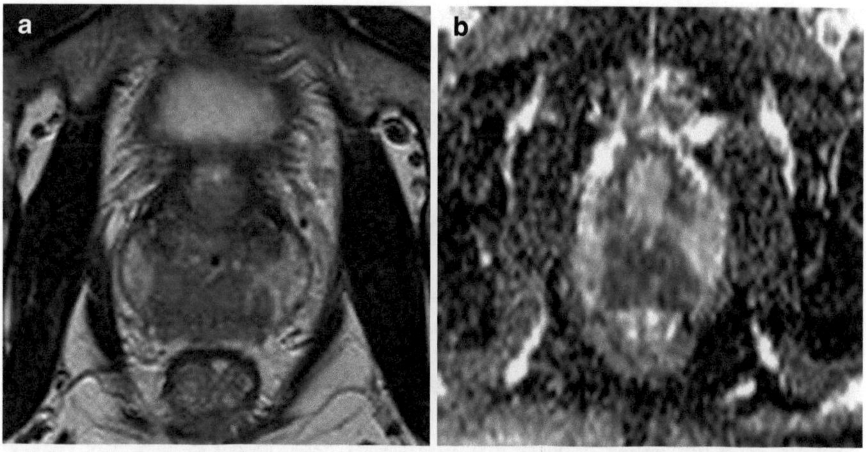

Fig. 3.11 Peripheral gland tumour on T2 MRI (**a**) and on DWI-MRI (**b**)

motion, is very sensitive to movement artefacts. Typically, artefacts occur in areas with large variations in magnetic susceptibility, such as in tissue-air interfaces (air in the rectum or endorectal coil) or in chemical shift in areas with water-fat interfaces.

Correlation of DW imaging results and histopathologic findings as well as to prognostic histological prostate cancer markers such as hypoxia-inducible factors is another area for future research. Of all the functional MRI techniques, DWI is the most practical and simple in its use and should be a core element of any multiparametric MR imaging examination. The combination of DWI and T2-weighted imaging significantly improves the accuracy of cancer detection beyond that achieved with T2-weighted imaging alone.

3.5 Proton MR Spectroscopy

MRS provides metabolic information about prostate tissue by analysing the relative concentration of various chemical compounds and metabolites in the prostate. In MRS, proton spectra are measured in two or three spatial dimensions and the specific resonance frequencies or chemical shifts are analysed for different metabolites present in the voxel of tissue sampled (Fig. 3.12).

In clinical prostate MRS, choline-to-citrate ratios are generally used as a metabolic biomarker for prostate cancer. Different anatomic zones of the healthy prostate have different spectral peaks for citrate, creatine, and choline, which are reflected in different choline-to-citrate ratios. High citrate concentrations are found in the glandular tissues of the prostate such as the peripheral zone, which contains epithelial cells and secretory ducts. Therefore, citrate concentrations are highest in the peripheral zone and lower in the central zone. In the transition zone, the citrate concentrations can vary depending on the degree of BPH present which will influence the extent of glandular and stromal proliferation present

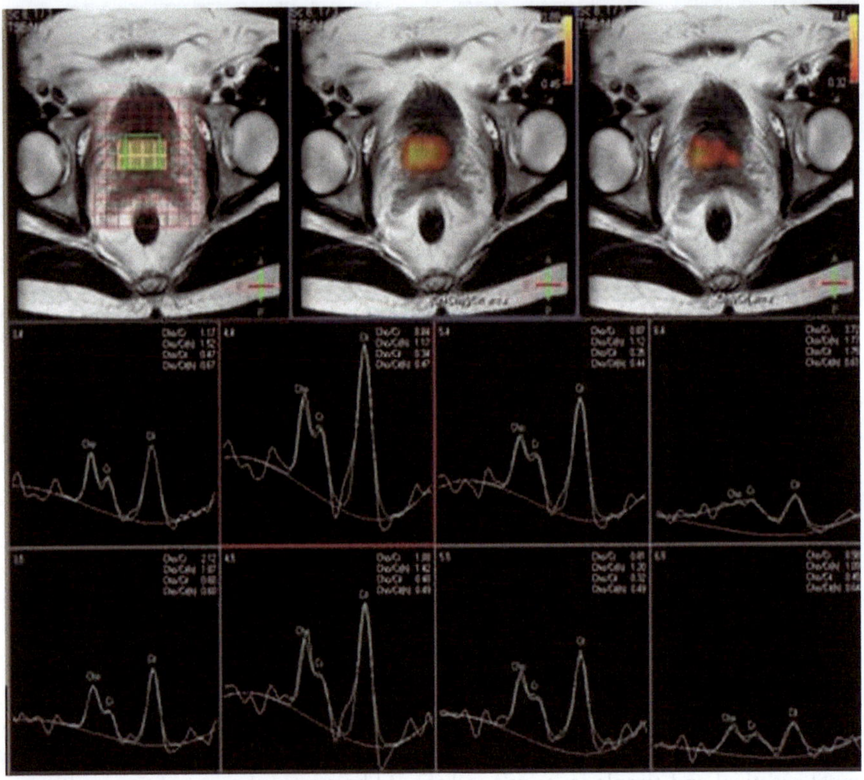

Fig. 3.12 MRS spectral analysis

(Mueller-Lisse and Scherr 2007). Compared with healthy peripheral gland tissue or BPH tissue, citrate signals are reduced and choline signals are often increased in prostate cancer tissue. Citrate is produced in epithelial cells as an intermediate product in the Krebs cycle and then accumulated in the luminal space of the prostate. The lower citrate peak in prostate cancer tissue may be caused by altered metabolism, as well as by a reduction of luminal space, which commonly occurs in prostate cancer. Choline compounds are involved in the biosynthesis and degradation of phospholipids, which are required for the build-up and maintenance of cell membranes. An increased cell turnover in prostate cancer results in an increased concentration of free choline-containing molecules within the cytosol and the prostate interstitial tissue.

In MRS the whole prostate is subdivided into a three-dimensional grid (Fig. 3.12) of multiple voxels. With the introduction of the endorectal coil for prostate MR examinations, it became possible to obtain in vivo MRS imaging spectra of small voxels in the prostate (less than 1 cm^3) with sufficient signal-to-noise ratios (Heerschap et al. 1997). The steps involved in MRS imaging, including the accurate placement of lipid saturation bands, spectral data acquisition, postprocessing, and interpretation, are labour intensive, and there is a significant learning curve with the implementation of a clinical MRS service. Since the prostate is relatively small and embedded in adipose tissue, much effort has been put into suppressing spectral contamination, not only of the high water signal but also of strong lipid signals. Three-dimensional MRS imaging sequences are currently preferred over two-dimensional sequences because of the possibility of complete coverage of the entire prostate gland (Scheenen et al. 2004). Three-dimensional acquisitions can be performed in approximately 10–15 min with a resolution as low as 0.4 cm^3 and with sufficient signal-to-noise ratio at 1.5 T.

The advantages of MRS are its generally accepted accuracy, its ability to depict possible cancer in the transitional zone, and its proven diagnostic performance. MRS has several limitations. The technique is hampered by a long acquisition time, possible variability in results dependent on postprocessing or shimming, and no direct visualisation of the periprostatic anatomy. Furthermore, a previous prostate biopsy may interfere with spectra making accurate interpretation of the metabolite ratios impossible. Considerable local magnetic field distortions may occur due to haemorrhage, which is why the examination should be performed with sufficient delay (6 weeks) from the time of biopsy. Spectral quality depends on magnetic field homogeneity, which must be optimised for each patient by coil shimming. Currently, the interpretation of MRS requires particular expertise and is time consuming. No consensus has been reached about the metabolite ratio that can definitively signify the presence of prostate cancer, and there may be individual variability in spectral analysis among patients. Automated measurement procedures, rapid display of examination results, and proper training of clinical users are still needed to fully implement MRS as a practical and widespread clinical tool for assessment of prostate cancer. The combined use of MRS and MRI has been shown to improve cancer detection and localisation in the peripheral zone and cancer volume measurement in the peripheral zone. Furthermore, on the basis of a strong correlation between the volume of prostate cancer and its extracapsular extension, investigators have shown that the combination of volumetric data from MRS and T2-weighted imaging may result in improved accuracy in determining extracapsular tumour extension.

In recent years, other merits of MRS have been noted (Dickinson et al. 2011). The results of several studies show that prostate biopsy directed with MRS may help increase the cancer detection rate in patients with an elevated prostate-specific antigen level and a previous negative TRUS biopsy. In addition, investigators have observed a trend towards increasing ratios of choline and creatine to citrate in association with higher Gleason grades, suggesting a potential role for MRS in the non-invasive estimation of prostate cancer aggressiveness. MRS may also be more useful than conventional MRI for detecting transitional zone cancer. However, the cancer metabolite ratio in the transitional zone varies broadly, and thus, there may be overlap in metabolite ratios between cancerous and benign tissues in the transitional zone.

3.6 Multiparametric MRI (mpMRI) for Brachytherapy

The various MRI techniques described all have their limitations and the optimum use of MRI in prostate cancer is likely to be a combined mpMRI prostate examination to maximise the overall accuracy and value of each of these techniques. An mpMRI prostate examination consists of T1- and T2-weighted imaging combined with one or more functional MRI techniques. Computer software programmes can register and fuse the information from anatomical and functional scans. Furthermore, the education, experience, and dedication of radiologists are essential for correct interpretation of findings from multiparametric MRI of the prostate. There is growing interest in integrating mpMRI into the prostate cancer diagnostic pathway ground (Sciarra et al. 2011).

Even before treatment, accurate definition of prostate cancer location helps improve cancer detection in targeting prostate biopsies with MRI guidance. Sampling the most abnormal area of the prostate on mpMRI is most likely to yield a positive diagnosis of cancer. Prostate cancer localisation is vital information to the surgeon in planning the type of radical prostatectomy, e.g. knowledge of the proximity or otherwise of the tumour to the neurovascular structures can help stratify the patient into nerve- or non-nerve-sparing surgery. The clinical oncologist is also likely to benefit from accurate tumour location in planning radiotherapy to the prostate, e.g. external beam and/or brachytherapy. Accurate definition of the prostate cancer location helps improve prostate cancer staging and may inform the decision as to whether LDR or HDR brachytherapy is most appropriate for an individual patient. Improved evaluation of prostate cancer location helps improve and support focused radiation therapy planning of the dominant prostatic lesion and improves guidance of minimally invasive focal therapies. Localisation accuracy with DCE-MRI increased to 72–91 %, as compared with 69–72 % for anatomic T2-MRI only (Fütterer et al. 2006). In other prospective studies (Kim et al. 2007, 2009), the addition of DWI to T2-MRI improved prostate cancer localisation performance. MRS has shown higher specificity (68–99 %) and lower sensitivity (25–80 %) for prostate cancer localisation, when compared with anatomic T2-MRI in prospective studies with prostatectomy specimens as reference standard. Lack of consensus on and multicenter validation of agreed thresholds for increased metabolic ratios as a criterion for malignancy as well as of a clear definition of tumour focus size are still unresolved issues for MRS in prostate cancer. By improving localisation, mpMRI techniques may also contribute to improved local staging accuracy. mpMRI techniques may also contribute in detection of transition zone prostate cancers. The combined use of DWI and DCE-MRI and T2-weighted MR imaging led to increased accuracy in detection of transition zone cancer. Of all the current clinical indications for mpMRI, localisation is perhaps the most relevant for brachytherapy. Accurate prostate cancer localisation results in more accurate prostate cancer staging and more choice of technique (Hoeks et al. 2011).

3.7 Determination of Prostate Cancer Aggressiveness: Implications for Brachytherapy Imaging and Patient Selection

Prostate cancer is graded according to the Gleason score, a combination of the two most prevalent Gleason grades (at prostatectomy), or the most prevalent and the highest grade (at prostate biopsy), based on architectural characteristics of prostate

Fig. 3.13 Multifocal cancer on T2-MRI at 3 T

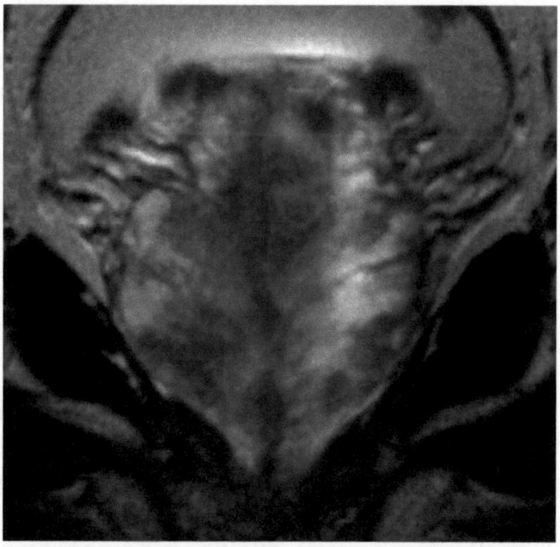

cancer tissue. Sampling error in biopsy specimens obtained at systematic TRUS occurs in approximately 64 % of procedures and results in a changed Gleason score at histopathologic evaluation of the prostatectomy specimen. This may result in incorrect evaluation of prostate cancer aggressiveness and subsequent inappropriate treatment including brachytherapy. MRI does offer some potential for evaluating the aggressiveness of a prostate cancer: on T2-weighted MRI signal intensity changes and detection rates for prostate cancer have been associated with its aggressiveness. At MRS, the choline-citrate ratios have been shown to be associated with Gleason score. Results for ADC as a possible marker of cancer aggressiveness are promising. DCE-MRI can yield parameters reflecting the grade of tumour. Suggested minimum requirements for an mpMRI study in the clinical workup of brachytherapy patients are T1- and T2-weighted MRI sequences in combination with DWI and possibly DCE-MRI. DCE-MRI can be used for optimal identification of potential prostate cancer foci. MRS may be added to improve specificity for different clinical indications including possible brachytherapy boosts or focal brachytherapy in the future.

MRI has considerable potential to improve the prostate cancer diagnostic pathway. Until fairly recently, the accuracy of morphologic MRI to detect, localise, and characterise prostate cancers was limited, and as a result, MRI has not been routinely incorporated into clinical care. However, evidence is accumulating that suggests an improved performance of MRI, provided that modern sequences are used and their outputs combined in a multiparametric MRI (mpMRI). MRI technology continues to evolve. Recent advances include increased magnetic field strengths, from 1.5 to 3.0 T and higher (Fig. 3.13), and the further development of multichannel receiver coils. Limitations associated with current functional MRI techniques will be largely resolved by these and other technical advances in the future. More clinical studies are needed to correlate pathologic findings with features observed at functional imaging. Standardised protocols and diagnostic criteria for functional imaging of the prostate should be established and implemented in clinical practice. The development of standardised reporting methods also should be encouraged, to facilitate the combination of anatomic and functional findings into an integral report. Finally, new functional and molecular imaging techniques, such as MR elastography and optical imaging, are on the verge of being used for clinical examinations, and no doubt eventually will be added to the list of functional imaging techniques available for evaluation of the prostate.

3.8 CT Scanning

Despite major developments in CT scanner technology over the past two decades, CT has virtually no role in prostate cancer detection or primary tumour staging (Hricak et al. 2007). CT has limited soft tissue contrast resolution, which prevents the accurate identification of the prostate margins from adjacent soft tissue structures. Prostate contour definition is poor even with modern multislice scanners particularly at the prostate apex and base. Zonal anatomy is difficult to discern on unenhanced scans but may be seen early after a bolus of iodinated contrast medium. The central gland generally enhances more than the peripheral gland. Nonetheless, tumours are very difficult to identify unless they are of large volume. CT may have a role in the assessment of locally advanced tumours, but these are generally not candidates for brachytherapy. CT scans and MRI can depict lymph node enlargement and have similar accuracy for the evaluation of lymph node metastases, generally based on size and morphology alone.

Arterial-phase multislice CT scanning may have a role in demonstrating intraprostatic pathology (Korporaal et al. 2010a, b). There is current interest in exploring the value of dynamic contrast-enhanced CT (DCE-CT) for detection of small prostate cancers, and this may have a future role for brachytherapy planning. DCE imaging of the prostate is primarily performed by using MRI. However, quantification of DCE- MRI data is difficult because of the complex relation between the signal intensity and the concentration of the contrast agent in tissue. In contrast, the quantification of DCE-CT data is more straightforward because of the linear relationship between Hounsfield units and contrast agent concentration. Although in the past only a limited scan volume in the craniocaudal direction was available, with modern multislice CT scanners, sufficiently large volumes can be imaged to make making DCE-CT of the entire prostate feasible. This is still work in progress and is not advocated for clinical practice at the present time.

CT is the current standard imaging technique for post-implant dosimetry recommended by the ESTRO-EAU-EORTC guidelines for post-implant evaluation.

3.9 Technetium Bone Scan

Screening potential brachytherapy patients for possible bony metastases may be required as part of the workup for brachytherapy. Not all patients with prostate cancer are at the same risk of bone metastases at diagnosis. Moreover, the effect of stage migration in recent years has led to a significant reduction in the prevalence of metastatic disease among patients with newly diagnosed prostate cancer. Despite this trend towards downstaging, a small number of patients will be found to have bone metastases at first diagnosis. Bone metastases follow the distribution of adult red bone marrow (skull, thorax, pelvis, spine, proximal long bones), followed by involvement of adjacent cortical bone (Fig. 3.14). Although these lesions appear radiographically sclerotic, they involve both osteoblastic and osteoclastic activity.

Technetium-99m (^{99m}Tc) bone scintiscanning is widely accepted as the most cost-effective imaging for the detection of bone metastases. ^{99m}Tc bone scans identify metastatic bone deposits by the increased osteoblastic activity they induce causing sclerotic bone lesions on plain radiographs and CT scanning. ^{99m}Tc bone scan findings are relatively nonspecific when solitary lesions are present, and such patients may require additional evaluation, e.g. MRI scanning, before a final treatment decision can be made. Bone scans furthermore have inherent poor spatial and contrast resolution, and in many patients, further imaging is also required to characterise such equivocal lesions. Despite the superior sensitivity of MRI compared with bone scintiscanning, bone scanning continues to be used as the initial screening investigation because of its relatively low cost, wide availability, and value in imaging the entire skeleton. Sensitivities of ^{99m}Tc bone scans are reportedly 62–89 %. In the evaluation of 1,403 patients with prostate cancer, bone scans were 28 % more sensitive than conventional radiographs in detecting metastatic bone lesions (Hricak et al. 2007).

Evidence-based clinical guidelines for the use of ^{99m}Tc bone scans in assessing the risk of distant spread of prostate cancer have been available since 1993, when it was proposed that routine bone scans should not be used for patients with PSA below 10 ng/mL. The role of the bone scan has been modified in recent years with the widespread introduction of the measurement of PSA levels. If the PSA level is less than 10 ng/mL, the risk of a positive bone scan is less than 1 %. When the PSA level is 10–50 ng/mL, the incidence of a positive bone scan increases to about 10 %, and with PSA level above 50 ng/mL, it increases to about 50 %. Staging bone scans

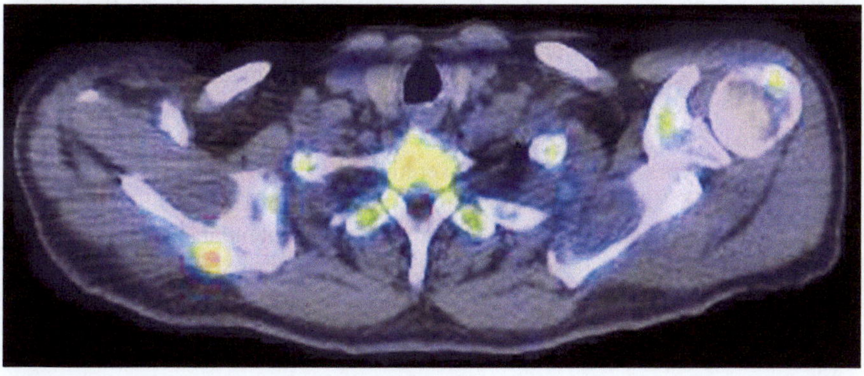

Fig. 3.14 Multiple bone metastases on ^{18}F-FCH PET scan

are best considered only for patients with a biopsy Gleason score >7 or with a PSA > 20 ng/ml and palpable disease (cT2/T3) prior to treatment.

Goerge……note this contrasts with your chapter suggesting bone scans only with PSA > 20 …… we need harmony??!!

3.10 SPECT Imaging

Single photon emission computed tomography (SPECT) studies of the skeleton have been shown to be more sensitive than planar images alone in the detection of bone metastases, but this technique is more expensive and less widely available than ^{99m} Tc scanning (Even-Sapir et al. 1993). Combined SPECT-CT, a recent development, adds anatomic information to SPECT, and its incremental value to SPECT is yet to be evaluated. It is not recommended for brachytherapy clinical practice.

3.11 Positron Emission Tomography (PET)

PET uses positron-emitting radioisotopes to identify tumour foci within tissue. Most clinical PET studies to date have been performed with the glucose analogue fluorine 18 (^{18}F) fluorodeoxyglucose (FDG) where the hydroxy function is replaced by the radioactive fluorine isotope. Glucose is an important cell nutrient playing an essential role in cellular energy metabolism. The accumulation of ^{18}F-FDG is based on enhanced glycolysis, which is associated with the growth rate and malignant potential of the tumour. Cancer cells have increased metabolism and utilise the less-efficient glycolytic pathway, both of which lead to increased glucose uptake. ^{18}F-FDG can therefore be used to study changes in glucose uptake. The magnitude of the elevated FDG uptake and its accumulation in tumours are commonly expressed by the standardised uptake value (SUV) defined as the ratio of activity per unit mass in the lesion to the administered activity per unit mass in the patient. Current PET spatial resolution is approximately 5 mm, thus limiting its capability in detecting small lesions. Prostate cancer is often characterised by multiple foci that may be smaller than current PET spatial resolution.

^{18}F-FDG has been widely shown to be clinically useful in many tumour sites, but the results in patients with prostate cancer have been disappointing and inconsistent (Liu et al. 2001). Prostate cancers have low metabolic glucose activity relative to other cancer types, and the urinary excretion of ^{18}F-FDG may obscure detail in the prostate itself. ^{18}F-FDG is also taken up in nodules of BPH, leading to lack of specificity. More recently, studies have emphasised the potential advantages of PET performed with radiotracers such as ^{11}C acetate, ^{11}Choline, ^{18}F fluoroethylcholine, and some other experimental tracers. ^{11}Choline is currently providing the most promising results in the assessment of prostate tumours. Two possible mechanisms have been suggested to explain the increased choline uptake in prostate cancer cells (Beheshti et al. 2010). Firstly, there is increased cell proliferation in tumours. Choline is a precursor for the biosynthesis of phosphatidylcholine and other phospholipids that are major components of the cell membrane. Choline uptake is a marker of cell proliferation in patients with prostate cancer, as malignancies are commonly characterised by increased proliferative activity. The second explanation is upregulation of choline kinase in cancer cells. Overexpression of choline kinase has been found in cancer cell lines, including human prostate cancer (Zheng et al. 2004). Elevated levels of choline and upregulated choline kinase activity have been detected in prostate cancer cells, and PET-labelled choline analogues can be used for targeted imaging of prostate cancer. ^{11}Choline is rapidly cleared from blood and accumulates in prostate tissue enabling imaging as early as 3–5 min after injection. ^{11}Choline has negligible urinary excretion. The short half-life of ^{11}Choline, however, is a major limitation for routine clinical application. Imaging prostate cancer with ^{11}Choline is possible in newly diagnosed patients but is not currently recommended mainly due to limited reported studies on sensitivity(71 %) and specificity (43 %) in localising tumour within the prostate gland (Picchio et al. 2010). It is clear that ^{11}C-choline PET has significant limitations for evaluation of the newly diagnosed prostate cancer. The degree tracer uptake by malignant tissue overlaps significantly with that of high-grade prostatic intraepithelial neoplasia, prostatitis, benign prostatic hypertrophy, and normal tissue. ^{11}C-choline PET is not routinely used for initial staging and is generally reserved for patients with biochemical recurrence including external beam radiation. For such restaging, ^{11}C-choline PET may be useful to screen for metastatic disease in patients who are potential candidates for salvage local treatment.

Newer radiotracers have been developed that label amino acids within malignant tissue. Acetate is a naturally occurring metabolite that is converted to acetyl-CoA, a substrate for the tricarboxylic acid cycle, and is incorporated into cholesterol and fatty acids. It is thought that acetate is involved in lipid synthesis and becomes incorporated into tumour cell membrane [50]. Because ^{11}C acetate (^{11}C AC) is excreted mainly via the pancreas and bowel, the pelvis can be imaged without urinary contamination. On the basis of these metabolic properties, ^{11}C AC PET may be able to detect and localise prostate tumours and hopefully monitor treatment response in patients with prostate cancer. The major limitation of ^{11}C AC PET is its short half-life (20 min), which requires that scanning be performed near a cyclotron; on the other hand, the short half-life of ^{11}C AC does provide the potential for multitracer PET imaging in a single session. A recently described choline analogue, ^{18}F-fluorocholine (^{18}F-FCH), has a longer half-life (approx. 110 min) but is excreted mainly in urine, therefore restricting its use in prostate cancer. Nonetheless, early imaging (before urinary excretion) with ^{18}F-FCH has successfully shown both localised and metastatic prostate cancer (Fig. 3.14). In radiotherapy dose escalation studies, ^{18}F-FCH has been used to delineate gross tumour volume and to generate planning target volumes within the prostate attempting to reduce irradiation dose to the bladder and the rectum (Weber et al. 2009). Methionine is an amino acid analogue that is rapidly cleared from the blood pool and primarily metabolised in the liver and pancreas with no significant urinary excretion; ^{11}C-methionine uptake reflects the increased protein synthesis associated with tumour cell proliferation and increased cell turnover. Clinical experience to date is limited with ^{11}C-methionine PET.

3.12 Radiolabelled Antibody Imaging

Radiolabelled monoclonal antibodies directed against specific cell surface antigens have been used for both imaging and therapy of several types of tumours. ^{111}In capromab pendetide (ProstaScint, Cytogen) is a radiolabelled murine monoclonal antibody to an intracellular component of the prostate-specific membrane antigen (PSMA). PSMA expression is normally low but is significantly increased in prostate cancer cells, and its expression correlates with tumour grade. However, because the target of capromab pendetide is intracellular, cell membranes must be disrupted (e.g. by apoptosis, hypoxia) for adequate binding to occur. This reduces the signal-to-noise ratio and also reduces the already poor spatial resolution. Anatomical localisation of ^{111}In capromab pendetide uptake has been difficult because of its nonspecific binding and high blood-pool activity. Imaging with capromab pendetide can be performed with both planar gamma cameras and cross-sectional SPECT cameras after administration of an infusion of ^{111}In-labelled antibody. Images are obtained several days after injection, which is possible because of its long half-life (2.8 days). Antibody imaging with ^{111}In capromab pendetide can allow detection of lymph node metastases, recurrence after prostatectomy, and occult metastatic disease. Although fusion with anatomic images and combined SPECT-CT improves specificity, the overall accuracy is still low. Because of its poor tissue penetration into bone, ^{111}In capromab pendetide imaging is suboptimal for detection of bone metastases; it is less sensitive than conventional bone scans. A number of other PSMA antibodies have been developed that target the external domain of the antigen and therefore may provide superior results for prostate cancer detection and staging (Schettino et al. 2004; Bander et al. 2005).

3.13 Imaging Focal Prostate Therapy: Potential Implications for Brachytherapy

The multifocal nature of prostate cancer has necessitated whole-gland therapy in the past. Since the widespread use of PSA screening, patients frequently present with less-advanced lower-grade disease and whole-gland treatment may represent overtreatment of such low-grade disease. The concept, therefore, of focal therapy for prostate cancer has recently been gaining in popularity owing to downward stage migration, improved biopsy and imaging techniques, and the prevalence of either

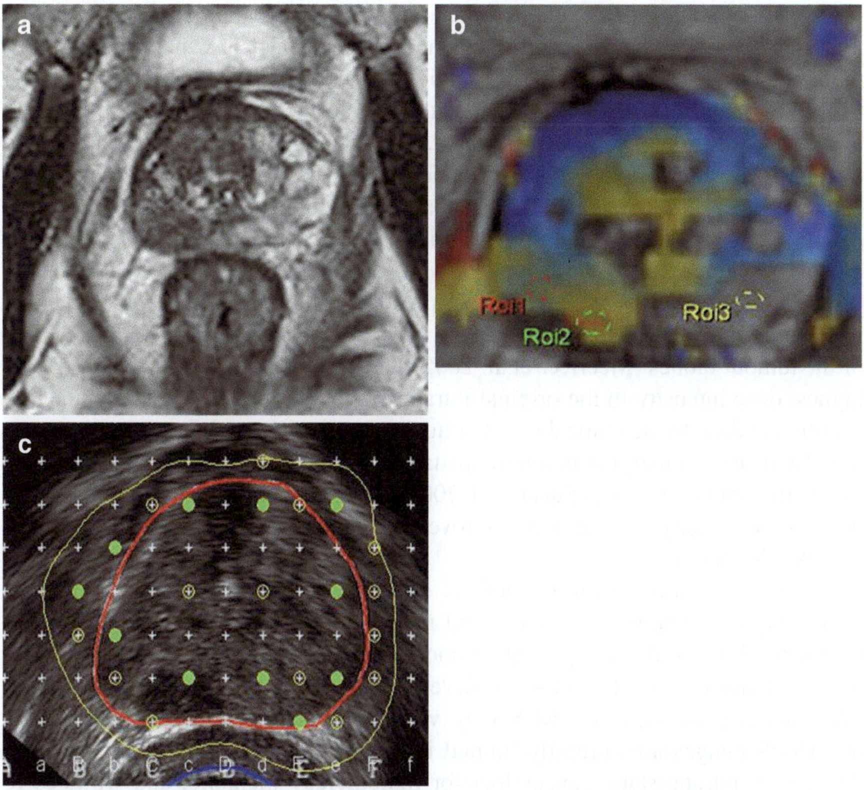

Fig. 3.15 (**a**) Tumour in right peripheral gland on T2-MRI. (**b**) Corresponding tumour on DCE-MRI. (**c**) Intraoperative TRUS during brachytherapy

unifocal cancer or a dominant cancer with secondary tumours of minimal malignant potential (Eggener et al. 2010). Multiple lines of evidence suggest that the index intraprostatic lesion, defined as the largest focus of cancer, is predominantly responsible for total tumour volume, risk of cancer recurrence, and Gleason grade. Many men with localised disease wish to avoid the morbidity of whole-gland therapy, and a range of minimally invasive ablative therapies, including cryotherapy, high-intensity focused ultrasound, vascular targeted photodynamic therapy, thermal laser ablation, and brachytherapy, could be used as alternatives for the treatment of low-risk prostate cancer (Fig. 3.15). These techniques are designed to destroy the tumour with less injury to periprostatic structures, aiming to achieve cancer control without the morbidity associated with whole-gland treatment.

Focal therapy for prostate cancer is in its infancy because of a number of unresolved problems: the main concern is the potential for suboptimum cancer control leading to poor outcomes compared with whole-gland therapies. With inappropriate patient selection and inaccurate mapping of multifocal disease, the potential for missed curative treatment exists. After an initial positive biopsy, it is still unclear how much further detailed imaging is needed or appropriate to fully and reproducibly assess the whole prostate gland. It is also unclear which ablative treatment will prove the best option for an individual patient. Brachytherapy focal boost treatments may be technically feasible with more accurate assessment of the overall burden and location of cancer in the gland. Focal therapy can encompass any degree of subtotal glandular ablation using a variety of devices or techniques derived from experience of whole-gland treatment. Focal brachytherapy could improve the therapeutic ratio by dose escalation to identified areas of aggressive cancer within the prostate; contemporary radiotherapy planning techniques are able to create the necessary dose distributions to do this. Intraprostatic failure usually occurs at the primary tumour location and is the result of intrinsic radiation resistance of a fraction of the tumour clones (Meerleer et al. 2005). It may be justified to focus the area of highest dose intensity to the original intraprostatic tumour site whilst maintaining a sufficient dose to the clinical target volume, planning target volume and organs at risk. Evidence is emerging that intraprostatic failures are mostly found at the location of the primary tumour (Pucar et al. 2007). This suggests that an additional boost dose to the primary tumour may improve clinical outcome and could be achieved with brachytherapy.

Adequate focal treatment is critically dependent on accurate imaging for diagnosis, staging, tumour localisation, and monitoring of treatment. CT scanning is unable to visualise the intraprostatic tumour site, and CT-based radiotherapy planning is obviously unsuitable for this development without image registration from other imaging modalities. TRUS may visualise the tumour within the prostate but TRUS images are currently limited in their ability to reliably and reproducibly depict intraprostatic cancer foci for radiotherapy planning, and we need to further investigate the role of the newer TRUS techniques such as elastography and HistoScanning™ for intraoperative brachytherapy planning. MRI can be used for local staging in patients who are candidates for whole-gland brachytherapy (Fig. 3.15). Patients with localised prostate cancer are generally candidates for total gland ablation regardless of the extent and distribution of cancer within the gland. On the other hand, for selection of patients who are appropriate candidates for focal ablation, the ability to accurately localise tumour within the prostate becomes relevant. Advances in functional MRI of the prostate have brought about improved tumour localisation, and it is likely that optimal patient selection will be achieved with a combination of mpMRI and extended TRUS biopsy. Techniques such as DWI and DCE-MRI can play an important role in the delineation of intraprostatic GTV for radiotherapy treatment planning. However, data about the sensitivity and specificity of small voxels are not available. This poses a problem for GTV delineation, since delineation implies that for each voxel, a decision is made whether or not that voxel is part of the target volume or not. Delineating a prostate tumour on any imaging essentially comes down to a voxel-level decision whether a voxel contains tumour or not. There are however some important issues to consider before we embark on this treatment pathway. First, the sensitivity and specificity of any imaging technique are not perfect, limiting the predictive power for the presence of tumour. Second, by definition no detailed spatial verification of imaging with pathology can be obtained from patients scheduled for brachytherapy. In the clinical practice of radiotherapy treatment planning, this means that there will never be a gold standard when delineating a prostate tumour for brachytherapy planning. The process of brachytherapy planning involves intrinsic uncertainties based on the limitations of the imaging technique used, and these are integral to current brachytherapy practice. The concept of adjacent intraprostatic regions of varying risk therefore needs consideration. Functional MRI techniques should help us define these regions of varying cancer risks within the prostate and adjust the radiation dose accordingly (Korporaal et al. 2010a, b). The decision whether or not to treat an area of the prostate need not be a binary process; brachytherapy may have an advantage in that dose painting within the gland would accommodate a gradation of risk from healthy tissue to poorly differentiated tumour tissue. The potential for dose painting with brachytherapy and boosting one or more volumes would be more appropriate than the all or nothing treatment offered by other ablative therapies.

3.14 Summary

Modern imaging has much to offer in the practice of prostate brachytherapy. Improved patient staging, better assessment of tumour aggressiveness, and localisation of tumour within the gland all offer the clinician the ability to more accurately identify men who are most likely to benefit from brachytherapy for their prostate cancer. It is important that the brachytherapy team have access to good imaging, understand its applications and limitations, and have core knowledge of the imaging anatomy of the prostate gland. TRUS and MRI have already demonstrated the ability to localise prostate cancers in many patients. Functional imaging investigations will help to differentiate low-grade from high-grade tumours; in the future this may help clinicians choose the most appropriate treatment for individual patients and select those who will benefit from more aggressive therapy. As imaging techniques improve, there is also the prospect of evaluating response to treatment and of early detection of tumour recurrence and potential salvage therapy. Although PET currently plays little role in prostate cancer management, newer radiopharmaceuticals demonstrate promise in tumour detection and staging although further research is required before they are introduced into routine clinical use.

References

Aigner F, Mitterberger M, Rehder P et al (2010) Status of transrectal ultrasound imaging of the prostate. J Endourol 24:685–691

Akin O, Sala E, Moskowitz C et al (2006) Transition zone prostate cancers: features, detection, localization, and staging at endorectal MR imaging. Radiology 239(3):784–792

Alonzi R, Padhani A, Allen C (2007) Dynamic contrast enhanced MRI in prostate cancer. Eur J Radiol 63(3):335–350

Ashley R, Inman B, Routh J et al (2008) Reassessing the diagnostic yield of saturation biopsy of the prostate. Eur Urol 53(5):976–981

Bander N, Milowsky M, Nanus D et al (2005) Phase I trial of 177lutetium-labeled J591, a monoclonal antibody to prostate-specific membrane antigen, in patients with androgen-independent prostate cancer. J Clin Oncol 23:4591–4601

Beerlage H, Aarnink R, Ruijter E et al (2001) Correlation of transrectal ultrasound, computer analysis of transrectal ultrasound and histopathology of radical prostatectomy specimen. Prostate Cancer Prostatic Dis 4:56–62

Beheshti M, Imamovic L, Broinger G et al (2010) ^{18}F Choline PET/CT in the preoperative staging of prostate cancer in patients with intermediate or high risk of extracapsular disease: a prospective study of 130 patients. Radiology 25:925–933

Bonekamp D, Macura K (2008) Dynamic contrast-enhanced magnetic resonance imaging in the evaluation of the prostate. Top Magn Reson Imaging 19(6):273–284

Braeckmann J, Autier Garbar C et al (2008a) Computer-aided ultrasonography (HistoScanning): a novel technology for locating and characterizing prostate cancer. BJU Int 101(3):293–298

Braeckmann J, Autier P, Soviany C et al (2008b) The accuracy of transrectal ultrasonography supplemented with computer-aided ultrasonometry for detecting small prostate cancers. BJU Int 102:1560–1565

Catalona W, Richie J, Ahmann F et al (1994) Comparison of digital rectal examination and serum prostate specific antigen in the early detection of prostate cancer: results of a multicenter clinical trial of 6,630 men. J Urol 151(5):1283–1290

Cornud F, Hamida K, Flam T et al (2000) Endorectal color doppler sonography and endorectal MR imaging features of nonpalpable prostate cancer: correlation with radical prostatectomy findings. Am J Roentgenol 75:1161–1168

Cruz M, Tsuda K, Narumi Y et al (2002) Characterization of low-intensity lesions in the peripheral zone of prostate on pre-biopsy endorectal coil MR imaging. Eur Radiol 12(2):357–365

Dickinson L, Ahmed H, Allen C et al (2011) Magnetic resonance imaging for the detection, localisation, and characterisation of prostate cancer: recommendations from a European consensus meeting. Eur Urol 59(4):477–494

Djavan B, Ravery V, Zlotta A et al (2001) Prospective evaluation of prostate cancer detected on biopsies 1, 2, 3 and 4: when should we stop? J Urol 66(5):1679–1683

Donovan J, Hamdy F, Neal D et al (2003) Prostate Testing for Cancer and Treatment (ProtecT) feasibility study. Health Technol Assess 14:1–88

Eggener S, Salomon G, Scardino P et al (2010) Focal therapy for prostate cancer: possibilities and limitations. Eur Urol 58(1):e1–e18

Eggert T, Khaled W, Wenske S et al (2008) Impact of elastography in clinical diagnosis of prostate cancer. A comparison of cancer detection between B-mode sonography and elastography-guided 10-core biopsies. Urologe A 47(9):1212–1217

Eggert T, Brock M, Noldus J et al (2010) Rising PSA level and negative prostate biopsy. Can prostate elastography help? Urologe A 49(3):376–380

Even-Sapir E, Martin R, Barnes D et al (1993) Role of SPECT in differentiating malignant from benign lesions in the lower thoracic and lumbar vertebrae. Radiology 187:193–198

Franiel T, Lüdemann L, Rudolph B et al (2008) Evaluation of normal prostate tissue, chronic prostatitis, and prostate cancer by quantitative perfusion analysis using a dynamic contrast-enhanced inversion-prepared dual-contrast gradient echo sequence. Invest Radiol 43(7):481–487

Fütterer J, Heijmink S, Scheenen T et al (2006) Prostate cancer localization with dynamic contrast-enhanced MR imaging and proton MR spectroscopic imaging. Radiology 241(2):449–458

Halpern E, Frauscher F, Strup S et al (2002) Prostate: high-frequency Doppler US imaging for cancer detection. Radiology 225:71–77

Halpern E, Ramey J, Strup S et al (2005) Detection of prostate carcinoma with contrast-enhanced sonography using intermittent harmonic imaging. Cancer 104:2373–2383

Heerschap A, Jager G, van der Graaf M et al (1997) In vivo proton MR spectroscopy reveals altered metabolite content in malignant prostate tissue. Anticancer Res 17(3A):1455–1460

Hoeks C, Barentsz J, Hambrock T et al (2011) Prostate cancer: multiparametric MR imaging for detection, localization, and staging. Radiology 261:46–66

Hricak H, Dooms G, McNeal J et al (1987) MR imaging of the prostate gland: normal anatomy. Am J Roentgenol 148(1):51–58
Hricak H, Choyke P, Eberhardt S et al (2007) Imaging prostate cancer: a multidisciplinary perspective. Radiology 243:28–53
Jacobs M, Ouwerkerk R, Petrowski K et al (2008) Diffusion-weighted imaging with apparent diffusion coefficient mapping and spectroscopy in prostate cancer. Top Magn Reson Imaging 19(6):261–272
Katahira K, Takahara T, Kwee T et al (2011) Ultra-high-b-value diffusion-weighted MR imaging for the detection of prostate cancer: evaluation in 201 cases with histopathological correlation. Eur Radiol 21(1):188–196
Kim C, Park B, Lee H et al (2007) Value of diffusion-weighted imaging for the prediction of prostate cancer location at 3T using a phased-array coil: preliminary results. Invest Radiol 42(12): 842–847
Kim H, Kim J, Kim K et al (2009) Prostate cancer: apparent diffusion coefficient map with T2-weighted images for detection-a multireader study. Radiology 250(1):145–151
Korporaal J, van den Berg C, Groenendaal G et al (2010a) The use of probability maps to deal with the uncertainties in prostate cancer delineation. Radiother Oncol 94(2):168–172
Korporaal J, van den Berg C, Jeukens C et al (2010b) Dynamic contrast-enhanced CT for prostate cancer: relationship between image noise, voxel size, and repeatability. Radiology 256: 976–984
Kuligowska E, Barish M, Fenlon H et al (2001) Predictors of prostate carcinoma: accuracy of grayscale and color Doppler US and serum markers. Radiology 2001(220):757–764
Langer D, van der Kwast T, Evans A et al (2008) Intermixed normal tissue within prostate cancer: effect on MR imaging measurements of apparent diffusion coefficient and T2-sparse versus dense cancers. Radiology 249(3):900–908
Lee S, Park K, Choi K et al (2010) Is endorectal coil necessary for the staging of clinically localized prostate cancer? Comparison of non-endorectal versus endorectal MR imaging. World J Urol 28(6):667–672
Liu I, Zafar M, Lai Y et al (2001) Fluorodeoxyglucose positron emission tomography studies in diagnosis and staging of clinically organ-confined prostate cancer. Urology 57:108–111
McNeal J (1981) The zonal anatomy of the prostate. Prostate 2(1):35–49
Meerleer G, Villeirs G, Bral S et al (2005) The magnetic resonance detected intraprostatic lesion in prostate cancer: planning and delivery of intensity-modulated radiotherapy. Radiother Oncol 75(3):325–333
Mueller-Lisse U, Scherr M (2007) Proton MR spectroscopy of the prostate. Eur J Radiol 63(3): 351–360
Noguchi M, Stamey T, McNeal J et al (2001) Relationship between systematic biopsies and histological features of 222 radical prostatectomy specimens: lack of prediction of tumour significance for men with nonpalpable prostate cancer. J Urol 66(1):104–109
Norberg M, Egevad L, Holmberg L et al (1997) The sextant protocol for ultrasound-guided core biopsies of the prostate underestimates the presence of cancer. Urology 50:562–566
Novara G, Boscolo-Berto R, Lamon C, Ficarra S et al (2010) Detection rate and factors predictive the presence of prostate cancer in patients undergoing ultrasonography-guided transperineal saturation biopsies of the prostate. BJU Int 105(9):1242–1246
Padhani A, Harvey C, Cosgrove D (2005) Angiogenesis imaging in the management of prostate cancer. Nat Clin Pract Urol 2(12):596–607
Partin A, Mangold L, Lamm D et al (2001) Contemporary update of prostate cancer staging nomograms (Partin Tables) for the new millennium. Urology 58(6):843–848
Picchio M, Gionannini E, Crivellaro C et al (2010) Clinical evidence on PET/CT for radiation therapy planning in prostate cancer. Radiother Oncol 96(3):347–350
Pucar D, Hricak H, Shukla-Dave A et al (2007) Clinically significant prostate cancer local recurrence after radiation therapy occurs at the site of primary tumour: magnetic resonance imaging and step-section pathology evidence. Int J Radiat Oncol Biol Phys 69:62–69

Qayyum A, Coakley F, Lu Y et al (2004) Organ-confined prostate cancer: effect of prior transrectal biopsy on endorectal MRI and MR spectroscopic imaging. AJR Am J Roentgenol 183(4): 1079–1108

Rifkin M, Zerhouni E, Gatsonis C et al (1990) Comparison of magnetic resonance imaging and ultrasonography in staging early prostate cancer: results of a multi-institutional cooperative trial. N Engl J Med 323:621–626

Salomon G, Graefen M, Heinzer H et al (2009) The value of real-time elastography in the diagnosis of prostate cancer. Urologe A 48(6):628–636

Scheenen T, Klomp D, Röll S et al (2004) Fast acquisition-weighted three-dimensional proton MR spectroscopic imaging of the human prostate. Magn Reson Med 52(1):80–88

Schettino C, Kramer E, Noz M et al (2004) Impact of fusion of indium-111 capromab pendetide volume data sets with those from MRI or CT in patients with recurrent prostate cancer. AJR Am J Roentgenol 183:519–524

Schröder F, van der Maas P, Beemsterboer P et al (1998) Evaluation of the digital rectal examination as a screening test for prostate cancer. Rotterdam section of the European Randomized Study of Screening for Prostate Cancer. J Natl Cancer Inst 90(23):1817–1823

Schröder F, Carter H, Wolters T et al (2008) Early detection of prostate cancer. Part 1: PSA and PSA kinetics. Eur Urol 53(3):468–477

Sciarra A, Barentsz J, Bjartelli A et al (2011) Advances in magnetic resonance imaging: how they are changing the management of prostate cancer. Eur Urol 59(6):962–977

Simmons L, Autier P, Zatura F et al (2012) Detection, localisation and characterisation of prostate cancer by Prostate HistoScanning. BJU Int 110(1):28–35

Spethmann J, Graefen M, Beckmann A et al (2010) Accuracy of computer-aided transrectal ultrasonography detection (HistoScanning) of prostate cancer in the predication of a negative margin in radical prostatectomy patients. Eur Urol Suppl 9(2):65, 103

Tomoaki M, Masakazu T, Takeshi T et al (2009) Real-time elastography for the diagnosis of prostate cancer: evaluation of elastographic moving images. Jpn J Clin Oncol 39(6):394–398

Trabulsi E, Sackett D, Gomella L et al (2010) Enhanced transrectal ultrasound modalities in the diagnosis of prostate cancer. Urology 76(5):1025–1033

Wang L, Mazaheri Y, Zhang J et al (2008) Assessment of biologic aggressiveness of prostate cancer: correlation of MR signal intensity with Gleason grade after radical prostatectomy. Radiology 246(1):168–176

Watanabe H, Kato H, Kato T et al (1968) Diagnostic application of ultrasonography to the prostate [in Japanese]. Nippon Hinyokika Gakkai Zasshi 59:273–279

Weber D, Wang H, Cozzi L et al (2009) RapidArc, intensity modulated photon and proton techniques for recurrent prostate cancer in previously irradiated patients: a treatment planning comparison study. Radiat Oncol 4:34

Zelhof B, Pickles M, Liney G et al (2009) Correlation of diffusion-weighted magnetic resonance data with cellularity in prostate cancer. BJU Int 103(7):883–888

Zheng Q, Gardner T, Raikwar S et al (2004) [11C]Choline as a PET biomarker for assessment of prostate cancer tumor models. Bioorg Med Chem 12(11):2887–2893

Patient Selection and Recommendations on Permanent Seed Implantation as Monotherapy for Localized Prostate Cancer

4

Jean-Michel Hannoun-Levi and Jean-Marc Cosset

4.1 Introduction

Transperineal permanent prostate brachytherapy is a safe and efficacious treatment option for selected patients with organ-confined prostate cancer (PC). Careful adherence to established brachytherapy standards has been shown to improve the likelihood of procedural success as well as to reduce the incidence of treatment-related morbidity. Candidates for treatment with prostate seed implant alone, i.e. as monotherapy, include those for whom there is a significant likelihood that their PC could be encompassed by an adequate dose distribution from permanent prostate seed implant alone. Currently, prostate seed implant results are reported with a long follow-up, and the key to success is based on the patient selection to identify implant candidates with a high probability of biochemical disease-free survival and a good functional outcome.

4.2 Pretreatment Investigations

4.2.1 History and Clinical Examination

Age, professional activity, marital status, comorbidity factors and paternity wishes represent some general factors not directly linked to PC but important to consider for treatment decision. Furthermore, investigation of urinary (with or without

J.-M. Hannoun-Levi, MD, PhD (✉)
Department of Radiation Therapy, Antoine Lacassagne Cancer Center,
33, avenue de Valombrose, 06189 Nice, France
e-mail: jean-michel.hannoun-levi@nice.fnclcc.fr

J.-M. Cosset, MD
Department of Oncology/Radiotherapy,
Institut Curie, Paris, France

G. Kovács, P. Hoskin (eds.), *Interstitial Prostate Brachytherapy*,
DOI 10.1007/978-3-642-36499-0_4,

α-blockers), digestive and sexual functions (considered in next chapters); common medications; alcohol and cigarette consumption; high levels of dietary fat; and family history of PC must be investigated.

4.2.2 Digital Rectal Examination (DRE)

Digital rectal examination is integral to staging PC. Treatment decisions are largely based on information gained from digital rectal examination. However, the accuracy of an abnormal digital rectal examination for predicting the location and extent of cancer remains highly variable and mainly related to the experience of the physician.

Some men clinically staged as "low risk" may have intermediate or even high-risk PC due to understaging based on the low level of accuracy for assessing actual T-category using the digital rectal examination alone (Obek et al. 1999).

4.2.3 Prostate-Specific Antigen (PSA)

4.2.3.1 Physiopathological Conditions

Age, recent prostate manipulation, ejaculation, prostatitis and prostate size are the main parameters which may influence the PSA levels. It is important to double-check this biochemical marker, especially in case of recent, rapid or unexpected rising PSA (Harnden et al. 2008).

4.2.3.2 PSA Variation and Common Medications

Medications including 5-alpha reductase inhibitors (5ARIs – finasteride, dutasteride) can dramatically and quickly decrease PSA (Etzioni et al. 2005). Recent studies suggest that other medications, including statins (Cyrus-David et al. 2005; Hamilton et al. 2008; Mondul et al. 2010) and nonsteroidal anti-inflammatory drugs (NSAIDs) (Singer et al. 2008), may also lower PSA levels.

NSAID, statin and thiazide diuretic intake was inversely related to PSA levels. Five years of NSAID, statin and thiazide diuretic use was associated with PSA levels lower by 6, 13 and 26 %, respectively (Chang et al. 2010).

4.2.3.3 PSA Velocity (PSAv)

PSAv represents the rate of change in the PSA level over time. Given two men with the same initial PSA level, the patient experiencing a rise of at least 2 points in the PSA level during the year before diagnosis (PSAv > 2 ng/ml/year) has been shown to have more advanced pathologic stage, grade and higher risk of recurrence after radical prostatectomy (RP) (Patel et al. 2005), prostate implant (Rossi et al. 2008) or external beam radiation therapy (EBRT) and androgen deprivation therapy (ADT) (Palma et al. 2008) when compared to men without at least a 2-point rise during the past year.

In addition, a higher risk of prostate cancer-specific mortality (PCSM) after RP (D'Amico et al. 2004) and EBRT with (Palma et al. 2008) or without (D'Amico et al. 2005) ADT has been observed for men with this rapid rise in PSA compared to those without such a rapid rise. For the specific case of men with low-risk PC managed with RP (D'Amico et al. 2006) at 7 years, PCSM estimates reached 5 % versus lower than 1 % for men whose PSA level rose by at least 2 points during the year prior to diagnosis as compared to those who did not. These respective estimates were 19 and 0 % for men undergoing EBRT (D'Amico et al. 2005). However, PSA velocity should be analyzed with the other prognostic factors (Vickers et al. 2011). Although not reported as a significant parameter after brachytherapy, it is tempting to extrapolate this experience with PSAv to permanent implants.

4.2.3.4 PSA Density (PSAd)

PSAd is defined as the ratio of the PSA divided by the prostate volume The use of PSAd less than 0.15 as a measure of potentially insignificant prostate cancer is well established (D'Amico et al. 2004). PSAd could be considered at an independent factor (multivariate analysis) for predicting tumour upgrading on repeat biopsy, whereas it showed a strong trend to predict PSAv on follow-up (Kotb et al. 2011). Again, no study has been reported to date about the value of PSAd after brachytherapy.

4.2.4 Prostate Biopsies

4.2.4.1 Gleason Score

Some men diagnosed with pathological low-risk PC on biopsies may have intermediate- or high-risk PC due to undergrading based on the sampling error associated with prostate biopsy technique (Zam et al. 2008). Furthermore, differences may exist between groups in reporting the biopsy Gleason score: for some authors, the Gleason primary pattern is defined by an involvement of at least 51 % of the specimen (Merrick et al. 2007), while for others, it is based upon the highest Gleason score present in any one core. Actually, details of how primary grade is assigned are not available in a significant number of studies.

4.2.4.2 Percent of Positive Prostate Biopsies

Evidence from surgical series shows that men who have PC found in more than 50 % of the biopsy specimens are at increased risk for undergrading on the basis of the biopsy Gleason score and understaging on the basis of the digital rectal examination compared with those with lower than 50 % (Cooperberg et al. 2004; Lotan et al. 2004). Data from specific EBRT series led to the same conclusion (D'Amico et al. 2001; Merrick et al. 2002). D'Amico et al. (2002) specifically examined the prognostic significance of this factor in men with low-risk PC and found that PCSM estimates were 9 % as compared with 0 % by 5 years in men with at least 50 % as compared with less than 50 % of the biopsy cores revealing Gleason score 6.

4.2.4.3 Percentage of Cancer on the Biopsy Core

The pathological prostate biopsy report should analyze different measures of tumour extent such as (1) number of cores positive for cancer in the number of cores examined, (2) percentage of needle core tissue affected by carcinoma and (3) linear millimetres of carcinoma present (Humphrey 2007). While the percentage of needle core tissue affected by carcinoma is described as a prognostic factor for biochemical relapse, its value remains controversial (Pe et al. 2009).

4.2.4.4 Prostatic Capsular Invasion (PCI)

Higher clinical stage, higher Gleason grade in the biopsy specimen and higher pretreatment serum PSA levels are all associated with increasing levels of prostatic capsular invasion (PCI). Wheeler et al. (1998) reported that in RP specimens increasing levels of PCI were significantly associated with increasing tumour volume ($p<.001$), Gleason grade ($p<.0001$), seminal vesicle involvement ($p<.001$) and lymph node metastases ($p<.001$). In a multivariate analysis, the authors noticed that the level of PCI was an independent prognostic factor ($p<.001$). There is a strong association between the level of invasion of cancer into or through the prostatic capsule and the volume, grade, pathological stage and rate of recurrence after radical prostatectomy.

4.2.4.5 Perineural Invasion (PNI)

The presence of perineural invasion on prostate biopsy in an otherwise low-risk patient is associated with a significant risk of upstaging and upgrading at RP (Lee et al. 2007) and a resulting higher risk of progression after EBRT (Beard et al. 2004) or RP (D'Amico et al. 2000). Specifically, for men with low-risk PC treated using EBRT, PNI was significantly associated with an increased risk of PSA recurrence (Beard et al. 2004), and this was also found to be true in an independent data set for men with low-risk PC undergoing RP where the time to PSA recurrence was significantly shorter in men with PNI in the prostate needle biopsy (D'Amico et al. 2000). Specifically, by 5 years after RP, estimates of PSA recurrence were 18 % as compared to 5 % in men with PNI versus none at biopsy. A confirmatory study (Yu et al. 2007) showed a 4.14-fold increase in the risk of PSA recurrence in men with low-risk PC undergoing EBRT (after adjusting for EBRT dose). Those data should probably be extrapolated to brachytherapy, with perineural invasion taken into account when selecting patients.

4.2.5 Prostate Volume and Prostate Staging

4.2.5.1 Imaging Assessment

Transrectal ultrasound (TRUS) has only moderate accuracy in the detection of prostate carcinoma but is useful in the estimation of prostate volume and thus calculation of PSA density as well as measuring the residual urine volume. The role of magnetic resonance imaging (MRI) in diagnosis and staging of prostate carcinoma is rapidly increasing (De Visschere et al. 2010). Morphologic T2-weighted magnetic resonance images (T2-WI) depict the prostatic anatomy with high resolution and can detect tumoral areas within the peripheral zone of the prostate. Addition of an endorectal probe, magnetic resonance spectroscopic imaging, dynamic contrast-enhanced MRI and diffusion-weighted imaging further increase the diagnostic performance of MRI.

The gold standard for diagnosis of PC is histological assessment currently obtained by transrectal ultrasound-guided systematic core needle biopsies. However, in the future, imaging-based targeted biopsies may improve the biopsy yield and decrease the total number of biopsy cores (Delongchamps et al. 2011).

Computed tomography (CT) and 5-FDG positron emission tomography (PET) have no value in early PC detection, and their indications are limited to lymph node staging and detection of distant metastases. The more recently introduced choline-PET scan could find a role in the staging of selected patients.

4.2.5.2 Large Prostates and Neoandrogen Deprivation

In case of prostate volume larger than 60 cc, ADT delivered during a 3- to 6-month period may be proposed to reduce the pre-implant volume of the gland (Rosenthal et al. 2011). The reduction of the prostate volume after ADT commonly ranges between 10 % (small glands) and 60 % (large glands) with a median decrease of 30 % (Whittington et al. 1999). However, it must be kept in mind that neoadjuvant ADT for volume reduction could increase the risk of brachytherapy-related urinary morbidity (Hinerman-Mulroy et al. 2004), leading some authors to consider that ADT should not be routinely given for low-risk patients (Gibbons et al. 2009), especially when an intraoperative planning technique is used for brachytherapy (Meyer et al. 2008).

In case of median lobes, Cosset at al. (2011) described a one-step customized transurethral resection of the prostate and permanent implant brachytherapy for selected PC patients. The authors considered this procedure as technically feasible but too toxic. Alike other authors, they now propose a two-step procedure, with the limited resection of the median lobe preceding by 3–6 months of the implantation.

4.2.6 Lower Urinary Tract Symptoms (LUTS)

The investigations of the LUTS must be performed and carefully analyzed, taking into account the eventual prescription of α-blocker medication for a pre-existing obstructive uropathy.

4.2.6.1 Individual Questionnaire

The *International Prostate Symptom Score* (IPSS) was designed to be self-filled in by the patient. IPSS provides subjective but informative information regarding the patient's urinary function. An IPSS of less than 13–15 is generally correlated with relatively mild lower urinary tract symptoms.

4.2.6.2 Urodynamic Studies

Urodynamic studies will provide more objective information regarding the urinary function but are highly dependent on the bladder status before the exam (Trachtenberg 2005). Residual urine volume less than 50 cc and urinary flow with a Qmax higher than 15 ml/s are usually required to consider the LUTS as compatible with interstitial brachytherapy (Kovacs et al. 2005).

4.2.7 Sexual Potency

The assessment of sexual function may orientate the treatment choice. Identifying the presence and severity of sexual concerns should be considered part of cancer treatment decision and follow-up care. Efforts to evaluate and characterize sexual problems in patients with prostate cancer have been hampered by a lack of consensus regarding valid outcome measures. However, the *International Index of Erectile Function* (IIEF5 – 5 items) questionnaire remains a simple, consistent and reproducible evaluation of sexual potency. A more extensive IIEF questionnaire with 15 items is also available.

A new flexible and psychometrically robust measure of sexual function "Supplement Sexual Function of the Patient-Reported Outcomes Measurement Information System (PROMIS) (CaPS-SF)" is also now available (Jeffery et al. 2009).

4.2.8 Bowel Function

Any history of bowel disorder such as chronic diarrhoea, blood in stools, constipation and anal sphincter dysfunction should be carefully investigated before treatment. Pre-existing digestive disorders are not usually considered as a contra-indication for permanent seed brachytherapy; however, in the case of digestive complication occurring after the implant, it will help define causality between irradiation and digestive symptoms.

4.3 Low-Risk Patients

4.3.1 Typical Indications

Patients with low-risk localized prostate cancer are classically considered as the cases most suitable for low-dose-rate brachytherapy as sole therapy (Table 4.1).

A number of different risk stratifications or classifications for localized prostate cancers do exist. The large majority of those systems divide PC patients into low-, intermediate- and high-risk groups according to pretreatment PSA level, Gleason score and clinical stage (D'Amico et al. 1998; Beyer et al. 2003). However, whatever the stratification system used, low-risk PC is constantly defined by the same three clinical parameters (D'Amico et al. 1998): a PSA level lower than 10 ng/ml, a biopsy Gleason score of 6 or lower (with no grade 4 or 5 disease) and American Joint Commission on Cancer staging tumour category (T) 1c or 2a (Greene 2002). However, there is additional information from the prostate biopsy and PSA history

Table 4.1 Criteria and limits to define most suitable patient for implant prostate brachytherapy as sole therapy

Criteria	Limits
Data related to the disease stage	
DRE	cT1b–T2a
MRI	No extracapsular extension
Pathological data	
Gleason score	≤6
Percentage of biopsy cores involved with cancer (%)	≤50
Percentage of cancer on the biopsy core (%)	≤50
Prostatic capsular invasion	No
Perineural invasion	No
Biochemical data	
Initial PSA (ng/ml)	<10
PSA velocity (ng/ml/year)	<2
PSA density (ng/ml/cc)	<0.15
Prostate volume (cc)	<50–60
Lower urinary tract symptoms	
IPSS	≤13–15
Residual urine volume (cc)	<50
Urinary flow (ml/s)	>15

DRE digital rectal exam, *MRI* magnetic resonance imaging, *PSA* prostate-specific antigen, *IPSS* International Prostate Symptom Score

that may also be considered. Table 4.1 summarizes the classical data provided by the standard classifications of low-risk PC and some additional factors aiming at refining the selection criteria for permanent implant brachytherapy (D'Amico 2011; Rosenthal et al. 2011).

Therefore, a patient with a low-risk PC according to the classical definition, and who respects the additional data regarding PSA and biopsies, and with few/no LUTS and acceptable sexual function can be considered as an excellent candidate for interstitial prostate implant (Rosenthal et al. 2011; EPROP 2011).

4.3.2 Exclusion Criteria

The following are potential exclusion criteria for permanent seed brachytherapy (Ash et al. 2007; Rosenthal et al. 2011):

- Life expectancy of less than 5 years
- Unacceptable operative risk
- Bleeding disorder or anticoagulation that cannot be stopped
- Poor anatomy which in the opinion of the radiation oncologist could lead to a suboptimal implant (e.g. narrow pubic arch, large or poorly healed transurethral resection of the prostate defect)
- Significant obstructive uropathy

- Pathologically positive lymph nodes
- Distant metastases

The following situations can be considered as relative contra-indications (to be discussed case by case) for permanent seed brachytherapy:

- Large median lobes (may benefit from limited and customized resection; see Chap. 2)
- Large gland size (may benefit from a volume reduction with ADT; see Chap. 2)
- Previous pelvic irradiation
- High AUA score (IPSS > 15) (but may improve with ADT and/or limited TURP)
- History of multiple pelvic surgeries

4.3.3 Special Situations

4.3.3.1 Men Under 60 Years of Age

Burri et al. (2010) retrospectively evaluated the biochemical outcomes of young men (<60 years) treated with low-dose-rate brachytherapy for PC. With a median follow-up of 68 and 66 months for men younger and older than 60 years, respectively, the 8-year freedom from biochemical failure rate was similar between the two groups (92 % vs. 87 % – p = NS). Merrick et al. (2008) observed equivalent results with an age cut-off fixed at 50 years for the "young man" definition – those authors suggested that young age should not be considered as a contra-indication when discussing brachytherapy as a primary treatment option for clinically localized PC.

4.3.3.2 Family History of Prostate Cancer

It is well established that there is an association between a family history of PC and increased risk of developing the disease. This risk increases with a greater proximity of relatedness, greater number of family members affected and/or earlier age at diagnosis of the family member (Madersbacher et al. 2011). However, the impact of the genetic risk of prostate cancer and the posttreatment clinical outcome remains questionable. In this setting, Szulkin et al. (2012) observed no evidence for an association between any of 23 established prostate cancer genetic risk variants and disease progression. Currently, there is no rational to consider family history of prostate cancer as a deleterious prognostic factor or a contra-indication for prostate seed implant.

4.3.3.3 Posttreatment Fertility

Because young men are often good candidates for prostate seed implant (low-risk PC and few LUTS), the question of fertility has to be clearly discussed with the patient. In most cases, brachytherapy irradiation makes patient infertile. However, although the therapy-related modifications of the semen reduce fertility, patients must be aware of the possibility of fathering children after such a permanent implantation, with a limited risk of genetic effects to the child (ICRP 2005). For young men with PC who wish to father children in the future, it is highly recommended that they consider sperm banking prior to brachytherapy (Mydlo and Lebed 2004).

4.3.4 Iodine Implant Brachytherapy Versus Other Techniques

Most patients with clinically localized PC can also benefit from RP, EBRT and active surveillance. The choice of the treatment requires a clear understanding by the patient of the risks and benefits attached to each option in order to make an informed decision.

4.3.4.1 Radical Prostatectomy

For a low-risk PC patient with significant LUTS or obstructive uropathy, radical prostatectomy is usually considered as more appropriate. However, the impact on LUTS should be carefully balanced against the urinary and sexual side effects observed after radical prostatectomy (Crook et al. 2011). In addition a very large prostate (>70–80 cc) will usually be preferentially treated with prostatectomy or EBRT as more adequate choices.

4.3.4.2 External Beam Radiation Therapy

For a low-risk PC patient, EBRT can be proposed as an alternative to interstitial seed implant. However, among low-risk PC patients, Zelefsky et al. (2011) reported that the 7-year biochemical tumour control was superior for brachytherapy compared with high-dose intensity-modulated radiation therapy (81 Gy), but there was a significant increase in grade 2 urinary and rectal symptoms for brachytherapy compared with EBRT.

4.3.4.3 Active Surveillance

For "very" low-risk PC defined with PSA less than 10 ng/ml, PSAd less than 0.15, T1c and Gleason score 6 (3 + 3) in at most two cores each having less than 50 % core involvement, active surveillance (AS) may be discussed with the patient as an alternative treatment (Mohler et al. 2010). The most mature AS study (8 years of follow-up) reported by Klotz (2008) for men with very low-risk PC and with up to a 20-year life expectancy suggests that AS is an acceptable treatment. However, even in this very good prognostic subgroup of patients, some patients will die from prostate cancer (D'Amico 2011). AS should therefore be carefully discussed with young men without comorbidity and presenting with a very low-risk PC.

Nevertheless, it has been stressed that men with low-risk PC and *significant comorbidities* are at risk of being overtreated (Daskivich et al. 2011). Measure of comorbidity (Charlson et al. 1987; Piccirillo et al. 2004) should be applied appropriately in the light of the data suggesting that overtreatment of low-risk PC mainly occurs in men with significant comorbidity (Piccirillo et al. 2004).

In terms of cost, Eldefrawy et al. (2011) reported the results of a cost comparison between AS and treatment for low-risk PC. The authors observed that AS was associated with a different cost distribution in which the initial cost of AS is low and the follow-up cost is higher than other therapies. However, despite the higher follow-up cost, AS logically remains the most cost-effective alternative for low-risk PC.

4.4 Intermediate-Risk Localized Prostate Cancer Patients

Those patients were not initially considered as "good" indications for prostate brachytherapy alone. However, most large historical series have included a proportion of intermediate-risk patients and even sometimes high-risk patients from which has emerged data strongly suggesting that selected intermediate-risk patients may also be adequately treated with brachytherapy as monotherapy. This remains an area of active investigation (Frank et al. 2007; Cosset et al. 2008; Stone et al. 2009; RTOG 2009).

Frank et al. (2007) reported on an American survey showing that most authors would propose brachytherapy as monotherapy to selected intermediate-risk patients with absence of PNI, GS 7 (3+4) *or* PSA 10–20 ng/ml, pT1c and positive cores <30 %. Among a series of 809 patients, Cosset et al. (2008) reported a 5-year relapse-free survival of 97 % for low-risk patients and 94 % for a subset of "favourable" intermediate-risk patients defined by either PSA between 10 and 15 ng/ml or Gleason 7 but with all the other criteria for low-risk present. These results suggest that selected patients in the intermediate-risk group of localized PCs can be safely considered for permanent implant brachytherapy as monotherapy. Those data have been recently confirmed by the same group with more patients and a longer follow-up (Wakil et al. 2010).

Currently, a phase III randomized trial is comparing for selected intermediate-risk PC, combined EBRT plus interstitial implant versus interstitial implant alone. Entry criteria are clinical stages T1c–2b, and either Gleason score = 7 with PSA less than 10 or Gleason score < 7 with PSA between 10 and 20 ng/ml [RTOG trial 0232].

4.5 Other Potential Indications and Investigational Treatments

4.5.1 Brachytherapy as a Boost After External Beam Radiation

In intermediate- or high-risk patients, interstitial seed implant can be performed as a boost after or before EBRT delivered to the whole pelvis or the prostatic fossa (Stock et al. 2004; Bittner et al. 2010). However, a brachytherapy boost performed with temporary high-dose-rate implant may be more cost-effective compared to a seed implant (Grimm et al. 1996).

4.5.2 Brachytherapy for Salvage After Failure of External Beam Radiation

Salvage prostate brachytherapy may be an attractive option for attaining disease control in patients with PC local failure after EBRT. The 5-year biochemical control rate after a salvage implant ranges between 50 and 60 % with acceptable and manageable urinary and digestive toxicities (Boukaram and Hannoun-Levi 2010).

4.5.3 Focal Therapy

Focal therapy aims to find a middle ground between active surveillance and radical therapies by treating the cancer alone, with a margin, and preserving as much tissue as is practical. Early feasibility studies have demonstrated an absence of rectal toxicity and preservation of genitourinary function in 80–90 % of men (Todor et al. 2011). Focal therapy for low-risk PC remains an experimental procedure which should only be performed in the context of prospective clinical trials for selected subgroups of patients.

Conclusion

Prostate seed implant is an efficient and well-tolerated treatment for low-risk PC and represents one of the three standard and validated treatments alongside RP and EBRT. The key to success is careful patient selection based on classical low-risk PC criteria as well as additional information from the prostate biopsy and PSA history not used in the original definition of low-risk PC. LUTS and comorbidity factors have to be carefully investigated. Taking all together, those data have to be clearly presented and discussed with the patient who requires an understanding of the risks and benefits of each therapeutic option in order to make an informed decision.

References

Ash D, Bottomley D, Al-Qaisieh B et al (2007) A prospective analysis of long-term quality of life after permanent I-125 brachytherapy for localised prostate cancer. Radiother Oncol 84(2):135–139

Beard CJ, Chen MH, Cote K et al (2004) Perineural invasion is associated with increased relapse after external beam radiotherapy for men with low-risk prostate cancer and may be a marker for occult, high-grade cancer. Int J Radiat Oncol Biol Phys 58:19–24

Beyer DC, Thomas T, Hilbe J et al (2003) Relative influence of Gleason score and pretreatment PSA in predicting survival following brachytherapy for prostate cancer. Brachytherapy 2(2):77–84

Bittner N, Merrick GS, Wallner KE et al (2010) Whole-pelvis radiotherapy in combination with interstitial brachytherapy: does coverage of the pelvic lymph nodes improve treatment outcome in high-risk prostate cancer? Int J Radiat Oncol Biol Phys 76:1078–1084

Boukaram C, Hannoun-Levi JM (2010) Management of prostate cancer recurrence after definitive radiation therapy. Cancer Treat Rev 36(2):91–100

Burri RJ, Ho AY, Forsythe K et al (2010) Young men have equivalent biochemical outcomes compared with older men after treatment with brachytherapy for prostate cancer. Int J Radiat Oncol Biol Phys 77(5):1315–1321
Chang SL, Harshman LC, Presti JC Jr (2010) Impact of common medications on serum total prostate-specific antigen levels: analysis of the National Health and Nutrition Examination Survey. J Clin Oncol 28:3951–3957
Charlson ME, Pompei P, Ales KL et al (1987) A new method for classifying prognostic comorbidity in longitudinal studies: development and validation. J Chronic Dis 40:373–383
Cooperberg MR, Broering JM, Litwin MS et al (2004) The contemporary management of prostate cancer in the United States: lessons from the Cancer of the Prostate Strategic Urologic Research Endeavor (CapSURE), a national disease registry. J Urol 171:1393–1401
Cosset JM, Flam T, Thiounn N et al (2008) Selecting patients for exclusive permanent implant prostate brachytherapy: the experience of the Paris Institut Curie/Cochin Hospital/Necker Hospital group on 809 patients. Int J Radiat Oncol Biol Phys 71:1042–1048
Cosset JM, Barret E, Castro-Pena P et al (2011) One-step customized transurethral resection of the prostate and permanent implant brachytherapy for selected prostate cancer patients: technically feasible but too toxic. Brachytherapy 10(1):29–34
Crook JM, Gomez-Iturriaga A, Wallace K et al (2011) Comparison of health-related quality of life 5 years after SPIRIT: Surgical Prostatectomy Versus Interstitial Radiation Intervention Trial. J Clin Oncol 29(4):362–368
Cyrus-David MS, Weinberg A, Thompson T et al (2005) The effect of statins on serum prostate specific antigen levels in a cohort of airline pilots: a preliminary report. J Urol 173:1923–1925
D'Amico AV (2011) Future of treatment for low-risk prostate cancer: for all, for some, or for none? J Clin Oncol 29(15):1940–1943
D'Amico AV, Whittington R, Malkowicz SB et al (1998) Biochemical outcome after radical prostatectomy, external beam radiation therapy, or interstitial radiation therapy for clinically localized prostate cancer. JAMA 280:969–974
D'Amico AV, Wu Y, Chen MH et al (2000) Perineural invasion as a predictor of biochemical outcome following radical prostatectomy for select men with clinically localized prostate cancer. J Urol 165:126–130
D'Amico AV, Schultz D, Silver B et al (2001) The clinical utility of the percent of positive prostate biopsies in predicting biochemical outcome following external-beam radiation therapy for patients with clinically localized prostate cancer. Int J Radiat Oncol Biol Phys 49:679–684
D'Amico AV, Keshaviah A, Manola J et al (2002) The clinical utility of the percent of positive prostate biopsies in predicting prostate cancer specific and overall survival following radiation therapy for patients with localized prostate cancer. Int J Radiat Oncol Biol Phys 53:581–587
D'Amico AV, Chen MH, Roehl KA et al (2004) Preoperative PSA velocity and the risk of death from prostate cancer after radical prostatectomy. N Engl J Med 351:125–135
D'Amico AV, Renshaw AA, Sussman B et al (2005) Pretreatment PSA velocity and risk of death from prostate cancer following external beam radiation therapy. JAMA 294:440–447
D'Amico AV, Chen MH, Renshaw AA et al (2006) Identifying men diagnosed with clinically localized prostate cancer who are at high risk for death from prostate cancer. J Urol 176: S11–S15
Daskivich TJ, Chamie K, Kwan L et al (2011) Overtreatment of men with low-risk prostate cancer and significant comorbidity. Cancer 117(10):2058–2066
De Visschere P, Oosterlinck W, De Meerleer G et al (2010) Clinical and imaging tools in the early diagnosis of prostate cancer, a review. JBR-BTR 93(2):62–70
Delongchamps NB, Rouanne M, Flam T et al (2011) Multiparametric magnetic resonance imaging for the detection and localization of prostate cancer: combination of T2-weighted, dynamic contrast-enhanced and diffusion-weighted imaging. BJU Int 107(9):1411–1418
Eldefrawy A, Katkoori D, Abramowitz M et al (2011) Active surveillance vs. treatment for low-risk prostate cancer: a cost comparison. Urol Oncol. Pubmed/21616691
Etzioni RD, Howlader N, Shaw PA et al (2005) Long-term effects of finasteride on prostate specific antigen levels: results from the prostate cancer prevention trial. J Urol 174:877–881

Frank SJ, Grimm PD, Sylvester JE et al (2007) Interstitial implant alone or in combination with external beam radiation therapy for intermediate-risk prostate cancer: a survey of practice patterns in the United States. Brachytherapy 6:2–8

Gibbons EP, Jacobs BL, Smith RP et al (2009) Dosimetric outcomes in prostate brachytherapy: is downsizing the prostate with androgen deprivation necessary? Brachytherapy 8(3):304–308

Greene FL (2002) Prostate. In: Greene FL, Page DL, Fleming ID et al (eds) AJCC cancer staging manual, 6th edn. Springer, New York, pp 309–316

Grimm PD, Blasko JC, Ragde H et al (1996) Does brachytherapy have a role in the treatment of prostate cancer? Hematol Oncol Clin North Am 10(3):653–673

Hamilton R, Goldberg KC, Platz EA et al (2008) The influence of statin medications on prostate-specific antigen levels. J Natl Cancer Inst 100:1511–1518

Harnden P, Naylor B, Shelley MD et al (2008) The clinical management of patients with a small volume of prostatic cancer on biopsy: what are the risks of progression? A systematic review and meta-analysis. Cancer 112(5):971–981

Hinerman-Mulroy A, Merrick GS, Butler WM et al (2004) Androgen deprivation-induced changes in prostate anatomy predict urinary morbidity after permanent interstitial brachytherapy. Int J Radiat Oncol Biol Phys 59(5):1367–1382

Humphrey PA (2007) Diagnosis of adenocarcinoma in prostate needle biopsy tissue. J Clin Pathol 60(1):35–42

International Commission on Radiological Protection (2005) Radiation safety aspects of brachytherapy for prostate cancer using permanently implanted sources. A report of ICRP Publication 98. Ann ICRP 35(3):iii–vi, 3–50

Jeffery DD, Tzeng JP, Keefe FJ et al (2009) Initial report of the cancer Patient-Reported Outcomes Measurement Information System (PROMIS) sexual function committee: review of sexual function measures and domains used in oncology. Cancer 115(6):1142–1153

Klotz L (2008) Active surveillance for prostate cancer: trials and tribulations. World J Urol 26: 437–442

Kotb AF, Tanguay S, Luz MA et al (2011) Relationship between initial PSA density with future PSA kinetics and repeat biopsies in men with prostate cancer on active surveillance. Prostate Cancer Prostatic Dis 14(1):53–57

Kovacs G, Potter R, Loch T et al (2005) GEC/ESTRO-EAU recommendations on temporary brachytherapy using stepping sources for localised prostate cancer. Radiother Oncol 74:137–148

Lee IH, Roberts R, Shah RB et al (2007) Perineural invasion is a marker for pathologically advanced disease in localized prostate cancer. Int J Radiat Oncol Biol Phys 68:1059–1064

Lotan Y, Shariat SF, Khoddami SM et al (2004) The percent of biopsy cores positive for cancer is a predictor of advanced pathological stage and poor clinical outcomes in patients treated with radical prostatectomy. J Urol 171:2209–2214

Madersbacher S, Alcaraz A, Emberton M et al (2011) The influence of family history on prostate cancer risk: implications for clinical management. BJU Int 107(5):716–721

Merrick GS, Butler WM, Galbreath RW et al (2002) Relationship between percent positive biopsies and biochemical outcome after permanent interstitial brachytherapy for clinically organ confined carcinoma of the prostate gland. Int J Radiat Oncol Biol Phys 52:664–673

Merrick GS, Galbreath RW, Butler WM et al (2007) Primary Gleason pattern does not impact survival after permanent interstitial brachytherapy for Gleason 7 prostate cancer. Cancer 110:289–296

Merrick GS, Wallner KE, Galbreath RW et al (2008) Biochemical and functional outcomes following brachytherapy with or without supplemental therapies in men<or=50 years of age with clinically organ-confined prostate cancer. Am J Clin Oncol 31(6):539–544

Meyer JP, Bell CR, Elwell C et al (2008) Brachytherapy for prostate cancer: is the pretreatment prostate volume important? BJU Int 102(11):1585–1588

Mohler J, Bahnson RR, Boston B et al (2010) NCCN clinical practice guidelines in oncology: prostate cancer. J Natl Compr Canc Netw 8:162–200

Mondul AM, Selvin E, De Marzo AM et al (2010) Statin drugs, serum cholesterol, and prostate specific antigen in the National Health and Nutrition Examination Survey 2001–2004. Cancer Causes Control 21:671–678

Mydlo JH, Lebed B (2004) Does brachytherapy of the prostate affect sperm quality and/or fertility in younger men? Scand J Urol Nephrol 38(3):221–224

Obek C, Louis P, Civantos F et al (1999) Comparison of digital rectal examination and biopsy results with the radical prostatectomy specimen. J Urol 161:494–498

Palma D, Tyldesley S, Pickles T et al (2008) Prostate Cohort Outcomes Initiative: pretreatment prostate-specific antigen velocity is associated with development of distant metastases and prostate cancer mortality in men treated with radiotherapy and androgen-deprivation therapy. Cancer 112:1941–1948

Patel DA, Presti JC Jr, McNeal JE et al (2005) Preoperative PSA velocity is an independent prognostic factor for relapse after radical prostatectomy. J Clin Oncol 23:6157–6162

Pe ML, Trabulsi EJ, Kedika R et al (2009) Effect of percentage of positive prostate biopsy cores on biochemical outcome in low-risk PCa treated with brachytherapy or 3D-CRT. Urology 73(6):1328–1334

Piccirillo JF, Tierney RM, Costas I et al (2004) Prognostic importance of comorbidity in a hospital-based cancer registry. JAMA 291:2441–2447

Expert Panel on Radiation Oncology–Prostate, Frank SJ, Arterbery VE, Hsu IC et al (2011) American College of Radiology Appropriateness Criteria permanent source brachytherapy for prostate cancer. Brachytherapy 10(5):357–362

Rosenthal SA, Bittner NH, Beyer DC et al (2011) American Society for Radiation Oncology (ASTRO) and American College of Radiology (ACR) practice guideline for the transperineal permanent brachytherapy of prostate cancer. Int J Radiat Oncol Biol Phys 79(2):335–341

Rossi PJ, Urbanic J, Clark PE et al (2008) Pretreatment prostate-specific antigen velocity is associated with freedom from biochemical recurrence of prostate cancer after low-dose-rate prostate brachytherapy alone. Brachytherapy 7:286–289

RTOG Trails/0232. http://www.rtog.org/ClinicalTrials/ProtocolTable/StudyDetails.aspx?study=0232. Accessed 07/02/2011

Singer EA, Palapattu GS, van Wijngaarden E et al (2008) Prostate-specific antigen levels in relation to consumption of nonsteroidal anti-inflammatory drugs and acetaminophen: results from the 2001–2002 National Health and Nutrition Examination Survey. Cancer 113:2053–2057

Stock RG, Cahlon O, Cesaretti JA et al (2004) Combined modality treatment in the management of high-risk prostate cancer. Int J Radiat Oncol Biol Phys 59:1352–1359

Stone NN, Potters L, Davis BJ et al (2009) Multicenter analysis of effect of high biologic effective dose on biochemical failure and survival outcomes in patients with Gleason score 7–10 prostate cancer treated with permanent prostate brachytherapy. Int J Radiat Oncol Biol Phys 73:341–346

Szulkin R, Holmberg E, Stattin P et al (2012) Prostate cancer risk variants are not associated with disease progression. Prostate 72(1):30–39. doi:10.1002/pros.21403

Todor DA, Barani IJ, Lin PS et al (2011) Moving toward focal therapy in prostate cancer: dual-isotope permanent seed implants as a possible solution. Int J Radiat Oncol Biol Phys 81(1): 297–304

Trachtenberg J (2005) Treatment of lower urinary tract symptoms suggestive of benign prostatic hyperplasia in relation to the patient's risk profile for progression. BJU Int 95(Suppl 4):6–11

Vickers AJ, Till C, Tangen CM et al (2011) An empirical evaluation of guidelines on prostate-specific antigen velocity in prostate cancer detection. J Natl Cancer Inst 103(6):462–469

Wakil G, Gobaux V, Hajage D et al (2010) Can intermediate-risk patients be safely treated with permanent implant prostate brachytherapy: long-term results of the first 1044 patients of the Paris Institut Curie/Cochin hospital/Necker hospital group. In: Abstract, American Brachytherapy Society (ABS), San Diego, 2010

Wheeler TM, Dillioglugil O, Kattan MW et al (1998) Clinical and pathological significance of the level and extent of capsular invasion in clinical stage T1-2 prostate cancer. Hum Pathol 29(8): 856–862

Whittington R, Broderick GA, Arger P et al (1999) The effect of androgen deprivation on the early changes in prostate volume following transperineal ultrasound guided interstitial therapy for localized carcinoma of the prostate. Int J Radiat Oncol Biol Phys 44(5):1107–1110

Yu HH, Song DY, Tsai YY et al (2007) Perineural invasion affects biochemical recurrence-free survival in patients with prostate cancer treated with definitive external beam radiotherapy. Urology 70:111–116

Zam NA, Tan PH, Sim HG et al (2008) Correlation between prostate needle biopsies and radical prostatectomy specimens: can we predict pathological outcome? Pathology 40:586–591

Zelefsky MJ, Yamada Y, Pei X et al (2011) Comparison of tumor control and toxicity outcomes of high-dose intensity-modulated radiotherapy and brachytherapy for patients with favorable risk prostate cancer. Urology 77(4):986–990

Patient Selection and Recommendations: HDR

5

György Kovács

Temporary prostate interstitial brachytherapy using remote afterloading technology was first established in routine clinical practice in the second half of the 1970s (Watanabe et al. 1974). Following the introduction of transrectal ultrasound imaging(TRUS) for the guidance of permanent transperineal seed implants by Holm in Denmark, the first prospective controlled clinical applications with high-dose-rate (HDR) stepping source technology were started in Kiel/Germany (Bertermann and Brix 1990) resulting in a series of publications describing clinical situations where patients with localized prostate cancer benefit from HDR brachytherapy (Kovács et al. 1995; Martinez et al. 1995; Borghede et al. 1997; Dinges et al. 1998; Demanes et al. 2000).

Patient selection criteria are very similar to that of LDR implants (see Chap. 6); therefore, only differences will be discussed here in more detail.

Pretreatment investigations include:

1. Patient history
2. Histology
3. Gleason sum
4. Initial PSA
5. Nodal staging
6. Bone scan if PSA was >15 ng/ml
7. IPSS score
8. Residual urine volume
9. Uroflow
10. Preimplant TRUS imaging
11. Eligibility for anesthesia
12. Quality-of-life assessment

G. Kovács
Interdisciplinary Brachytherapy Unit,
University of Lübeck, Lübeck, Germany
e-mail: kovacsluebeck@gmail.com

G. Kovács, P. Hoskin (eds.), *Interstitial Prostate Brachytherapy*,
DOI 10.1007/978-3-642-36499-0_5, © Springer-Verlag Berlin Heidelberg 2013

Patient History: All patients should have complete medical history as well as an actual general physical examination to clear their suitability to the treatment. This should contain eligibility investigations for the planned anesthesia.

Histology: All patients need to have proven histology as a result of a 12-core TRUS-guided biopsy procedure. The percentage of positive biopsies, the proportion of each core involved, and the presence of perineural infiltrations are factors that are important influences on outcome (de la Rosette et al. 2009).

Gleason Sum Score: Gleason sum score has a strong prognostic relevance, and there is a large variety in the quality of different pathology reports due to differences in the level of experience of the pathologist. Second opinion by an expert could be advantageous if scientific evaluation of patients from different institutions will be performed (Egevad et al. 2002; Kronz et al. 2003).

Initial PSA: It should be recorded.

Nodal Staging: Computed tomography (CT), magnetic resonance imaging (MRI), and 5-FDG positron-emission tomography (PET) are the most common imaging methods for lymph node staging and detection of distant metastases. The more recently introduced choline-PET scan may also find a role in the staging of selected patients. The usefulness of surgical lymph node staging is controversial, but in selected patients with high risk of nodal involvement, lymph node sampling will give additional staging information in patients. Patients with stage T2 or less, PSA < 10 ng/ml, a Gleason score 6, and <50 % positive biopsy cores have <10 % likelihood of having node metastases and can be spared invasive nodal evaluation. Sentinel lymph node sampling is a promising option (Heidenreich et al. 2011).

Bone Scan: This may not be indicated in asymptomatic patients if the serum PSA level is <20 ng/ml in the presence of well-differentiated or moderately differentiated tumors (Heidenreich et al. 2011).

International Prostate Symptom Score (IPSS): Lower-urinary-tract function should be documented by this scoring system. Alternatively, the AUA (American Association of Urologists) scoring system can be used.

Post-voidal Urine Volume: It is an important indicator for existing obstructive symptoms if the value is over 100 ml.

Urinary Flow Rate (Qmax): It is another indicator for obstruction if the flow value is low.

Preimplant TRUS Imaging: The investigation should be performed by en expert at all patients before treatment decision. It is the most powerful and common imaging method for the guidance of prostate interstitial brachytherapy, both LDR and HDR. Clearance of pubic arch incompatibility, volume study, and imaging basis of dose calculations are the most important advantages. Other imaging modalities may also be used to compliment TRUS including CT and MR.

Anesthetic Assessment: During HDR brachytherapy, pain management and monitoring of vital functions are obligatory. General anesthesia is usual, but other groups have good experience with spinal or local anesthesia ± sedation (Wallner 2002).

Quality-of-Life Assessment: Quality-of-life measurement (including sexual function) results in the follow-up period are important information. Validated scoring systems should be used, e.g., EORTC QLQC30, FACT, IIEF, and PROMIS, including a baseline pretreatment.

Table 5.1 Patient selection criteria for a curative combined HDR brachytherapy and EBRT treatment

Inclusion criteria	Stages T1b–T3b
	Any Gleason score
	Any iPSA without proven distant metastases
Exclusion criteria	Pubic arch interference
	Rectal wall infiltration
	TURP within 6 months
	Infiltration of the external sphincter of the bladder neck
	Significant urinary obstructive symptoms
	Lithotomy position or anesthesia not possible

5.1 Patient Selection for HDR Brachytherapy in Combination with External Beam Radiation

HDR brachytherapy provides high-radiation-dose delivery; therefore, it is ideal for local dose escalation to limited volumes. Adaptation of stepping source dwell locations and dwell times within the target with no impairment from edema, source migration, or prostate movement during the short time of the boost delivery are important factors. BT boost with a stepping source in combination with EBRT can be recommended for a selected group of patients (Table 5.1). Although dose escalation is not necessary in low-risk cases (risk definition by at least initial PSA, Gleason sum, and T stage [Vicini et al. 2002]), the use of 50 Gy external beam dose combined with one or two HDR implantations shortens the total treatment time from the usual 8 to 6 weeks and results in significant dose reduction to normal tissues around the prostate. In the case of intermediate- or high-risk cases, this effect is even more prominent.

5.1.1 Potential Indications and Investigational Treatments

Limited numbers of patients with limited follow-up have been treated with temporary interstitial brachytherapy for localized prostate cancer without EBRT. Although this experience is promising, there is not yet enough evidence in the literature to advise this approach outside controlled clinical trials.

5.1.1.1 HDR Monotherapy

In patients with localized prostate cancer (T1b, T2a) presenting with favorable prognostic factors (iPSA ≤10 ng/ml, Gleason ≤6) who have an 80 % or greater probability of localized disease, temporary BT alone may be performed using appropriate fractionation to reach the optimal therapeutic ratio (Yoshioka et al. 2006; Demanes et al. 2007; Mark et al. 2007; Martinez et al. 2010). For patients with higher Gleason scores and higher PSA values, the risk of disease outside the prostate capsule increases. In these situations, temporary BT alone may also be effective but is not indicated out of controlled clinical trials (Hoskin et al. 2011).

5.1.1.2 Focal Therapy

Focal therapy aims to eradicate known cancer within the prostate and at the same time preserve uninvolved prostatic tissue, preserving genitourinary function. HDR brachytherapy is an excellent and proven method to concentrate high dose to a well-defined small volume and may be an ideal method for this approach. However, as yet there is no consensus on adequate target definition for focal therapy of prostate cancer, and this should only be delivered within strict controlled study circumstances (Polascik and Mouraviev 2008; de la Rosette et al. 2010).

5.1.1.3 HDR Brachytherapy for Salvage After Failure of Surgery, Primary Hormonal Treatment, or EBRT

HDR brachytherapy with or without adjuvant EBRT can be considered for salvage after failure of surgery, primary hormonal treatment, or EBRT. In locally recurrent disease, EBRT has been shown to modify the course of the disease. Local recurrence should be proven by biopsy and has to be visible by TRUS for target definition. The risk of side effects is significantly higher than for brachytherapy as first-line treatment and as salvage treatment after radical prostatectomy (RPE). Very limited experiences are published regarding salvage BT, but early reports demonstrate the feasibility of salvage temporary BT with or without complementary EBRT. Once again, however, such procedures should be applied only within prospective clinical trials (Lee et al. 2007; Allen et al. 2007).

5.1.2 Role of Antiandrogen Treatment

Androgen-deprivation hormonal treatment (ADT) has a significant role in reducing prostate volume before treatment ("downsizing") by reducing benign prostate hyperplasia contributing to the volume of the gland. The role of short-course neoadjuvant ADT combined with EBRT and temporary BT is under investigation. So far no significant advantage of short hormonal treatment has been observed in dose escalation studies (total biologic effective dose > 72 GyEQD2) with regard to long-term results (Krauss et al. 2011). On the other hand, when using EBRT combined with HT and applying much lower total radiation doses, ADT showed a significant benefit for such combinations (Bolla et al. 1997; Hinkelbein 1998; Lawton et al. 2001).

5.2 Clinical Team

An experienced team of different specialists is needed to perform treatment planning and delivery, as well as to control all issues necessary for a successful clinical treatment. The interdisciplinary team should be experienced in prostate interventional procedures, in TRUS (urologist, radiologist, or radiation oncologist), and in interstitial HDR brachytherapy. Urological assessment should evaluate clinical tumor stage, prostate volume, IPSS, urinary flow, and residual urine volume and exclude together with the radiation oncologist contraindications for transperineal TRUS-guided brachytherapy. These findings need to be discussed within the interdisciplinary team. Documentation of the brachytherapy must be performed according to national standards. It is helpful, when starting with this new treatment modality, to have a radiotherapist experienced in prostate temporary brachytherapy on-site during the first 3–5 implant procedures. In addition, a center undertaking such procedures and courses on TRUS brachytherapy should have been visited (Kovács et al. 2005).

5.3 Equipment

In contrast to LDR treatments, HDR brachytherapy requires stronger radiation protection during radiation delivery. When real-time planning is used (Kovács et al. 2007), the implantation room needs to have full radiation protection in order to irradiate the patient without movement post-implant. Required equipment includes:

- Operating room or brachytherapy suite suitable for sterile procedures and access to anesthetic support.
- HDR afterloader.
- TRUS unit with template; the ultrasound should be capable of both transaxial and sagittal image acquisition.
- TRUS fixation and stepping unit.
- Interstitial implant catheters of a suitable design compatible with the TRUS-based template; they should also be CT or MR compatible if this imaging method is to be used.
- Appropriate planning software to enable importation of post-implant imaging, implant reconstruction, and three-dimensional dosimetry.
- A brachytherapy suite with adequate shielding to perform the HDR treatment, according to national radiation protection rules.

5.4 Recording and Reporting

The ICRU recommendations (ICRU 1997) for recording and reporting brachytherapy applications should be followed as far as possible. Technical developments in treatment planning software solutions and quality assurance have introduced additional factors for improved description of the treatment, such as the dose non-homogeneity ratio (DNR) and the dose to organs at risk (bladder base, urethra, rectal wall) which should be part of the report. The dose should be related to fixed points and/or fixed volumes. It is suggested that the dose to 2 cm^3 (D_{2cc}) for the rectum and bladder, and D_{10} (dose to 10 %) and D30 (dose to 30 %) of the contoured prostatic urethra is used. Additionally, time-dose pattern should include dose rate and dose per fraction of the target dose (D100, D90) for CTV 1, CTV 2, and CTV 3, number and duration of the fractions, time interval between fractions, and the overall time

(Kovács et al. 2005). Furthermore, parameters which have been proven to be useful for reporting permanent prostate implants should be applied as they seem to be valid and reliable and also enable a better comparison between the different brachytherapy procedures: D90, D100, V100, V150, and V200 (Ash et al. 2000; Salambier et al. 2007).

5.5 Follow-Up

Extended follow-up is necessary because the yearly incidence rates for biochemical failure do not seem to plateau until at least 10 years following treatment and because a significant percentage of failures occur well beyond 5 years. PSA nadir and nadir time represent potential factors to be analyzed for future predictors of biochemical control (Vicini et al. 2011).

References

Allen GW, Howard AR, Jarrad DF et al (2007) Management of prostate cancer recurrences after radiation therapy – brachytherapy as a salvage option. Cancer 110:1405–1416

Ash D, Flynn A, Battermann J et al (2000) ESTRO/EAU/EORTC recommendations on permanent seed implantation for localized prostate cancer. Radiother Oncol 57:315–321

Bertermann H, Brix F (1990) Ultrasonically guided interstitial high dose rate brachytherapy with Ir-192: technique and preliminary results in locally confined prostate cancer. In: Martinez AA, Orton CF, Mould RF (eds) Brachytherapy HDR and LDR: remote afterloading state of the art. Nucletron International BV, Leersum, pp 281–303

Bolla M, Gonsales D, Warde P et al (1997) Improved survival in patients with locally advanced prostate cancer treated with radiotherapy and goserelin. N Engl J Med 337:295–300

Borghede G, Hedelin H, Holmang S et al (1997) Combined treatment with temporary short-term high dose rate iridium-192 brachytherapy and external beam radiotherapy for irradiation of localized prostatic carcinoma. Radiother Oncol 44:237–244

De la Rosette JJMCH, Wink MH, Mamoulakis C et al (2009) Optimization of prostate cancer detection in a clinical population: results from an eight versus 12-core transrectal ultrasound guided biopsy protocol. J Urol 182(4):1329

De la Rosette JJMCH, Ahmed HJ, Barentsz J, Barentsz T (2010) Focal therapy in prostate cancer—report from a consensus panel. J Endourol 24(5):775–780

Demanes DJ, Rodrigues RR, Altieri GA (2000) High dose rate prostate brachytherapy: the California endocurietherapy (CET) method. Radiother Oncol 57:289–296

Demanes DJ, Gilhezan M, Schour L et al (2007) High dose rate brachytherapy (HDR-BT) as monotherapy for favorable prostate cancer: excellent 5-year control rates and low toxicity. Int J Radiat Oncol Biol Phys 69:S83

Dinges S, Deger S, Koswig S et al (1998) High dose rate interstitial brachytherapy combined with external beam irradiation for localized prostate cancer—a prospective study. Radiother Oncol 48:197–203

Egevad L, Granfors T, Karlberg L et al (2002) Prognostic value of the Gleason score in prostate cancer. BJU Int 89(6):538–542

Heidenreich A, Bellmunt J, Bolla M et al (2011) EAU guidelines on prostate cancer. Part 1: screening, diagnosis, and treatment of clinically localized disease. Eur Urol 59:61–71

Hinkelbein W (1998) Adjuvant or therapeutic androgen suppression in locoregional advanced prostatic carcinoma (RTOG 85–31) [Article in German]. Strahlenther Onkol 174:385–386

Hoskin P, Rojas A, Lowe G et al (2011) High-dose-rate brachytherapy alone for localized prostate cancer in patients at a moderate or high risk of biochemical recurrence. J Radiat Oncol Biol Phys. doi:10.1016/j.ijrobp.2011.04.031

http://cancer.beaumont.edu/beaumont-launches-prostate-cancer-study-single-dose-radiation-treatment. Accessed 25 Aug 2011

ICRU Report 58 (1997) Dose and volume specification for reporting interstitial brachytherapy, International Commission on Radiation Units and Measurements, ICRU, Bethesda

Kovács G, Galalae R, Wirth B et al (1995) Optimization of interstitial brachytherapy by a new implant technique in the treatment of prostate cancer (ger). Strahlenther Onkol 171:685–688

Kovács G, Pötter R, Loch T et al (2005) GEC/ESTRO-EAU recommendations on temporary brachytherapy using stepping sources for localized prostate cancer. Radiother Oncol 74:137–148

Kovács G, Melchert C, Sommerauer M et al (2007) Intensity modulated high-dose-rate brachytherapy boost complementary to external beam radiation for intermediate- and high-risk localized prostate cancer patients – how we do it in Luebeck/Germany. Brachytherapy 6:142–148

Krauss D, Kestin L, Ye H et al (2011) Lack of benefit for the addition of androgen deprivation therapy to dose-escalated radiotherapy in the treatment of intermediate- and high-risk prostate cancer. Int J Radiat Oncol Biol Phys 80(4):1064–1071

Kronz JD, Milord R, Wilentz R, Weir EG, Schreiner SR, Epstein JI (2003) Lesions missed on prostate biopsies in cases sent in for consultation. Prostate 54(4):310–314

Lawton CA, Winter K, Murray K et al (2001) Updated results of the phase III Radiation Therapy Oncology Group (RTOG) trial 85–31 evaluating the potential benefit of androgen suppression following standard radiation therapy for unfavourable prognosis carcinoma of the prostate. Int J Radiat Oncol Biol Phys 49:937–946

Lee B, Shinohara K, Weinberg V et al (2007) Feasibility of high-dose-rate brachytherapy salvage for local prostate cancer recurrence after radiotherapy: the University of California-San Francisco experience. Int J Radiat Oncol Biol Phys 67(4):1106–1112

Mark RJ, Akins RS, Anderson PJ et al (2007) Interstitial high dose rate (HDR) brachytherapy as monotherapy for early stage prostate cancer: a report of 206 cases. Int J Radiat Oncol Biol Phys 69:S329

Martinez AA, Gonzalez J, Stromberg J et al (1995) Conformal prostate brachytherapy: initial experience of a phase I/II dose-escalating trial. Int J Radiat Oncol Biol Phys 33:1019–1027

Martinez AA, Demanes J, Vargas C et al (2010) High-dose-rate prostate brachytherapy: an excellent accelerated-hypofractionated treatment for favorable prostate cancer. Am J Clin Oncol 33:481–488

Polascik TJ, Mouraviev V (2008) Focal therapy for prostate cancer. Curr Opin Urol 18:269

Salambier C, Lavagnini P, Nickers P et al (2007) Tumour and target volumes in permanent prostate brachytherapy: a supplement to the ESTRO/EAU/EORTC recommendations on prostate brachytherapy. Radiother Oncol 83:3–10

Vicini FA, Martinez AA, Hanks G et al (2002) An interinstitutional and interspecialty comparison of treatment outcome data for patients with prostate carcinoma based on predefined prognostic categories and minimum follow-up. Cancer 95:2126–2135

Vicini FA, Shah C, Kestin L et al (2011) Identifying differences between biochemical failure and cure: incidence rates and predictors. Int J Radiat Oncol Biol Phys. doi:10.1016/j.ijrobp.2011.05.017

Wallner K (2002) Prostate brachytherapy under local anesthesia: lessons from the first 600 patients. Brachytherapy 1:145–148

Watanabe H, Igari D, Tanahasi Y et al (1974) Development and application of new equipment for transrectal ultrasonography. J Clin Ultrasound 2:91–98

Yoshioka Y, Konishi K, Oh RJ et al (2006) High-dose-rate brachytherapy without external beam irradiation for locally advanced prostate cancer. Radiother Oncol 80:62–68

Prostate Brachytherapy Implantation Treatment Techniques

6

Jan J. Battermann

6.1 Introduction

Modern permanent prostate brachytherapy (PPB) is based on imaging of the prostate by transrectal ultrasound (TRUS) and by MRI, using different sequences, blood perfusion and spectroscopy. With these techniques, accurate volume measurement can be done, as well as exclusion of extracapsular growth. But up till now, only TRUS is used during the implant procedure to guide the needles at the right position into the prostate.

6.2 History of Prostate Brachytherapy

Historically prostate brachytherapy was already performed in 1913 by Pasteau and Degrais using radium needles (Pasteau and Degrais 1914). Flocks performed prostate implants with radioactive gold (Au-198), first with a colloidal solution, but due to technical difficulties, gold seeds were developed for brachytherapy alone or in combination with external beam irradiation (Flocks et al. 1954; Lannon et al. 1993). This technique, however, was not very popular, because of the high energy of Au-198. The introduction of low-energy Iodine-125 seeds in the beginning of the 1980s renewed interest in prostate brachytherapy.

Hilaris and Whitmore at the Memorial Sloan-Kettering Cancer Center reported the largest series of 606 patients treated with the retropubic technique (Fig. 6.1) resulting in a 5-year survival of 79 % (Whitmore et al. 1972). However, about half of the patients experienced local recurrence after longer follow-up. This was the result of poor distribution of the seeds over the prostate, especially at the apex.

J.J. Battermann
Department of Radiation Oncology, University Hospital Utrecht, Utrecht, The Netherlands
e-mail: j.j.battermann@umcutrecht.nl

G. Kovács, P. Hoskin (eds.), *Interstitial Prostate Brachytherapy*,
DOI 10.1007/978-3-642-36499-0_6, © Springer-Verlag Berlin Heidelberg 2013

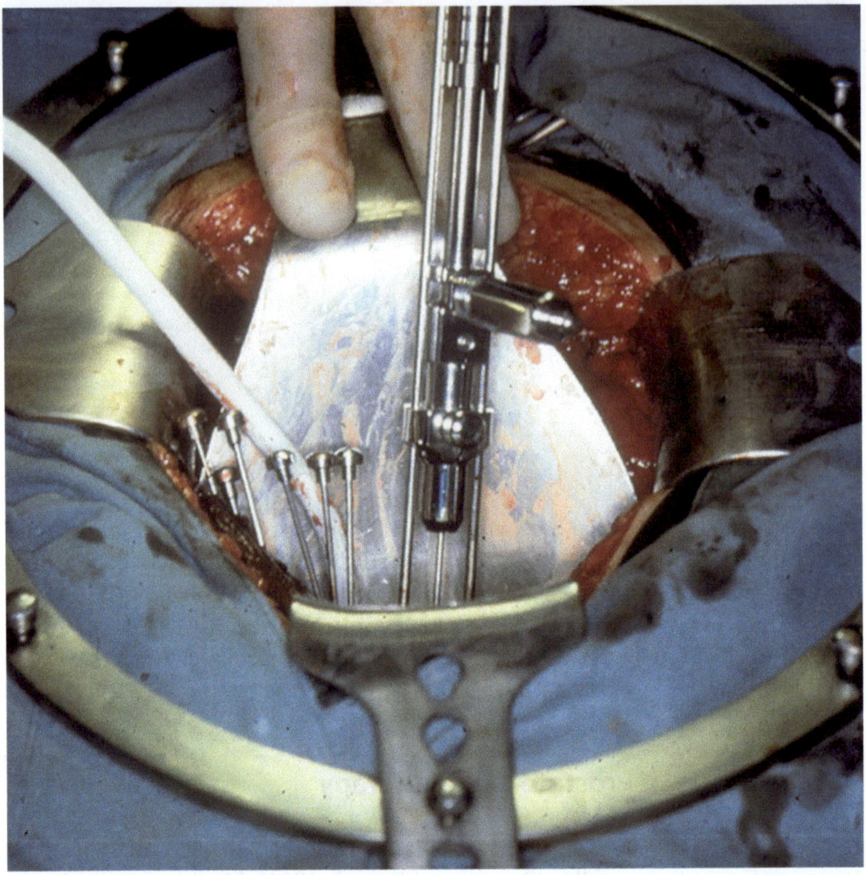

Fig. 6.1 Retropubic technique, using Mick applicator This is an open abdominal technique with guidance of the needles through a finger in the rectum. The Mick applicator has a cartridge of 10 or 15 seeds that can be pushed into the prostate with a stylet

Furthermore, the retropubic technique, combined with lymph node dissection, resulted in a postoperative complication rate of 23 %.

6.2.1 Perineal Implantation

In 1980 Charyulu described a perineal approach, combining EBRT and radon seeds in patients with advanced tumours (Charyulu 1980). Kumar improved this technique with the use of a C-arm to guide the needles (Kumar and Bartone 1981), but the greatest improvement came from Holm with the introduction of TRUS-guided needle placing (Fig. 6.2) (Holm et al. 1983). The TRUS probe was attached to a Perspex plate with holes at 1 cm apart, both in horizontal and vertical direction. This technique was refined over the following years, especially in Seattle, with the introduction of the support system with stepping unit, improved ultrasound equipment and planning systems in place of the nomogram (Blasko et al. 1987).

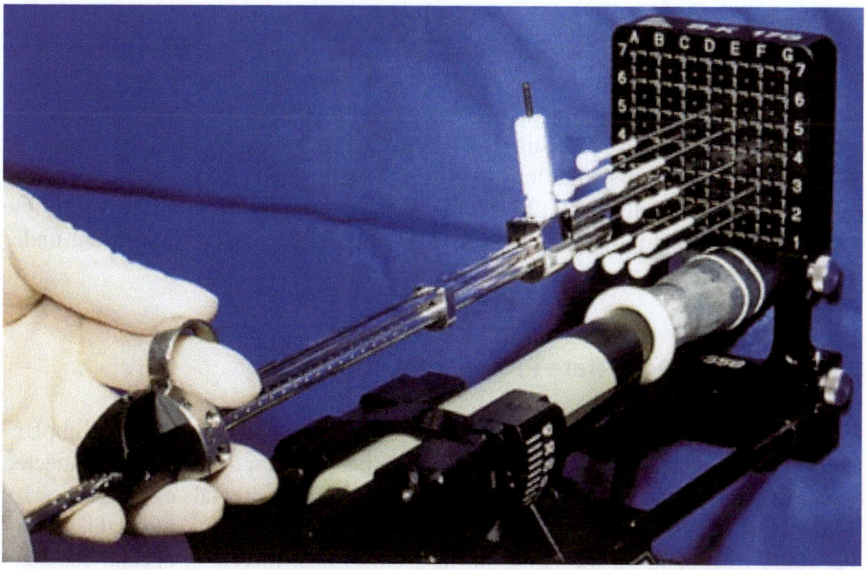

Fig. 6.2 Perineal technique, using Mick applicator Here the Mick applicator is placed on the template to insert the seeds with a similar applicator as in Fig. 6.1

6.3 Current Implantation Techniques

Current implantation techniques make use of permanent low-dose-rate brachytherapy, using I-125 or Pd-103 seeds; high dose rate using Iridium-192; and pulsed dose rate, using Iridium as well.

6.3.1 Permanent Low-Dose-Rate Prostate Brachytherapy

The longest experience of prostate brachytherapy is with PPB, using Iodine-125 or Palladium-103 seeds. Pd-103 is not popular in Europe, but in the USA still a substantial number of patients are treated with this isotope. The advantage is a shorter half-life of 17 days, versus 60 days for I-125, with similar energy (27 KeV versus 21). The shorter half-life has been proposed as an advantage in the treatment of aggressive cancers with Gleason scores of 8–10, but this has not been substantiated in clinical studies.

6.3.2 High-Dose-Rate Prostate Brachytherapy

Since the concept of the low αβ ratio in prostate cancer, there has been interest in the use of fractionated high-dose-rate (HDR) radiation (Brenner and Hall 1999). HDR brachytherapy has been used for many years for cervical cancers and has proven at least as good as low-dose-rate (LDR) brachytherapy. In prostate HDR was introduced in Germany by Bertermann and Kovacs in the management of advanced prostate cancers (Kovacs et al. 1996). These patients received EBRT, HDR brachytherapy and androgen deprivation (AD) with promising results. The group in Detroit showed similar good results, even without AD (Martinez et al. 2010). However, for HDR several fractions are needed, and that means either several hospitalisations or one hospitalisation for several days, where the needles remain in the patient until the final fraction is given. The first paper on HDR monotherapy was published by Grills et al. (2004). She showed excellent results, with very low complications and similar cancer control as with PPB. Several studies are under way to confirm these results.

6.3.3 Pulsed-Dose-Rate Prostate Brachytherapy

A third technique is with pulsed dose rate in place of HDR. The techniques of implantation are identical for all three methods, but in this situation the patient is connected to an afterloading machine, to give a radiation fraction at intervals every hour or so. With this technique it is essential that the needles remain in place and stable during the whole treatment. At the AMC in the Netherlands, therefore, special catheters were developed that are anchored in the prostate tissue (Pieters et al. 2006).

6.4 Transperineal Technique for LDR, HDR and PDR Brachytherapy

As mentioned above, the technique of needle placement is identical for PPB, HDR and PDR. Details of the technique in different centres may vary, depending on the use of TRUS, CT scan and fluoroscopy, but the majority of centres performing PPB will use TRUS to guide needle placement.

6.4.1 Preplanning

After preoperative workup and patient selection, a preplan can be made based on TRUS imaging or CT/MRI imaging. In general TRUS will be used for preplanning.

The patient is positioned in the same lithotomy position as during the treatment procedure (Fig. 6.3). The template on the ultrasound monitor should be placed in such a way that the largest prostate slice is in the middle of the template and the dorsal side of the prostate is a little bit lower than the first row of the holes in row 1 to avoid needles having to be placed where no holes are present on the template (Fig. 6.4). Usually the urethra is located in the D line (the midline on the template), but when the urethra is not in the D line, the prostate should still be centralised and a careful note taken of the alternative row defining the urethra. Using the stepping unit, the ultrasound probe can be moved with increments of 5 mm through the prostate from base to apex. To define base and apex, fusion of TRUS and MRI can be very helpful. With a computer planning program, the outline is introduced

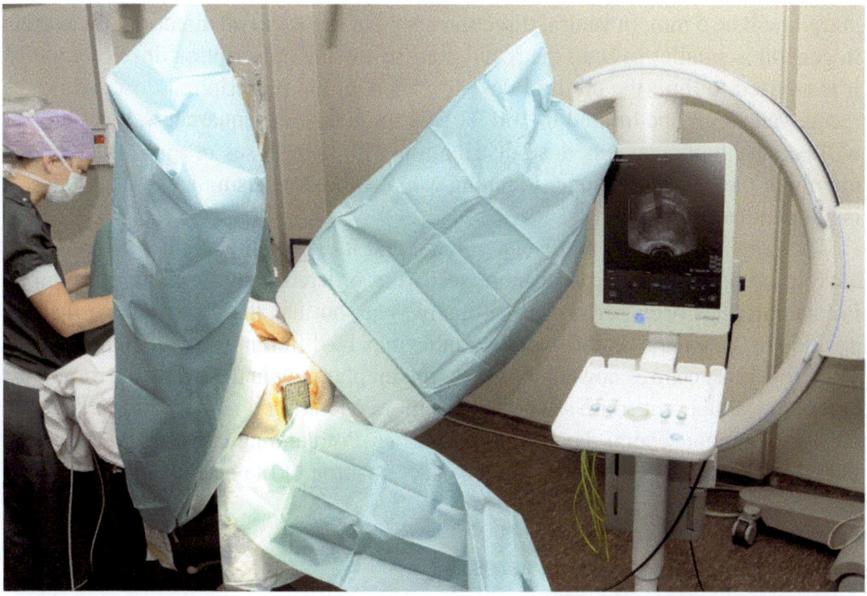

Fig. 6.3 Patient set-up in lithotomy position The patient is placed in the same position as during the preplanning with the legs in stirrups and the knees in about 90°. If the prostate is large the knees may be bent somewhat more to the nose

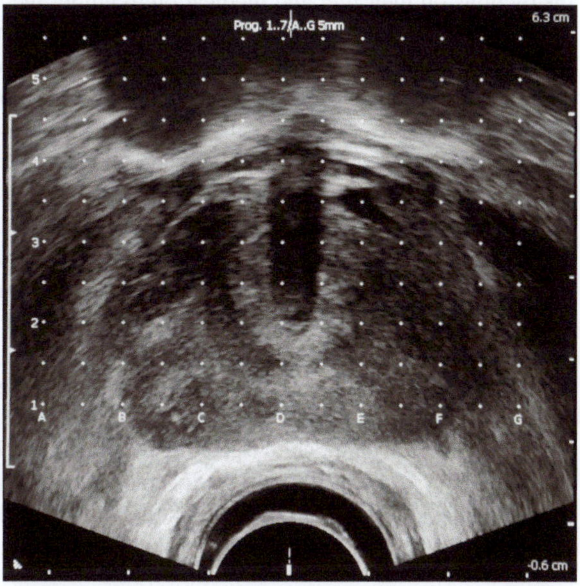

Fig. 6.4 Template with contour of the prostate in the middle of the template. The largest contour of the prostate is placed in the middle of the grid. In general the urethra will be in the *vertical D line*, and the dorsal side of the prostate should be a few millimetres under row 1 to keep the rectal dose below the prescription dose

in the system. The target volume should be the whole prostate with some margin, according to the GEC ESTRO guidelines (Salembier et al. 2007). The urethra and rectal wall are critical organs and should also be defined. In lateral directions, the margin will be 5 mm, in ventral direction also, but in the dorsal direction, the margin in general is smaller to keep the rectal dose below the prescription dose of 144 Gy. It is advised to limit the urethral dose to 150 % of the prescription dose, while the rectal wall should be limited to 100 %. Due to the inverse-square law, there always will be an inhomogeneous dose distribution with high doses of more than 200 % mainly in the peripheral zone. Planning can also be done using a CT scan, but this is not popular in Europe.

The preplan will give a prostate volume and indicates the number of seeds that have to be ordered. In general more seeds are ordered than the required number. Preplanning has been performed for many years but has the disadvantage that the position of the patient on the operation table may be different from the initial TRUS. Therefore most centres nowadays will do the preplanning in the operation room, just before the actual procedure starts. It takes more operating time but is more accurate than preplanning several days or weeks earlier. Preplanning can also be combined with intraoperative planning, using the exact position of the needles inserted.

6.4.2 Pretreatment Measures

Patients should be reviewed by an anaesthetist some days before the implant. Either general or spinal/saddle block anaesthesia can be used. For centres beginning with prostate implants, it is advised to use general anaesthesia. Patients will have had instructions on their diet and medication to empty the lower digestive tract. A few hours before the procedure, an enema is given. Instructions must also be given on stopping of medication, such as anticoagulants. Some centres give an alpha blocker some weeks before the treatment to improve urine flow. Antibiotics may also be given, either for several days orally before or as a bolus during the procedure.

Before the procedure starts, a final check will be done to be sure that all equipment is present and working. This includes the support system (Fig. 6.5) with stepping unit, a template, the ultrasound system with stand-off for the probe, C-arm if available, planning system and disposables (Fig. 6.6) such as Foley catheter, bevel-shaped implantation needles with stylet and locking needles (Fig. 6.7). The support system can be mounted on the operating table or can be a system placed on the ground. Locking needles (generally 2) are used for fixation of the prostate but are not used universally. The C-arm can be used during the procedure to check the needle position in relation to bladder or balloon of the catheter and afterwards for checking the total number of seeds in the prostate. New developments enable a C-arm to take CT slices through the prostate, and this can be used to check whether there are cold spots in the prostate with the option to add one or more seeds.

6.4.3 Implant Procedure

The actual implant procedure is carried out in a dedicated room for brachytherapy or in a general operating room. After the patient is anesthetised, he is placed in lithotomy position with the legs in stirrups and the femur in a vertical position.

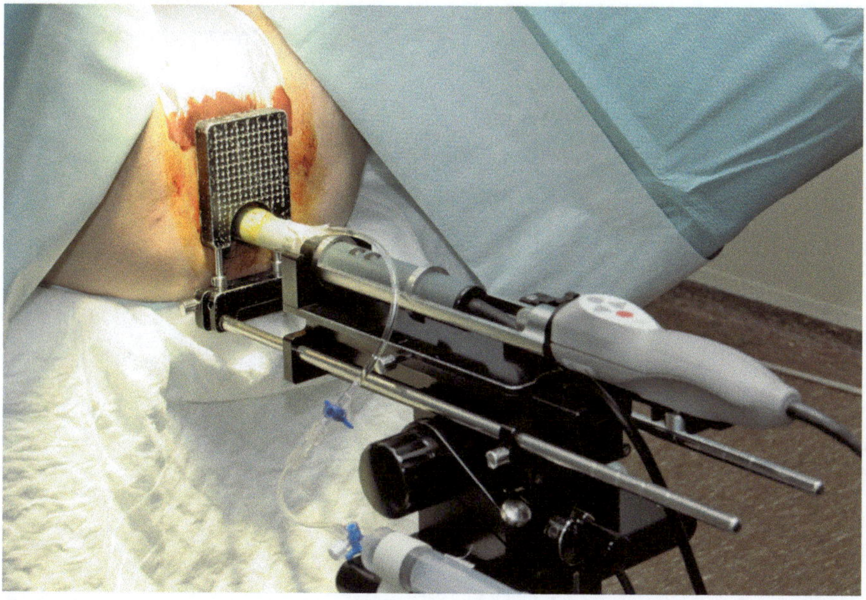

Fig. 6.5 Support system with stepping unit, template, ultrasound system and probe with 3D capability

In case of a large prostate volume, hyperextending the legs will widen the pubic arch. A Foley catheter is introduced to drain the bladder and to visualise the urethra. Some centres use aerated gel (half lubricating gel, half air mixed in a syringe to produce small bubbles that are visible on TRUS) for better urethral visualisation. The scrotum is fixed away from the perineum and the perineal area is disinfected.

The next step is to introduce the probe in the rectum and fix it on the stepping unit, reproducing the position of the preplanning if done earlier. The template is then mounted on the cradle of the stepping unit and checked for alignment of the prostate in the centre of the template (D line and row 1, Fig. 6.4). If no preplan has been performed, a set of outlines is taken, using the stepper with 5 mm increments, and increasingly a 3D acquisition will be performed also. The data are transferred to the planning computer and a dose plan can be generated (Fig. 6.9).

Stabilising needles are inserted under TRUS vision to monitor their entry both in transaxial (as an ultrasound shadow) or sagittal (whole needle visible) positions (Fig. 6.8). The treatment needles can be inserted one by one and directly loaded, or all needles can be placed first and loaded afterwards. Preloaded needles are also available and may be preferred. Needle placing is guided by TRUS using the transversal position and in the sagittal direction needles when they can often be better guided to the required position (Fig. 6.8).

If preloaded needles are not used, seeds are loaded after the needle is placed and checked for correct position. One can also insert all necessary needles first, followed by a second 3D scan, and check the position of each individual needle with sagittal TRUS. This gives also the opportunity to make an intraoperative plan based on the actual position of the needles.

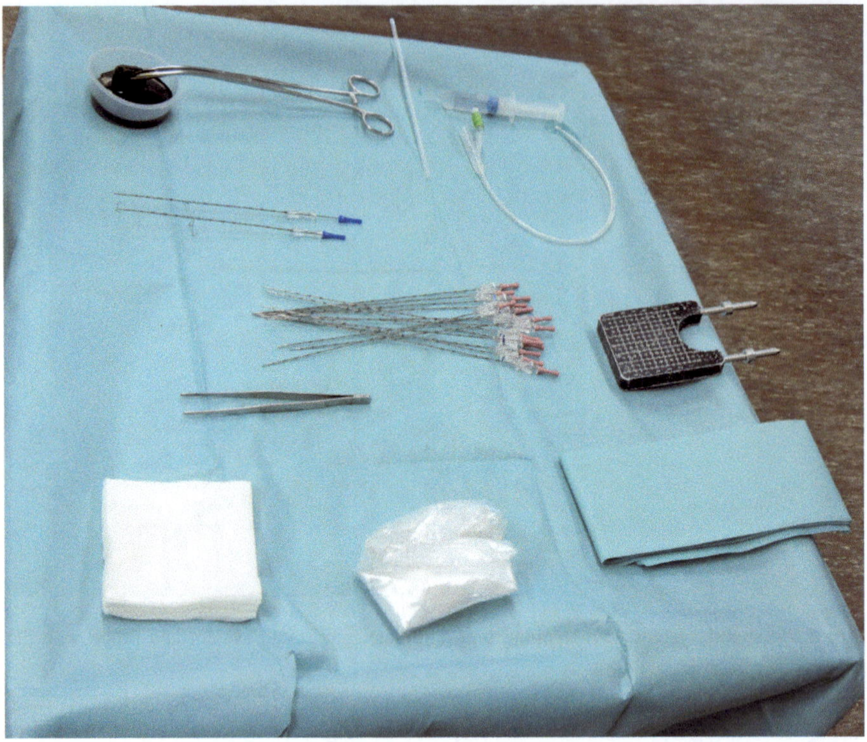

Fig. 6.6 Table with disposables Foley catheter, lubricating gel, implantation needles with stylet, locking needles and disinfectants. The needles have a bevel-shaped tip, and therefore, the needle can be directed in the desired direction. With the tip up, it will go upwards, etc.

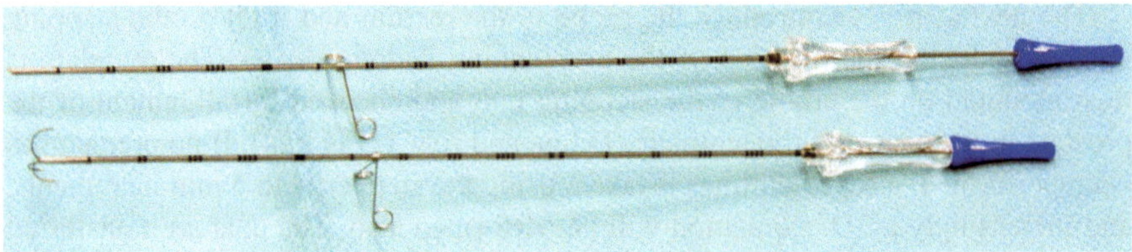

Fig. 6.7 Locking needles. These needles are used by many centres to fixate the prostate to the template. Some have a single hook, others three. Especially in combination with intraoperative planning, these needles are useful

6.4.4 Implantation Devices

There are many vendors in the seed market. All of them are using the similar seeds, typically 4.5 mm length and 0.8 mm diameter. Seeds can be introduced with the Mick applicator (Best Industries, USA), a simple device with a cartridge of 10 or 15 seeds, where with a stylet one seed can be pushed into the prostate (see Fig. 6.2).

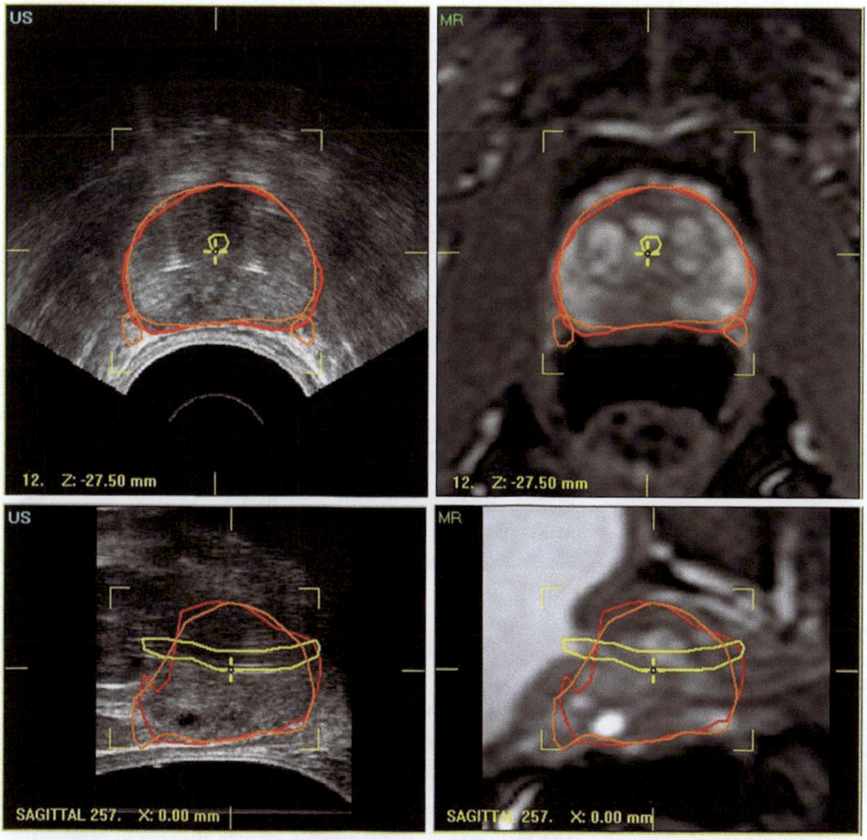

Fig. 6.8 Contoured outline of the prostate in transversal and sagital direction. The *red* line is the prostate contour using TRUS. The *orange* line is the contour outlined with MRI. With MRI the neurovascular bundle can also be depicted. *Yellow* is the outlined position of the urethra. Both the transversal and the sagital view should be used when placing the needles. Ultrasound systems may have a revolving probe to make 3D images of the prostate. This is used to optimize the position of the needles.

Then, the device is retracted for 5, 10 or more millimetres and the next seed is placed, and so on. This is a fast system with flexibility in the placing of seeds to spare urethra or rectum wall. Seeds can also be stranded in a suture of polyglactin (a biodegradable material), typically with 10 seeds per strand (Fig. 6.10). During the procedure the strands can be cut into the required lengths (e.g. 3 or 4 seeds) (Fig. 6.11). There are so-called strand holders available to reduce radiation exposure and to facilitate the introduction of the strands into the needles (Fig. 6.12). The disadvantage of strands is that differential loading (placing seeds further or closer to each other to improve the dose distribution) is more difficult, but their advantage is less chance of migration of seeds and straighter lines of seeds in the prostate. Bard offers the Quick link system to connect loose seeds and spacers to make strands with the advantage of differential loading within a strand (Fig. 6.13). Nucletron

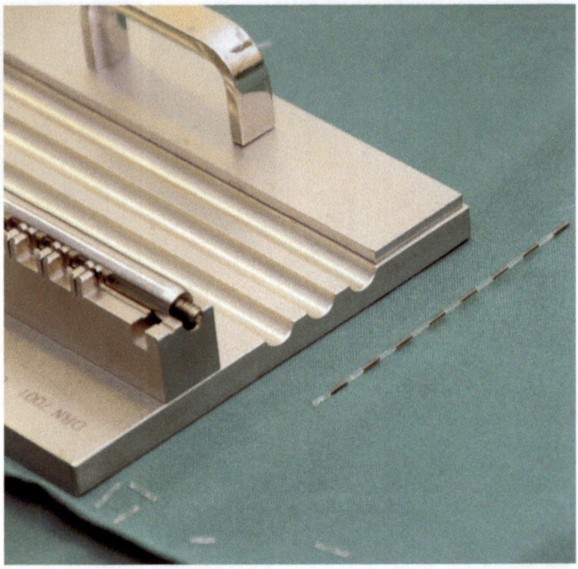

Fig. 6.9 Dose planning More and more intraoperative planning is used to get a dose plan as accurate as possible. With the semi-3D planning, both the dose in transversal and in sagittal direction can be depicted

Fig. 6.10 Strand with 10 seeds at 0.55 mm between active seeds. Typically the seeds are 4.5 mm length and 0.8 mm in diameter. This means that when several spacers are used in a needle, the distance between the first and last seed will increase

offers a complete system for remote afterloading of the seeds, the FIRST system (Fig. 6.14). Customers order a cartridge of the desired number of seeds (10 or multiples of 10 up till 120) and a standard cartridge of 100 spacers. The machine makes the desired configuration without any human support and brings the train of seeds and spacers into the prostate. The brachytherapist only has to connect the cable to the consecutive needles and, after inserting the train of seeds and spacers, retract the needle from the patient. The advantages of this system are no radiation exposure to

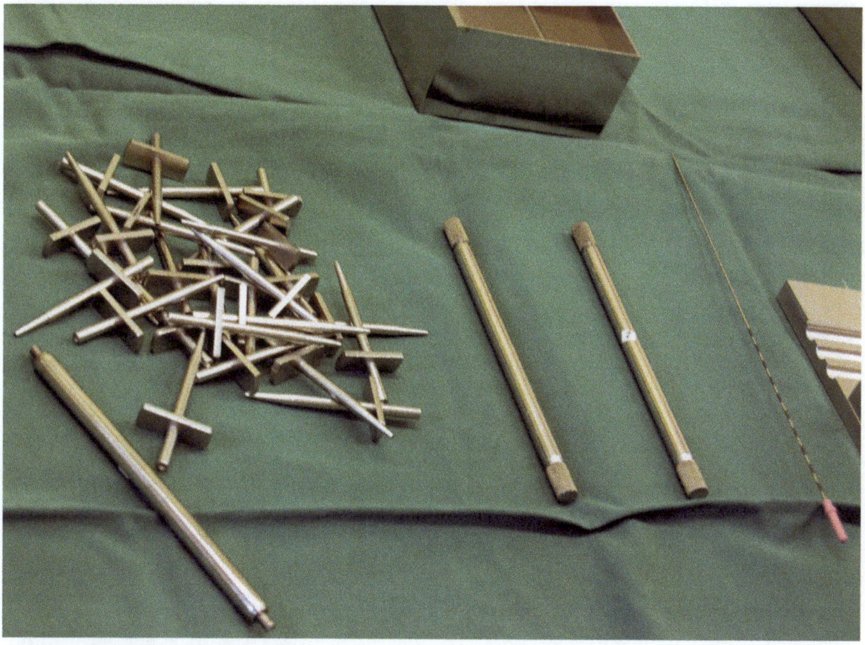

Fig. 6.11 Preparing strand in desired pieces for loading into needle. From a strand seeds can be cut with a knife, or it can be broken

personnel, full flexibility in loading configuration and fewer personnel needed as there is no preparation of seeds.

6.4.5 End of Procedure

After the insertion of all seeds and removal of the needles, the procedure ends with fluoroscopy using a C-arm or CT-C-arm to count the number of seeds in the body of the patient (Fig. 6.15). Seed migration is rare and occasionally seeds may be exposed through the catheter. Migration can occur after 1 or more months up to 1 year; it is often to the lung, but seeds can also be found in pelvic lymph nodes and more often seeds may be passed in the urine. With loose seed techniques, more seeds are lost than with stranded seeds (Hinnen et al. 2010).

Treatment can be undertaken as an outpatient procedure, or the patient can stay for one or two nights in the hospital. Before discharge from the hospital, it is checked that the patient can urinate and he will receive information on radiation safety and side effects that might occur.

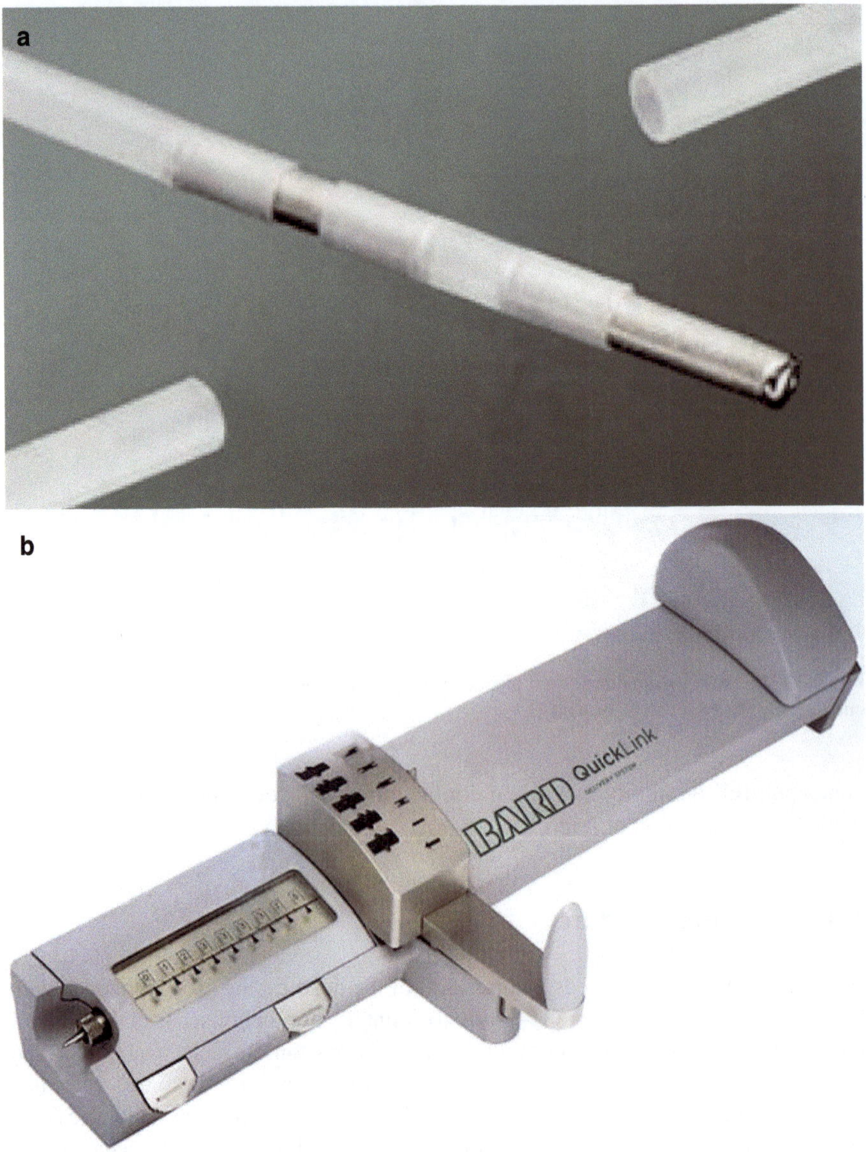

Fig. 6.12 (**a**) Utrecht strand holders in a box with loading of similar number of seeds to be introduced in the needles. The holder is placed on the hub of the needle and the composed train of seeds and spacers is pushed into the needle. At the desired position the needle is retracted over the obturator. (**b**) Needles are visible; a number of needles were already retracted, showing the piles of seeds deposited

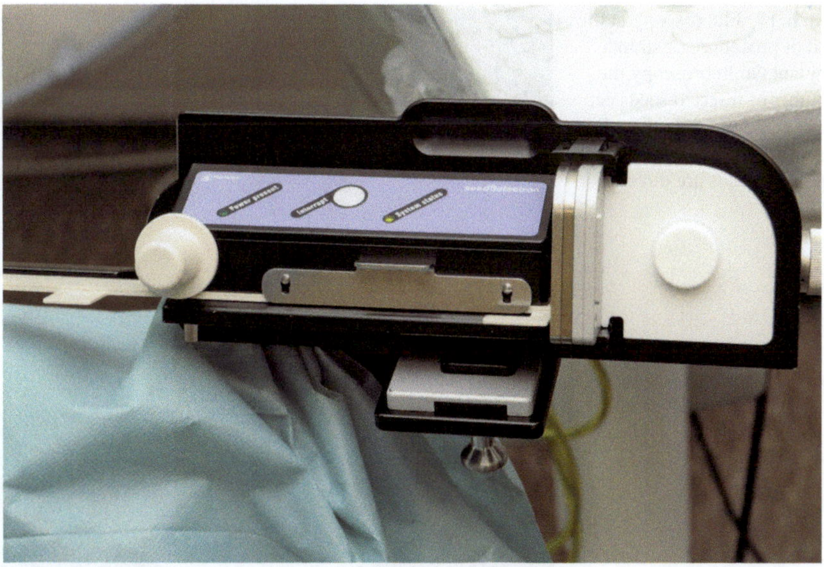

Fig. 6.13 Quick link system (Bard) With this system loose pieces of a strand can be connected at desired configuration to make a whole strand

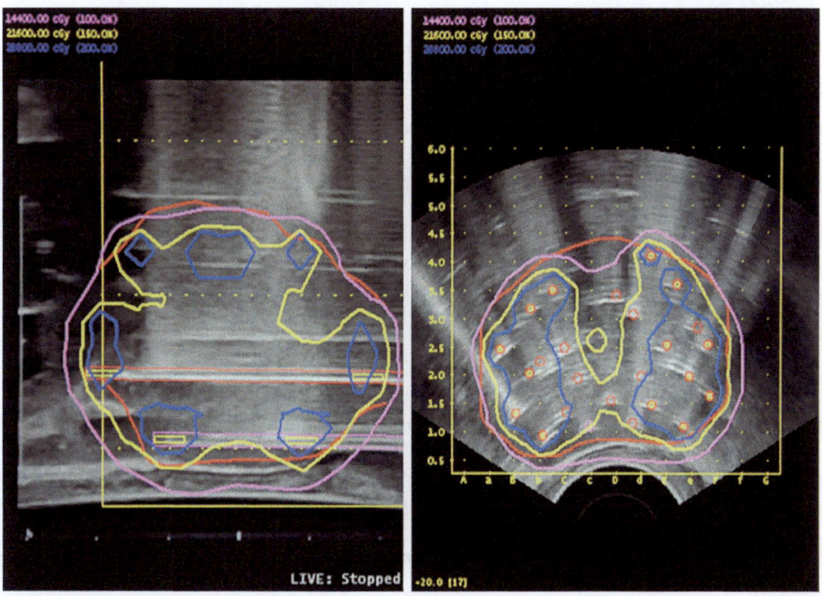

Fig. 6.14 FIRST system (Nucletron) The FIRST system with seedSelectron, seed cartridge, spacer cartridge, drive wire and compose element. This is the first afterloading system for the introduction of seeds. The desired composition of a train of seeds and spacers is composed by the computer, and the train is pushed into the prostate to position. Then, the needle is retracted till just outside the prostate and further retracted by the operator. Intra-operative dose planning in transversal and sagital direction. Colored lines show the isodose lines, *purple* line is the 100 % isodose, *yellow* line is 150 % and *blue* is 200 %

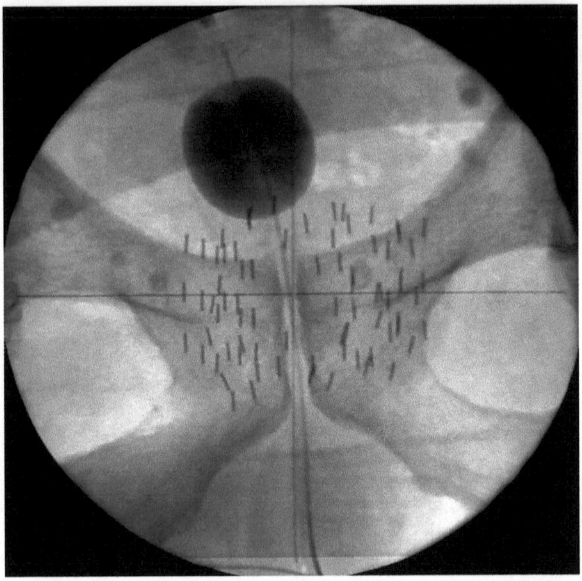

Fig. 6.15 Fluoroscopy at the end of procedure of strand implant On fluoroscopy the number of inserted seeds can be counted to verify that all seeds are in the prostate. Stranded seeds are directed in line

References

Blasko JC, Radge H, Schumacher D (1987) Transperineal percutaneous iodine-125 implantation for prostatic carcinoma using transrectal ultrasound and template guidance. Endocurie/ Hypertherm Oncol 3:131–139

Brenner DJ, Hall EJ (1999) Fractionation and protraction for radiotherapy of prostate carcinoma. Int J Radiat Oncol Biol Phys 43:1095–1101

Charyulu KKN (1980) Transperineal interstitial implantation of prostate cancer: a new method. Int J Radiat Oncol Biol Phys 6:1261–1266

Flocks RH, Kerr HD, Elkins HB et al (1954) Treatment of carcinoma of the prostate by interstitial radiation with radioactive gold (Au-198): a follow-up report. J Urol 71:628–633

Grills IS, Martinez AA, Hollander M et al (2004) High dose rate brachytherapy as prostate cancer monotherapy reduces toxicity compared to low dose rate palladium seeds. J Urol 171:1098–1104

Hinnen KA, Moerland MA, Battermann JJ et al (2010) Loose seeds versus stranded seeds in I-125 prostate brachytherapy: differences in clinical outcome. Radiother Oncol 96:30–33

Holm HH, Juul N, Pedersen JF et al (1983) Transperineal ^{125}iodine seed implantation in prostatic cancer guided by transrectal ultrasonography. J Urol 130:283–286

Kovacs G, Wirth B, Bertermann H et al (1996) Prostate preservation by combined external beam and HDR brachytherapy at nodal negative prostate cancer patients – an intermediate analysis after ten years experience. Int J Radiat Oncol Biol Phys (Abstr) 36(Suppl 1):198

Kumar PP, Bartone FF (1981) Transperineal percutaneous I-125 implant of prostate. Urology 17:238–240

Lannon SG, el-Araby AA, Joseph PK et al (1993) Long-term results of combined interstitial gold seed implantation plus external beam irradiation in localised carcinoma of the prostate. Br J Urol 72:782–791

Martinez A, Demanes J, Vargas C et al (2010) High-dose-rate prostate brachytherapy. An excellent accelerated-hypofractionated treatment for favourable prostate cancer. Am J Clin Oncol 33:481–488

Pasteau O, Degrais D (1914) The radium treatment of cancer of the prostate. Arch Roentg Ray 28:396–410

Pieters BR, van der Grient JNB, Blank LECM et al (2006) Minimal displacement of novel self-anchoring catheters suitable for temporary prostate implants. Radiother Oncol 80:69–72

Salembier C, Lavagnini P, Nickers P et al (2007) Tumour and target volumes in permanent prostate brachytherapy: a supplement to the ESTRO/EAU/EORTC recommendations on prostate brachytherapy. Radiother Oncol 83:3–10

Whitmore WF Jr, Hilaris B, Grabstald H (1972) Retropubic implantation of iodine-125 in the treatment of prostatic cancer. J Urol 108:918–920

7 HDR Technique

Peter Hoskin

7.1 Introduction

High dose-rate (HDR) prostate brachytherapy is a transrectal ultrasound-guided transperineal procedure similar to the procedure used for low-dose rate seed brachytherapy, but instead of depositing low-dose-rate seeds, afterloading applicators are placed in position to direct the high-dose-rate source once the implant is completed and dosimetric calculations have been approved. To achieve a good implant, three basic principles must be applied, meticulous technique, individualised dosimetry and good quality assurance. Training both in the theory and in the practical aspects of high-dose-rate afterloading brachytherapy is essential together with a period of mentorship from an experienced centre.

7.2 Equipment

The transperineal technique is performed under anaesthesia. It is therefore important to have access to an environment where this can be carried out safely with appropriate support. Many units will use general anaesthesia, whilst others prefer spinal anaesthesia for its more rapid recovery and the few systemic effects. Local anaesthetic procedures have also been described, and even where general or spinal anaesthesia is used, local infiltration of the perineum is of value in improving pain relief after implantation before treatment delivery.

- Transrectal ultrasound is essential for guidance of the applicators into the prostate in an appropriate pattern. This should have transaxial and sagittal crystal arrays allowing imaging in both planes. Access to Doppler and elastography may also enhance the value of imaging for the implant procedure.

P. Hoskin
Mount Vernon Cancer Centre,
University College London, London, UK
e-mail: peterhoskin@nhs.net

G. Kovács, P. Hoskin (eds.), *Interstitial Prostate Brachytherapy*,
DOI 10.1007/978-3-642-36499-0_7, © Springer-Verlag Berlin Heidelberg 2013

- A stepper unit is required with a cradle for the ultrasound probe. This should allow adjustment in all three planes for optimal positioning of the probe. It may be either floor mounted or mounted on the couch according to preference and design.
- A template mounted on the stepper and calibrated so that positions in the template correspond precisely to those displayed on the ultrasound image will be required. The template design will vary, with a range of spacing from 3 to 10 mm to define applicator position. The mechanism for fixation of the catheter within the template to prevent movement once in position should be considered, and a means of fixation to the perineum typically by sutures in each corner may be required.
- Applicators compatible with the HDR afterloader are available in different forms, the two main types being either steel needles or flexible plastic tubing stiffened with a trocar for insertion. It is important that these are CT compatible producing as little artefact as possible, and if MR is to be used for imaging, then steel needles are clearly inappropriate, and MR-compatible metal will be required. The manufacturers all provide suitable applicator kits which will be chosen on personal preference.

7.3 Procedure

The procedure can be broken down into a number of individual steps as follows:

1. Patient setup
2. Catheter insertion
3. Catheter fixation
4. Post-implant imaging
5. Outlining and dosimetry
6. Quality assurance
7. Treatment delivery

7.3.1 Patient Setup

The initial setup of the patient is critical to achieving a good implant with adequate coverage of the planned clinical target volume (CTV). The patient will be placed in an extended lithotomy position and the transrectal ultrasound probe placed in position to provide serial images of the prostate from base to apex, including seminal vesicles if these are to be included in the CTV. It is critical that the margins of the CTV are within the limits of the template positions, and in particular, the inferior border is on or above the lowest row of catheter positions. If this is not the case, then it will be impossible to achieve an adequate implant. The gland should be centralised and urethra lined up along the central row of catheter positions, row D.

A urinary catheter will be passed to confirm the urethral position and enable urinary drainage during the procedure. Sterile drapes will be placed over the surrounding

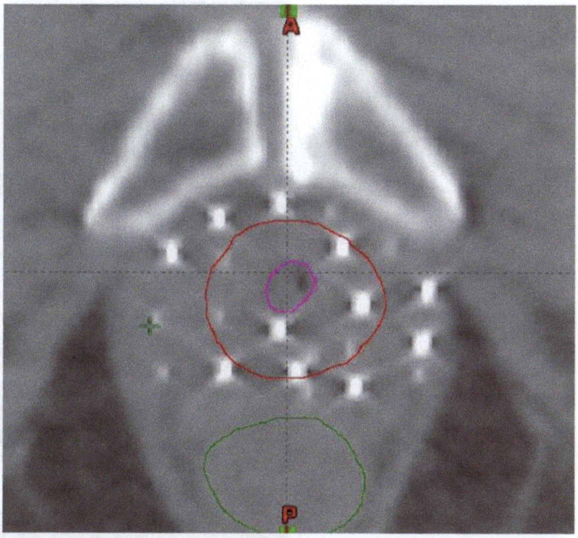

Fig. 7.1 CT image of HDR implant at region of apex to show catheter distribution and role of *inner ring* of catheters avoiding urethra but providing dose to the apical segment [CTV in *red*, urethra in *blue*, rectum in *green*]

areas, leaving the perineum open. The skin is sterilised with appropriate preparation, and the scrotum held outside the implant area with drapes and tape.

Local anaesthetic may be infiltrated into the area to be implanted at this stage.

7.3.2 Catheter Insertion

Catheters are passed through the template into the adjacent perineal skin and into position in the prostate gland. This is performed whilst observing their passage on the transrectal ultrasound image to ensure their correct placement. The most common approach is to treat the entire prostate gland aiming for homogenous cover. This will be achieved by catheters placed at approximately 10 mm distance around the periphery of the gland, close to the capsule, allowing for a CTV expansion of 3 mm to define the final PTV. Catheters on row D should be placed at the extreme periphery of the gland to avoid urethral damage. An inner row of applicators will also be required both to allow better dose control around the urethra and also to provide dose to the periphery of the apex as the gland tapers. These should, however, avoid row D to avoid the urethra as shown in Fig. 7.1. If the seminal vesicles are to be treated, then these should be included in the most inferior row of applicators as defined at the initial setup. Occasionally the seminal vesicles will droop posteriorly and not be encompassed by the template positions. If this is the case, then freehand placement of additional catheters may be of value to reach more peripheral areas shown in Fig. 7.2.

It is important during applicator insertion to scroll the ultrasound probe from base to apex and ensure that the position is appropriate for the entire length of the gland. The final and important part of applicator insertion is to define the *Z*-axis co-ordinate which should be beyond the prostate base and may encroach into the

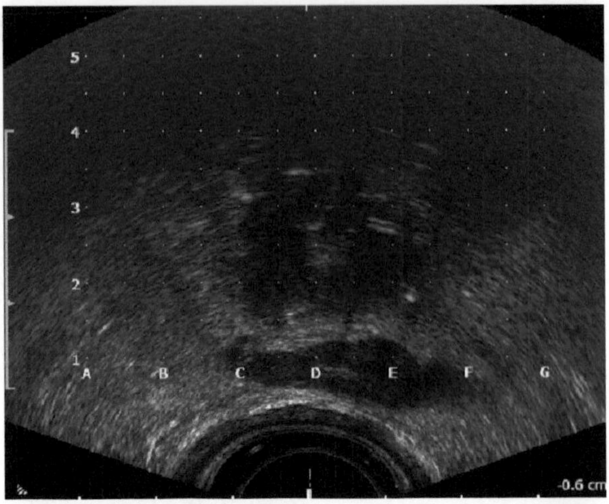

Fig. 7.2 Transaxial image of HDR implant with catheters placed to treat seminal vesicles

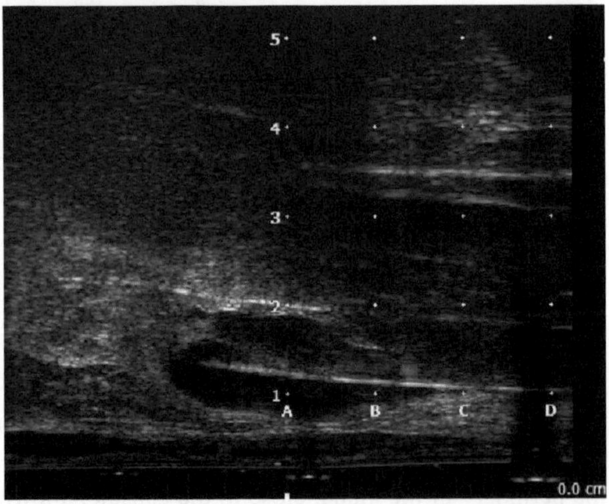

Fig. 7.3 Sagittal ultrasound image to show catheters placed within seminal vesicles

bladder in order to ensure adequate dose is delivered to the base. Where the seminal vesicles are to be treated, these should also be included in the *Z*-axis positioning, noting that they will often extend further cranially than the prostate base shown in Fig. 7.3.

7.3.3 Catheter Fixation

Once the applicators are all in a satisfactory position, it is critical that they remain there until treatment has been delivered. If the patient is to remain under anaesthetic at the site of implantation for imaging and dosimetry, the problem is less than if the patient is to be moved to an imaging facility and afterloader elsewhere. Two main systems exist:

1. The rigid template will have a fixation device so that the applicators are fixed in the template, unable to move through it, and the template is then sutured to the skin of the perineum. This is entirely adequate if the patient is not to be moved.
2. An alternative form of skin device is to use a flexible template which reproduces the applicator positions on the ultrasound template and through which the applicators will have been passed on their way to the skin. This is then adhered to the skin, and rubber 'O' rings will hold the applicators in position to prevent movement.

7.3.4 Post-implant Imaging

Once the implant is in place and secure, the next step is to undertake imaging which will be used for definition of the CTV and subsequent dosimetry. This may use ultrasound, CT or MR, and image fusion may enable more than one modality to be used. Ultrasound is most appropriate when the patient is kept anaesthetised in the lithotomy position with the ultrasound probe in situ following implantation. A 3–5 mm volume study is then acquired on which the CTV can be defined.

The alternative approach is for the patient to recover from the anaesthetic and be transferred to a CT or MRI scanner. Transaxial 3 mm slices will be taken through the CTV with the catheters in situ. The patient is usually imaged with their legs down rather than in the lithotomy position, and this will be the position for treatment also. Similarly MRI can be used at this point and will provide better soft tissue definition although reconstruction of the applicator positions may be more difficult.

These images are then imported into the planning system for volume definition.

7.3.5 Outlining and Dosimetry

Modern planning software enables the operator to define both the CTV and the organs at risk, using an appropriate drawing tool. Currently the GEC-ESTRO guidelines define three possible CTVs to be drawn:

- CTV 1 which is the entire gland with the margin at the prostate capsule and including any extracapsular disease identified on imaging seminal vesicles where appropriate. This will be the volume that most HDR teams will use.
- CTV 2, the peripheral zones.
- CTV3, any identifiable tumour reflecting a radiological gross tumour volume (GTV).

In addition rectum and urethra should be outlined as a minimum to allow normal tissue constraints to be introduced to the planning procedure.

The planning target volume (PTV) is defined with a 3 mm margin on the prostate capsule and any other microscopic tumour areas identified, constrained posteriorly to the rectum and superiorly to the bladder base.

Each applicator must be identified and tracked for its entire length to define the tip from which the first HDR dwell position will be related. Planning may be undertaken using manual techniques or dose optimisation or a combination of the two to achieve appropriate cover of the planning target volume achieving a D90 of at least 100 % and V100 greater than or equal to 95 %. Tolerance doses must also be defined for the rectum and urethra which should take into account combined dose delivery from external beam and brachytherapy and individual dose per fraction using an appropriate radiobiological correction and using D_{2CC} as the primary parameter for the rectum and D_{10} and D_{30} for the urethra.

7.3.6 Quality Assurance

This covers all the processes involved in the brachytherapy procedure. There will therefore be important equipment quality assurance checks on the afterloader to ensure accurate source position and source activity, the imaging to ensure good correlation between ultrasound template positions and actual positions, CT and MR machine QA and on the day of implant individual quality assurance of each applicator to ensure patency and safety.

There are other important considerations for an HDR implant. Following implantation, there will be haemorrhage and oedema causing displacement of the prostate gland and swelling of the perineal tissues. This will vary and may or may not have an impact on the application position relative to the prostate gland. This is however a critical parameter since the dosimetry will be defined based on the position of each applicator tip and first dwell position. Where an ultrasound-based general anaesthetic technique is used, checking is relatively simple, since the ultrasound probe remains in situ and there is a short time from image capture to treatment. Simple checks on the position of the template to the perineum can be undertaken using a ruler or more complex scale.

When the patient is moved from the place of implant to CT or MR and then to the afterloader for treatment, there is more opportunity for applicator displacement, and a longer period typically ensues during which oedema may occur with displacement of the prostate cranially. It has been shown that over an 18 h period, significant moves with a median of 11.5 mm cranial-caudal displacement will be seen having a fundamental impact on the dose delivered to the CTV if not corrected. Thus, repeated imaging is required prior to dose delivery in this situation to ensure that catheter tips are in a constant position in relation to the most cranial part of the PTV. If movement is detected, then an adjustment to the first dwell position may be adequate correction, or there may be a case for reinsertion of an individual applicator to ensure it is sufficiently cranial in its extent. Significant movement has been identified over a 3-fraction treatment spanning 30 h. Correction of any changes, however, can enable the dosimetric parameters to accurately reproduce those of the initial plan and ensure adequate and effective dose delivery.

As a general principle therefore, it is critical that any implant is imaged within a short time of the dose delivery, and certainly treatment should not be given any more than an hour or two after the most recent imaging verification.

7.3.7 Treatment Delivery

The treatment plan is transferred to the afterloader and the applicators connected using the appropriate source transit tubes. Again it is vital at this stage that careful quality assurance and double checking are used to connect the appropriate channel to its corresponding catheter since dosimetry will be critically dependent on this.

Following treatment delivery and disconnection of the source guide tubes, the implant can be removed. In most instances this will be a simple procedure, whether the patient remains under anaesthesia or not. At the same time, the urinary catheter is removed and the patient returns to a recovery environment.

Pulsed-Dose Rate Brachytherapy in Prostate Cancer

8

Bradley R. Pieters

8.1 Introduction

Very similar to high-dose-rate (HDR) brachytherapy, pulsed-dose-rate (PDR) brachytherapy is delivered as an afterloading technique using a temporary implant. Several catheters or needles are implanted into the prostate according to a pretreatment plan. PDR brachytherapy is given as a boost to external beam radiotherapy. Treatment usually last about 48 h. Because catheters/needles can move during treatment, careful attention should be paid. Anchoring catheters in the prostate with specially designed anchoring catheters can be useful for this purpose. Because the treatment is delivered without the rectal ultrasound probe in, it is advisable to perform definitive treatment planning on CT scan or MRI.

8.2 Pulsed-Dose Rate

For pulsed-dose-rate (PDR) brachytherapy, a single radioactive source is used for dose delivery, as is also the case with high-dose-rate (HDR) brachytherapy, but the source activity is approximately ten times weaker for PDR, in the range of 18.5–74 GBq. Both PDR and HDR brachytherapy are based on a single stepping source modality. This means that one source passes along all the needles or catheters for dose delivery, and by varying the dwell time per source position, optimization of the dose distribution can be obtained for both prostate coverage and limiting the dose to the organs and structures at risk. Another advantage of the stepping source modality is that treatment is given in fractions with an interval during which no radiation is delivered. For HDR this interval is at least 6 h, and for PDR the interval is between 1 and 3 h during which time medical personnel and relatives can visit patients without being exposed to radiation.

B.R. Pieters
Radiation Oncology, Academic Medical Center/University of Amsterdam,
Meibergdreef 9, 1105 AZ Amsterdam, Netherlands
e-mail: b.r.pieters@amc.uva.nl

G. Kovács, P. Hoskin (eds.), *Interstitial Prostate Brachytherapy*,
DOI 10.1007/978-3-642-36499-0_8, © Springer-Verlag Berlin Heidelberg 2013

Brachytherapy is delivered as either a sole treatment or as a boost to external beam radiotherapy. The experience of PDR brachytherapy for prostate cancer as sole treatment until now only has only been published as a meeting abstract (Geiger et al. 2008); however, there are several full paper publications on PDR as a boost (Lettmaier et al. 2012; Pieters et al. 2006, 2010, 2011).

PDR differs from HDR mainly in the different time dose pattern of dose delivery. In PDR multiple low-dose pulses are given separated by 1–3 h, whereas with HDR fraction doses are high. With HDR few fractions are given, 1 or 2 a day, and sometimes separated by a week. There is no evidence that the difference in dose delivery will result in a different clinical outcome. It is hypothesized that the α/β-ratio of prostate cancer cells using the linear quadratic radiobiological model is about 1.5 Gy (Brenner and Hall 1999; Brenner et al. 2002; Fowler et al. 2001; Wang et al. 2003). With a lower α/β-ratio of tumor cells compared to surrounding organs at risk, a hypofractionated treatment with HDR would be in advantage. For equal late complication risk, a higher radiobiological dose is expected with HDR hypofractionation compared to PDR hyperfractionation. A treatment planning model study of external beam radiotherapy and brachytherapy boost has shown that if the same radiobiological dose is prescribed on the periphery of the prostate, for an α/β-ratio of prostate cancer of 1.5 Gy, only small volumes in the prostate are treated to a higher equivalent radiobiological dose in case of HDR compared to PDR (Pieters et al. 2008). HDR also achieves a physical dose reduction so that the equivalent radiobiological dose to the rectum and bladder will be less for HDR compared to PDR. However, these differences are small and critically dependent on the value of the α/β-ratio for prostate cancer which is still a matter of debate. Several studies have suggested that the α/β-ratio is in the order of 2–4 Gy (van Kal and Gellekom 2003; Nahum et al. 2003; Valdagni et al. 2005); in which case, no difference will be expected between HDR and PDR.

8.3 Specialities of PDR Implantation Technique

Implantation for PDR prostate brachytherapy is performed like all modern prostate implantation under transrectal ultrasound guidance. Before placement of the needles in the prostate, it is advisable to undertake pretreatment planning to guide the positioning of needle placement. This should preferably be performed intraoperatively with the patient in the lithotomy position to reproduce the prostate position during the implantation itself. The prostate gland, rectum, and urethra are contoured. In case of intermediate- or high-risk disease, the base of the seminal vesicles is also contoured. A pretreatment plan can be constructed with adequate coverage of the prostate and if necessary the base of the seminal vesicles avoiding high dose to the rectum and urethra.

The number of needles to be used is dependent on the prostate volume. The larger the volume, the more needles are needed. At the Academic Medical Center in Amsterdam (AMC), the positioning of the needles is standardized. The distance between the needles is between 10 and 12 mm. Two planes of needles are planned between rectum and urethra to allow for optimization of dose distribution by taking into account the dose constraints for rectum and urethra (Fig. 8.1). The rest of the needles are placed at the lateral and ventral periphery of the prostate.

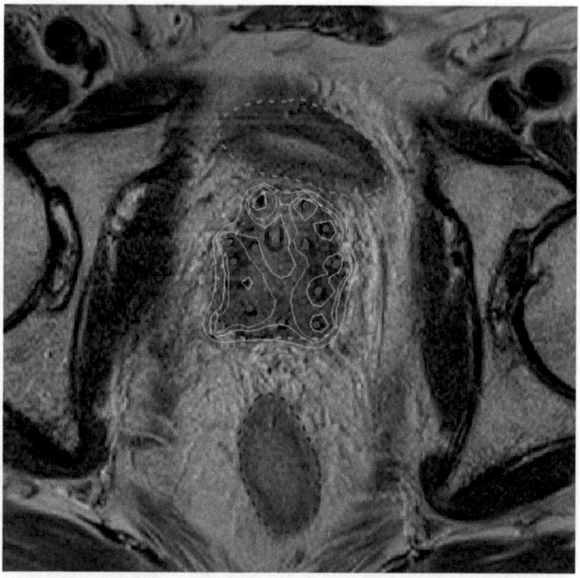

Fig. 8.1 Dose distribution of a PDR brachytherapy prostate implant. Isodose lines from the outside to the inside are 80 % (*pink*), 100 % (*blue*), 120 % (*yellow*), 140 % (*green*), and 200 % (*white*) of the reference dose

8.4 Catheter Displacement

The major problem encountered while performing a temporary implant in the prostate is the displacement of needles. Displacement up to 40 mm has been described (Damore et al. 2000; Galalae et al. 2002; Martinez et al. 2001; Mullokandov and Gejerman 2004). Such large deviations will inevitably cause modifications in the planned dose distribution. For this reason needle positions should be checked before each treatment fraction, if the same implant is used for several HDR treatments. Needle-positioning checks can be performed with fluoroscopy and fiducial markers into the prostate or with a CT scan, but in the case of PDR, multiple pulses are given separated by 1–2 h, making the positioning checks an impracticable solution. Self-anchoring catheters are available for PDR brachytherapy to prevent movement during therapy. These self-anchoring catheters have an umbrella-like mechanism at the tip of the catheter that can be unfolded when introduced into the prostate. At first a 7F needle covered by a synthetic splitsheath is inserted in the desired position, guided by a template. In contrast to common templates (a block with multiple holes in it), this template has an arm that can be moved and placed in any desired position. The needle can easily be released from the moving arm through an opening (Fig. 8.2). After the needle is removed, the 6F self-anchoring catheter can be placed into the splitsheath which is removed afterwards, and the catheter is fixed into the prostate by unfolding the anchor (Fig. 8.3). When all catheters are placed accordingly, the implantation is finished. By preventing movement of catheters, alterations of dose-volume parameters from the definitive treatment plan are prevented (Pieters et al. 2006).

The situation in which the patient is treated with PDR, which is in bed and without an ultrasound probe in, is different from the situation during the pre-planning

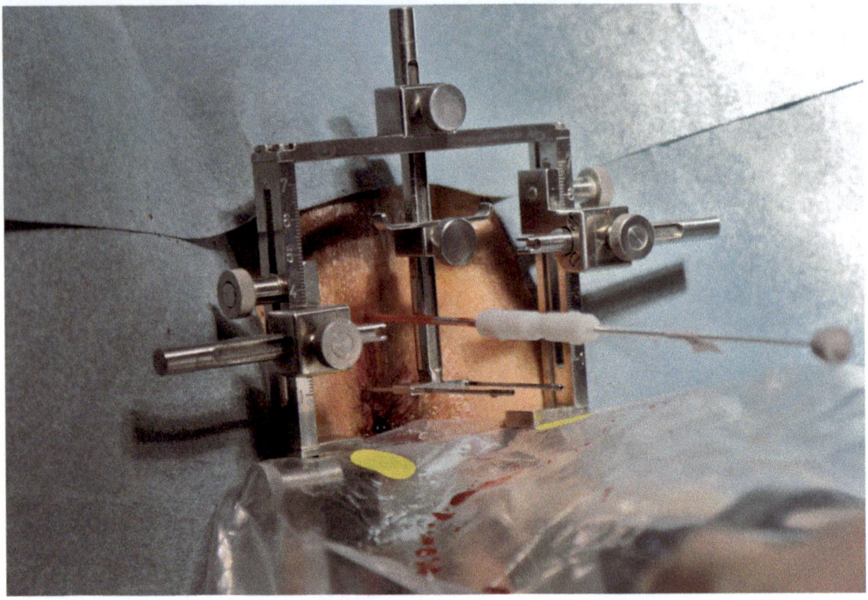

Fig. 8.2 Open template with two moving arms. The arms can be moved in ventro-dorsal direction and in medio-lateral direction. The brachytherapy catheters can be released from the arm

Fig. 8.3 Brachytherapy catheter with unfolded anchoring tip

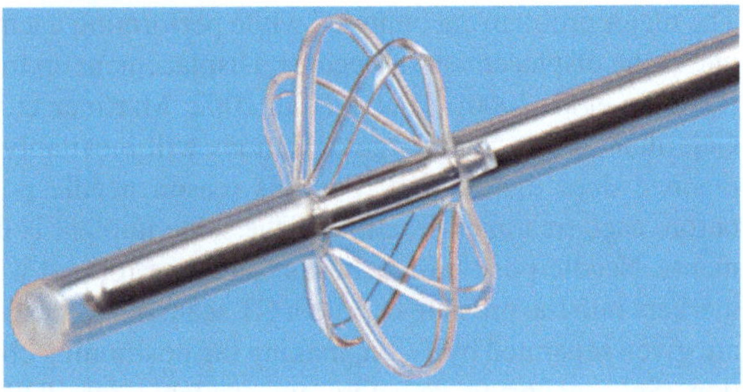

and implantation; in particular, the shape of the prostate and the position of the rectum relative to the prostate can be different. Because of this, a definitive treatment plan should be acquired after completion of the implant and removal of the rectal probe visualizing the definitive position of the organs on a CT or MRI scan following the implantation.

8.5 Dose Schedule

In the AMC, PDR prostate brachytherapy is given as a boost to external beam radiotherapy of 46 Gy in 2 Gy fractions to the prostate and base of the seminal vesicles. PDR dose escalation has been undertaken with schedules from 24 pulses of 1.04 Gy to 24 pulses of 1.2 Gy (Pieters et al. 2011). The interval time between the start of two pulses is 2.0 h. The EQD2 for an α/β-ratio of 3 Gy and T1/2 of 1.5 h with this schedule is 78 Gy. The dose is prescribed to the periphery of the prostate, and consequently large areas of the prostate will receive a dose of approximately EQD2 100 Gy.

Dose constraints for organs at risk are D2cc of 70 Gy EQD2 for the rectum (0.97 Gy/pulse) and maximal dose to the urethra of 140 %.

At the University Hospital Erlangen, an external beam dose of 50.4 Gy in 1.8 Gy fraction was followed by 50 hourly pulses of 0.7 Gy (35 Gy) PDR brachytherapy boost (Lettmaier et al. 2012).

8.6 Patient Care

Patients are treated in bed and connected to the PDR afterloader treatment machine. To prevent bending of the catheters, a mattress with a cutaway between the legs can be used (Fig. 8.4). A special fluid diet and anti-diarrhea medication are prescribed.

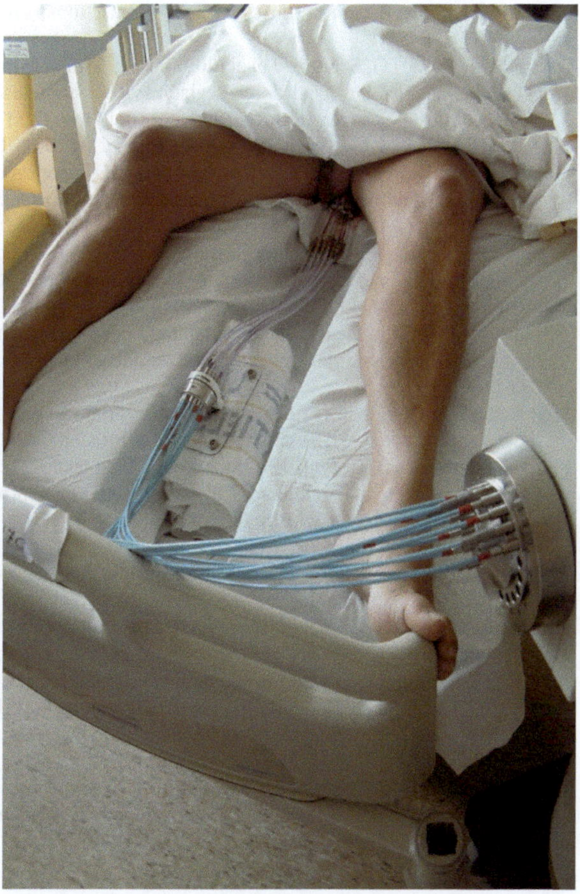

Fig. 8.4 Patient attached to the treatment machine on a mattress with a cut-away between the legs

Urine is drained via a Foley-balloon catheter that was inserted at the time of implantation.

At termination of the treatment, the catheters are removed. The catheter anchor is unfolded before withdrawal. An opiate, e.g., fentanyl, is given before removal.

PDR brachytherapy for prostate cancer has proven to be feasible. To prevent catheter movement during a 2-day treatment, the use of anchoring catheters is essential. Because the treatment is given without a rectal ultrasound probe in situ, it is advisable to perform a definitive treatment plan with CT or MRI scan.

References

Brenner DJ, Hall EJ (1999) Fractionation and protraction for radiotherapy of prostate carcinoma. Int J Radiat Oncol Biol Phys 43:1095–1101

Brenner DJ, Martinez AA, Edmundson GK, Mitchell C, Thames HD, Armour EP (2002) Direct evidence that prostate tumors show high sensitivity to fractionation (low α/β ratio), similar to late-responding normal tissue. Int J Radiat Oncol Biol Phys 52:6–13

Damore SJ, Syed N, Puthawala AA, Sharma A (2000) Needle displacement during HDR brachytherapy in the treatment of prostate cancer. Int J Radiat Oncol Biol Phys 46:1205–1211

Fowler J, Chapell R, Ritter M (2001) Is α/β for prostate tumors really low? Int J Radiat Oncol Biol Phys 50:1021–1031

Galalae RM, Kovács G, Schultze J et al (2002) Long-term outcome after elective irradiation of the pelvic lymphatics and local dose escalation using high-dose-rate brachytherapy for locally advanced prostate cancer. Int J Radiat Oncol Biol Phys 52:81–90

Geiger MH, Lotter M, Seeger AR, Sauer R, Strnad V (2008) Die alleinige interstitielle Pulse-Dose-Rate Brachytherapie in der Therapie des lokal begrenzten Prostatakarzinoms. Erfahungen hinsichtlich Durchführbarkeit und Toxizität. Strahlenther Onkol 184(Suppl 1):172

Lettmaier S, Lotter M, Kreppner S, Strnad A, Fietkau R, Strnad V (2012) Long term results of a prospective dose escalation phase-II trial: interstitial pulsed-dose-rate brachytherapy as a boost for intermediate- and high-risk prostate cancer. Radiother Oncol 104:181–186

Martinez AA, Pataki I, Edmundson G, Sebastian E, Brabbins D, Gustafson G (2001) Phase II prospective study of the use of conformal high-dose-rate brachytherapy as monotherapy for the treatment of favorable stage prostate cancer: a feasibility report. Int J Radiat Oncol Biol Phys 49:61–69

Mullokandov E, Gejerman G (2004) Analysis of serial CT scans to assess template and catheter movement in prostate HDR brachytherapy. Int J Radiat Oncol Biol Phys 58:1063–1071

Nahum AE, Movsas B, Horwitz EM, Stobbe CC, Chapman JD (2003) Incorporating clinical measurements of hypoxia into tumor local control modeling of prostate cancer: implications for the a/b ratio. Int J Radiat Oncol Biol Phys 57:391–401

Pieters BR, van der Grient JN, Blank LE, Koedooder K, de Hulshof MC, Reijke TM (2006) Minimal displacement of novel self-anchoring catheters suitable for temporary prostate implants. Radiother Oncol 80:69–72

Pieters BR, van de Kamer JB, van Herten YRJ et al (2008) Comparison of biologically equivalent dose–volume parameters for the treatment of prostate cancer with concomitant boost IMRT versus IMRT combined with brachytherapy. Radiother Oncol 88:46–52

Pieters BR, Rezaie E, Geijsen ED et al (2011) Development of late toxicity and IPSS resolution after external beam radiotherapy combined with PDR brachytherapy for prostate cancer. Int J Radiat Oncol Biol Phys 81:758–764

Pieters BR, Geijsen ED, Koedooder K et al (2011) Treatment results of PDR brachytherapy combined with external beam radiotherapy in 106 patients with intermediate- to high-risk prostate cancer. Int J Radiat Oncol Biol Phys 79:1037–1042

Valdagni R, Italia C, Montanaro P et al (2005) Is the alpha-beta ratio for prostate cancer really low? A prospective, non randomized trial comparing standard and hypofractionated conformal radiation therapy. Radiother Oncol 75:74–82
van Kal HB, Gellekom MPR (2003) How low is the α/β ratio for prostate cancer? Int J Radiat Oncol Biol Phys 57:1116–1121
Wang JZ, Guerrero M, Allen Li X (2003) How low is the α/β ratio for prostate cancer? Int J Radiat Oncol Biol Phys 55:194–203

Imaging for Post-implant Dosimetry

9

Brendan M. Carey

Prostate brachytherapy is a treatment that needs a dedicated, properly trained and coordinated team approach to achieve good clinical outcomes. A successful implant is one in which there is adequate coverage of the prostate gland whilst limiting the dose to the rectum, urethra and neurovascular structures. Many factors influence these outcomes, including the skill of the operator, the expertise of the medical physicists and the support care from the entire team. Technical factors such as the optimisation of TRUS imaging and the quality of the radiotherapy planning software will also impact on patient outcomes. The lessons learned from the retropubic era of prostate brachytherapy in the 1960s and 1970s when source placement was often less than ideal emphasised that implant quality can greatly affect treatment outcomes both in terms of treatment-associated morbidity and local control (Fuks et al. 1991). Because patients differ in their pelvic anatomy, some implants are technically more difficult than others. Hence, a variation in implant quality may occur, even for experienced teams. Post-implant assessment of implant quality should be an integral part of the overall treatment process (Salembier et al. 2007). Every brachytherapy centre should have a well-designed, effective and adequately monitored post-brachytherapy quality assurance programme that should review every aspect of the implant process including post-implant dosimetry. This should include an agreed pathway for post-implant imaging of the prostate in order to provide regular and quality data for dosimetric analysis.

In those early days of prostate brachytherapy, there was no useful imaging technique for evaluating prostate implant quality. Poor source distribution throughout the gland was often a combination of poor TRUS imaging during the procedure itself, suboptimal treatment planning and inexperience of the operator. Plain radiographs of the pelvis were obtained after the procedure to provide a basic count of the number of implanted sources and in order to compute radiation dose distributions

B.M. Carey
Consultant Radiologist, Institute Of Oncology,
St. James Hospital, Leeds, UK
e-mail: brendan.carey@btinternet.com

G. Kovács, P. Hoskin (eds.), *Interstitial Prostate Brachytherapy*,
DOI 10.1007/978-3-642-36499-0_9, © Springer-Verlag Berlin Heidelberg 2013

around these sources even when there was no way of relating those calculated doses to the actual anatomy of the prostate since the gland itself was not imaged.

Computed tomography (CT)-based post-implant dosimetry became available in the 1980s and relied on CT images to depict source location in relation to prostate anatomy. This was a significant improvement over plain radiographs and provided the implant team with the first realistic data on implant quality. Developments in CT and also in MRI over the past two decades have further improved our knowledge of the requirements for and the importance of good implant technique. TRUS has been advocated as a technique for assessing prostate implants both interactively during the implant procedure itself and following completion of the implant. Various image fusion techniques have been described all attempting to combine the differing advantages of each individual imaging technology. Improvements in TRUS imaging technology and in computer-based dose planning have facilitated more reproducible, safer and better quality implants with consequent better outcomes. Post-implant dosimetry has allowed us to assess potential improvements in our implant techniques and helped identify systematic errors in the procedure should they occur.

Post-implant dosimetry is by definition performed after the procedure has finished. The implant cannot be reversed and additional radiation either with further implanted sources or addition of external beam radiation is limited by the technical issues involved with matching doses and potential treatment-related toxicity. Fundamentally, the post-implant calculated dosimetry is a measure of the overall performance of the implant team. Such a multidisciplinary group of health professionals will all have some contribution to the outcome and it is important that a team approach is used when assessing results as well as performing the implant itself. Many different dosimetric indices have been used and studied: the purpose of this chapter is to look at the imaging techniques used and how these might be optimised to yield reliable and accurate post-implant dosimetry data. The ultimate aims of post-implant dosimetry are:

- To obtain information about dose parameters to the prostate
- To assess dose to critical structures at risk
- To correlate with toxicity and clinical outcomes
- To compare results with other centres
- To assess modifications to existing implant techniques

Post-implant evaluation provides the foundation upon which we can compare results, identify problems with technique and evaluate technical improvements and changes to technique over time. In an age where there are competing treatments for early-stage prostate cancer and comparative outcomes are under close scrutiny, it is vital that each centre has a comprehensive programme for post-implant dosimetry assessment. Discussion of post-implant imaging should consider:

- What imaging techniques should be routinely used?
- What is the optimal timing for post-implant dosimetry?
- What are the limitations of each of these imaging techniques?
- How can we use the results to improve subsequent implant quality?

The clinical results and interpretation of post-implant dosimetry are discussed elsewhere in this book. Poor prostate dosimetry merely reflects that an imaging technique has identified a region of the prostate considered underdosed according to the treatment plan; different imaging techniques may identify different areas of underdosage and these underdosed areas may or may not actually contain prostate cancer. The measured dosimetry therefore will have varying degrees of clinical correlation for an individual patient. To date, post-implant imaging has not concentrated on this aspect: dosimetric parameters are correlated to the whole of the gland and we have little data on whether or not the area(s) of cancer has been appropriately treated or otherwise. Furthermore, dose calculation is usually based on a single imaging session, ignoring potential geometrical changes in the implanted prostate during the months of dose accumulation.

9.1 Identification of Implanted Sources on Imaging

The implanted sources have to be identified, registered and then correlated with prostate anatomy in order to calculate the various dosimetric parameters that are used to assess implant quality. Different imaging techniques achieve these requirements to a varying degree and this is a factor when interpreting any post-implant results.

The potential variation in the calculated dosimetry as a function of seed detection rates was investigated for iodine-125 implants with seed activities commonly used (Su et al. 2004, 2005). A total of 108,000 complete sets of post-implant dose volume statistics were computer modelled. The results demonstrated that although the average D90 differed from the true value by less than 5 % when 70 % or more sources were identified, the D90 of an individual case could deviate by up to 13 %. The 95 % confidence interval of estimated D90 values differed by less than 5 % from the actual value when 95 % or more sources were detected. The authors concluded that 95 % or more sources need to be confidently identified in order to provide an accurate estimation of dose parameters for contemporary iodine-125 permanent prostate brachytherapy.

Source visibility on TRUS has been investigated (Al-Qaisieh et al. 2007) for different sources from different manufacturers. Such visibility may also potentially bias any form of TRUS-based post-implant dosimetry. The ultrasound signal intensity detected by the TRUS probe may depend upon a number of factors. The surface preparation of each source will determine the level and direction of reflected ultrasound. The materials from which a source is constructed will also impact upon the proportion of ultrasound signal which is reflected and transmitted because of different acoustic impedances. These authors concluded that until new developments in TRUS technology and implant source manufacturing allow more precise source segmentation on imaging, the use of TRUS alone for intraoperative source localisation could not be advocated. Interestingly, the CT greyscale beam profiles were similar for all sources and there were only minor variations in the MRI signal voids observed in this study. The detection rates for various CT-based algorithms have also been studied (Holupka et al. 2004). Source visualisation on MRI is a factor to be considered when interpreting the data from MRI-based post-implant dosimetry

(Thomas et al. 2009; De Brabandere et al. 2006). It should be remembered that each imaging technique has its own inherent limitations in the identification of implanted sources in the prostate based on the laws of physics and some of these will not be overcome by future improvements in imaging technology.

9.2 Plain Radiography

The earliest attempts at assessing implant quality relied simply on a radiograph of the pelvis to provide a source count for the records without any attempt at dosimetric analysis. The orientation of the sources and any clustering could be recognised but little further useful information was available from plain radiographs alone. Unfortunately, however accurate this source count is, there is no anatomical correlation with the soft tissue structure of the prostate and no useful information regarding dosimetry to the prostate possible. Developments such as automatic source reconstruction from plain radiographs have been described by various authors over the years (Todor et al. 2002; Zhang et al. 2004). In those early days of brachytherapy, most treatment-planning systems had the capacity to localise points and sources from a paired film set. Using this system to identify and localise a prostate implant with on average 100 sources was possible though hugely time-consuming and of very limited value without any anatomical reference to the prostate itself. Orthogonal couch films and gantry stereo-shift films can still be used but are far from satisfactory. Various authors have proposed enhancements of the plain radiograph assessment such as paired or multiple-projection film localisation as well as image fusion of plain radiographs with other imaging techniques such as TRUS, CT and MRI. The optimum use plain radiography in the post-brachytherapy setting lies with such fused data sets that can depict soft tissue as well as the implanted sources. Despite its limitations, plain radiographs still provide a basic source count following the implant and can be used for image registration and fusion when in conjunction with other imaging techniques.

9.3 Computed Tomography

The first published use of CT for post-implant dosimetry was in the early 1990s at Memorial Sloan–Kettering Cancer Center (Roy et al. 1993). This was the first display of organ and dose distribution together for post-implant prostate dosimetry. Post-implant CT evaluation quickly achieved its intended purpose: to provide enough feedback to the implant team to allow them to modify or correct implant technique and furthermore to evolve acceptable planning dose constraints to the structures at risk in order to minimise toxicity. Sources and soft tissue structures are localised on the axial CT images and this information entered into the treatment-planning systems in order to calculate dosimetric indices as required. By far, the majority of the post-implant imaging performed today is CT based and is widely recommended for routine post-brachytherapy dosimetry (Salembier et al. 2007;

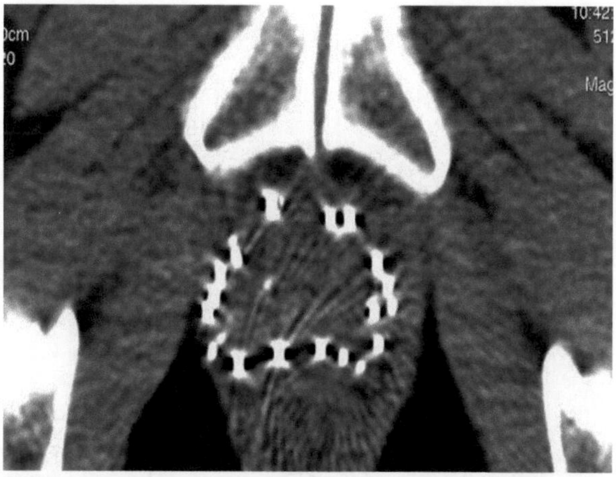

Fig. 9.1 Axial post-implant CT showing sources peripherally located in prostate

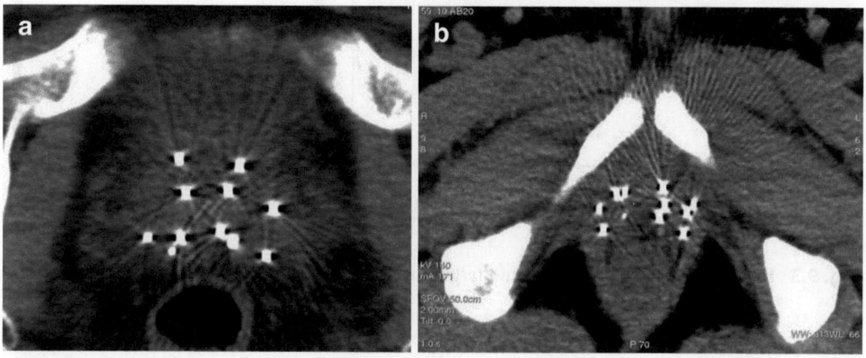

Fig. 9.2 Axial CT images at prostate base (**a**) and apex (**b**) showing poor identification of prostate margins

Nag 2000). CT imaging is relatively inexpensive and the source locations are easily visualised (Fig. 9.1). CT however does have significant limitations in definition of the prostate margins due to the inherent lack of contrast between adjacent soft tissue structures with this imaging technique. This lack of ability to visualise the prostate borders clearly, particularly in the oedematous gland, remains a significant problem for accuracy and reproducibility when using CT for post-implant dosimetry. Large uncertainties exist in the delineations of the prostate margins (Dubois et al. 1998; Al-Qaisieh 2002, 2003). Resolution of different soft tissue structures of similar CT attenuation is a particular problem at the prostate base and apex (Fig. 9.2), and it is often impossible to be certain of where the boundaries of the gland should be drawn in these locations. There is a degree of subjectivity based on experience and perhaps some clinical bias. Small changes in the contoured prostate can have a significant impact on the measured prostate volume (Fig. 9.3) and this of course affects the calculated dosimetric indices. Choice of CT technique is important: scan parameters are usually chosen based upon the ability to best determine source locations.

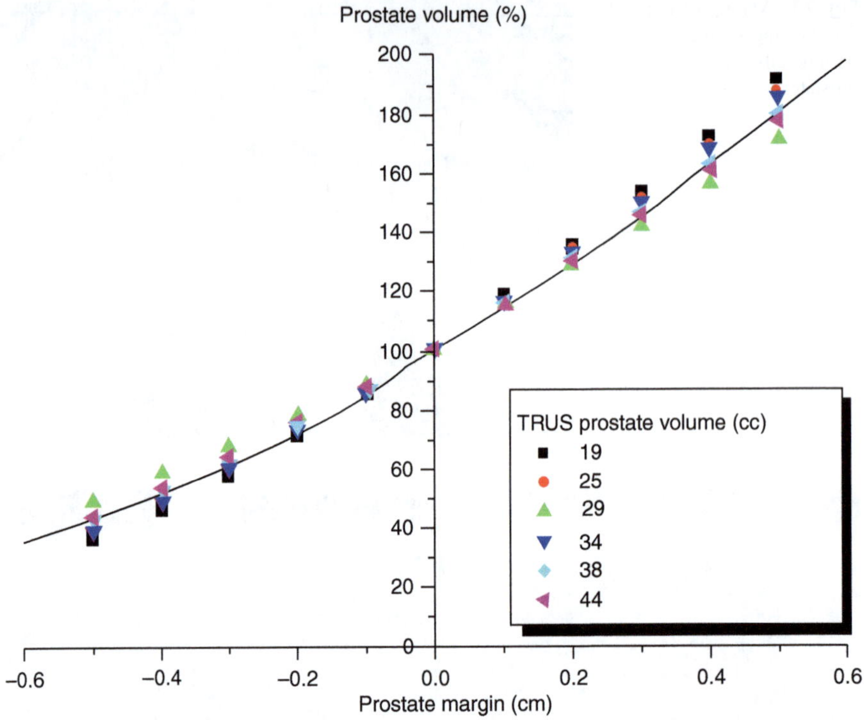

Fig. 9.3 Influence of contoured prostate margin on prostate volume

This involves a reduced field of view, usually about 15 cm, just large enough to encompass the meaningful dose regions and the critical structures that need to be contoured also. Slice spacing greater than 3 mm is not recommended (Nag et al. 2000). The margins of the prostate remain poorly defined, however, even with modern state-of-the-art helical CT scanners. There is no advantage in using intravenous contrast and the inherent limitations with CT imaging of the prostate have to be accepted. Accurate dose clouds can be constructed around the seed locations based on the volumetric CT data (Fig. 9.4), but errors in accurately defining prostate margins still occur especially at the prostate base and apex. Initially, there is always a temptation to "draw around the sources" but this is inevitably flawed. It may yield good dosimetry results but may equally have no correlation with what has been adequately treated or otherwise.

Although post-implant CT-based dosimetry with determination of the dose delivered to 90 % of the prostate gland (D90) has become a standard tool for assessing implant quality for prostate brachytherapy (Potters et al. 2001), the volume derived from the CT scan is generally different from the TRUS-derived volume (Smith et al. 2007). The differences in prostate volume as measured with different imaging techniques has been well documented and remains a confounding factor in the interpretation of post-implant dosimetry results. As TRUS is used for pre- or intra-operative planning of the implant, the data from any post-implant CT-based dose

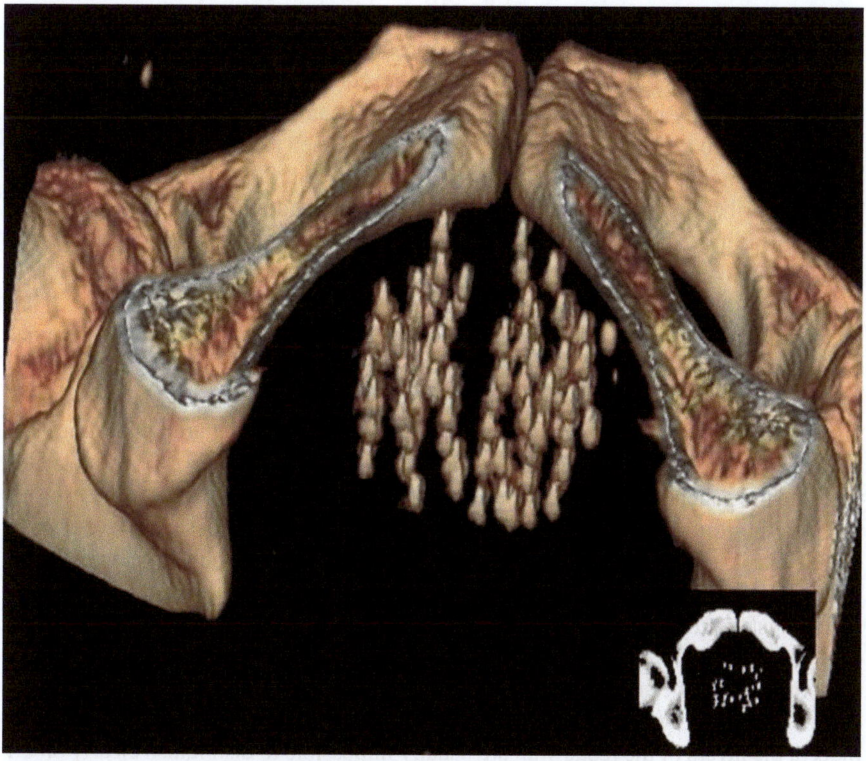

Fig. 9.4 Volumetric CT to show 3D source cloud

distribution may well be based on a different measured volume and so the results of post-implant dosimetric analysis will be will be subject to this bias. CT generally overestimates the volume of the prostate compared with TRUS and therefore the percentage of the prostate covered by the prescribed dose will apparently be different from that planned with TRUS. Intra-observer and inter-observer variations in evaluation of the prostate gland on CT scans have also been investigated and well documented in different studies. Different observers can outline the prostate differently whilst keeping the total prostate volume constant. The D90 is therefore not only a function of dose distribution and prostate volume but is also a function of prostate shape and position relative to the source position. It is essential that some form of source location method (source sorting) based on nearest neighbours be employed because the sources may appear on more than one CT section. Software algorithms have been developed to locate sources, but the exact number of sources in the prostate at the time of dosimetry evaluation needs to be inputted into the system. Once the actual number of sources in the prostate is known, introduction of this number into the treatment-planning system allows localisation of the sources by the automatic source finder software. Source position detection can then be performed and each source delineated. Scatter around the sources can complicate this evaluation on CT, but the use of specific filters may decrease this scatter. Ultimately, however, post-implant dosimetry based on CT imaging alone is a subjective process (Crook et al. 2002) and will always be prone to systematic error, random error and bias. Newer CT developments for image guidance may have a role in prostate brachytherapy in the future. Flat-panel cone-beam CT is a versatile volumetric image-guidance technique that is an emerging CT-based technique to improve intraoperative imaging for brachytherapy (Siewerdsen et al. 2005; Jaffray et al. 2002). This C-arm-based cone-beam CT system could be used to image and localise the prostate during the implant allowing intraoperative planning of source placement, providing real-time fluoroscopic monitoring during the source placement and providing cone-beam CT verification distribution.

Identifying and contouring the critical structures at risk on a CT scan may also present difficulties and limit the accuracy of CT-based post-implant dosimetry. The urethra is best visualised on CT following catheterisation; not necessarily a problem if the post-implant imaging is performed immediately following the procedure with the urinary catheter still in place, but at significant discomfort to the patient if performed at a later time and the patient has to be recatheterised just for the CT scan. In practice, this is not routinely performed and so the vast majority of post-implant CT scans have poor or absent visualisation of the urethra. CT definition of the neurovascular structures is also poor making it very difficult to correlate dosage to these structures with post-implant sexual dysfunction. The rectum has a varying appearance on CT scans and is dependent on the degree of rectal filling at the time of the CT examination. This may well be quite different to the appearances on the intraoperative TRUS, upon which the intraoperative dosimetry was planned, and therefore valid comparison is again difficult. Indeed, the layers of the rectal wall are usually impossible to distinguish clearly on CT particularly if the rectum is empty at the time of the scan.

9.4 MRI

Magnetic resonance imaging has better soft tissue resolution and is acknowledged as a better imaging technique for the prostate and periprostatic structures. MRI-based prostate contours correlate better with TRUS-based imaging, thus making it an attractive image technique for post-implant prostate dosimetry. MRI also provides superior delineation of the structures at risk. MRI improves definition of the prostate apex (Fig. 9.5) decreasing overestimation of volume in this region of the implant, and it also improves delineation of the prostate base and helps differentiate the prostate itself from the bladder neck as these adjacent soft tissue structures look very similar on CT scans. Furthermore, at the prostate base, MRI may clearly show the true extent of any enlargement of the transition zone and median lobe into the bladder neck area (Fig. 9.6) and help differentiate normal prostate from thickened bladder neck muscle. This is a particularly difficult area on CT scans where it is not easy to differentiate normal prostate from bladder tissue. Despite its excellent soft tissue resolution, there are some unresolved issues with source localisation on MRI. Sources in vascularised areas or outside the prostate can be difficult to identify if there is MR signal drop-off in

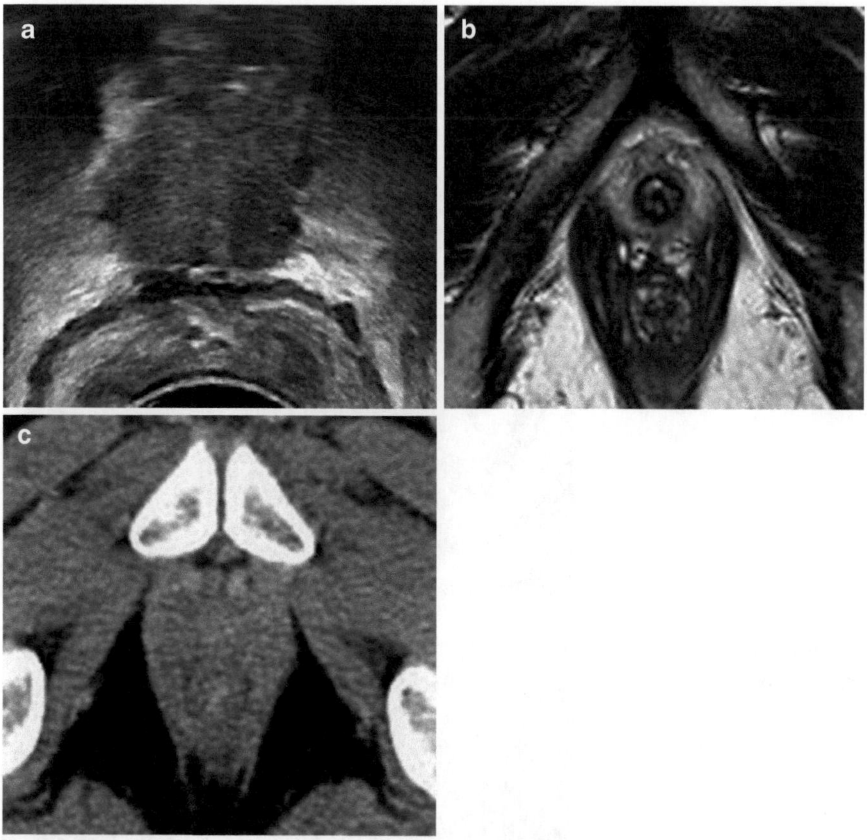

Fig. 9.5 Prostate apex as seen on TRUS (**a**) MRI (**b**) and CT (**c**) in the same patient

these areas. Sources essentially show as signal voids (Figs. 9.7 and 9.8) on MRI and may be confused with small areas of calcification which show similar characteristics on MRI scans. Dosimetry based upon MRI alone, therefore, may be inadequate due to this inability to reliably identify all sources. There are various suggestions in the literature as to the optimum sequences for MRI-based post-implant dosimetry (McLaughlin et al. 2002), but generally, a phased array pelvic coil is recommended (not an endorectal coil because of prostate distortion) and a combination of T1-weighted, T2-weighted and proton-density sequences are adequate. If the data is to be used for CT-MRI fusion (Fig. 9.9), then care needs to be taken to address the specific registration requirements of the software being used for the image fusion such as fields of view, slice thickness and slice spacing. Scans should preferably be 3 mm section thick with no intersection gap, have a 15 cm field of view and ideally be within 30 min of each other. This sequential timing of the scans is important to try to minimise changes in rectal filling between scans and also to eliminate changes in oedema-dependent volume if the scans are acquired on different dates.

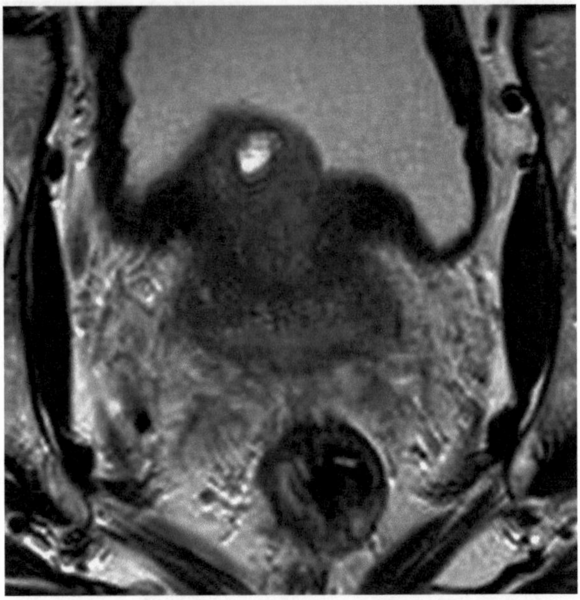

Fig. 9.6 Axial MRI at bladder base with prominent median lobe distinguished from the low-signal bladder wall

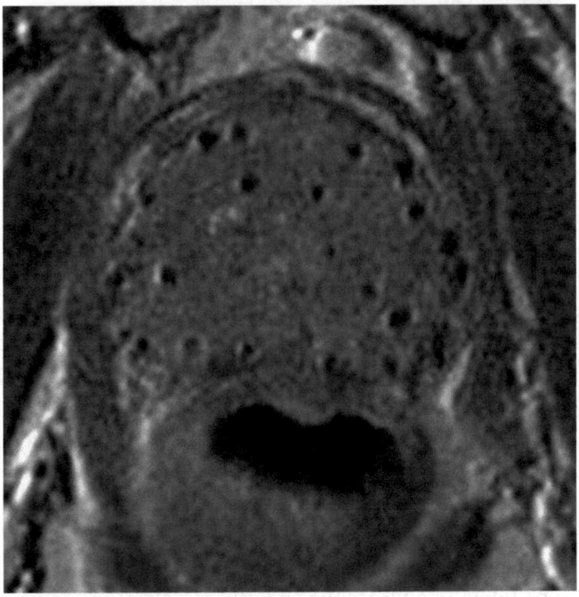

Fig. 9.7 Axial MRI scan showing sources in prostate

9.5 TRUS

Prostate brachytherapy is based on TRUS imaging and techniques have evolved considerably since the original Seattle 2-step implant technique. Nearly all modern prostate LDR brachytherapy uses TRUS to plan the location of sources in the gland

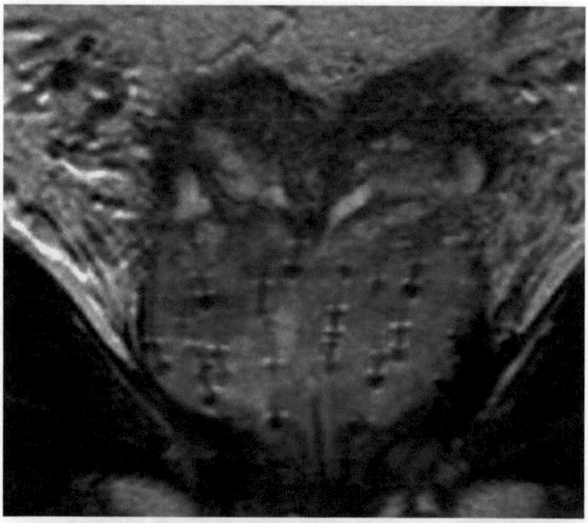

Fig. 9.8 Coronal MRI scan demonstrating linked sources in prostate

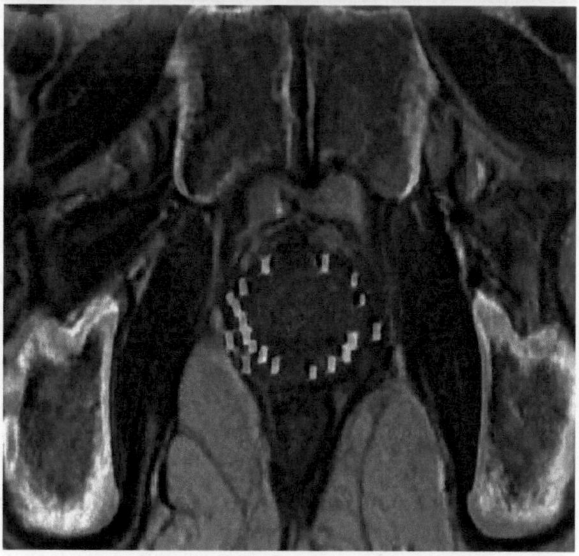

Fig. 9.9 Axial CT-MRI fused image

with some variation on the pre- or intraoperative TRUS-based planning technique. It seems intuitive therefore to consider using TRUS also for post-implant assessment of the implant. Interactive implant techniques incorporate some form of TRUS-based dosimetry to monitor and adapt the implant as it progresses. The major problem with regard to using TRUS for post-implant dosimetry, whether interactive or post-procedure, is the inability to accurately and reproducibly locate all implanted sources on TRUS imaging. Defining the prostate and marking source positions intraoperatively on TRUS are inevitably subjective with image quality deteriorating as the implant progresses because of prostate oedema and haemorrhage. Needle and source artefact may also hinder accurate source identification. Calcification within

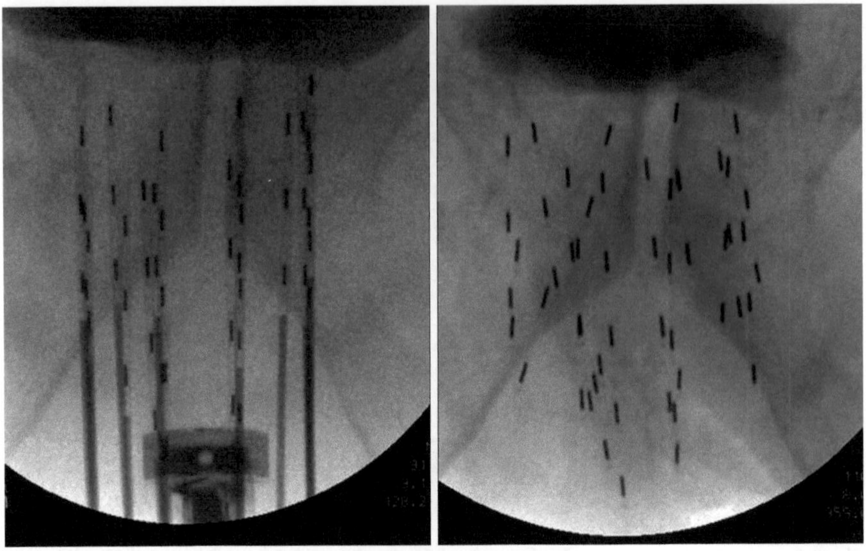

Fig. 9.10 Radiographs before and immediately after source insertion

the gland can obscure detail on TRUS images and result in major difficulties with source identification. Needle location may be used as a reasonable surrogate but may not truly reflect final source location within the prostate (Fig. 9.10) as sources may alter position or orientation slightly after deposition in the gland. The dynamic intraoperative TRUS-based assessment of dosimetry may therefore be hindered by inaccurate recognition of actual source placement. Repeating the TRUS examination for post-implant dosimetry some time after the procedure is potentially uncomfortable and inconvenient for the patient and so in general the use of TRUS for post-implant imaging has been limited to immediate TRUS acquisition following the procedure while the patient is still anaesthetised. 3D TRUS can be used both intra- and perioperatively but the same issues remain regarding identification of all sources in the gland (Fig. 9.11).

There are additional confounding factors when comparing TRUS-based intraoperative dosimetry to post-implant CT-based dosimetry (Steggerda et al. 2007). The intraoperative patient set-up is different: the patient is in a lithotomy as opposed to supine position; anaesthesia induces relaxation of pelvic musculature; and the gland is deformed by the TRUS probe. The TRUS images are generally segmented at 5 mm as compared to 3 mm CT slices. Various researchers have tried to segment the sources from TRUS images by linking sources with spacers, using radiographs to initialise segmentation, using elastography, or segmenting them directly. Even when carefully hand segmented, up to 25 % of the sources may remain undetected on ultrasound (Paulo et al. 2010). The underlying assumption has been that the coordinates for all the sources are known allowing intraoperative dosimetric modification and constant updating of the dose plan on an individual basis. However, segmentation algorithms are only accurate in segmenting sources that are fully visible,

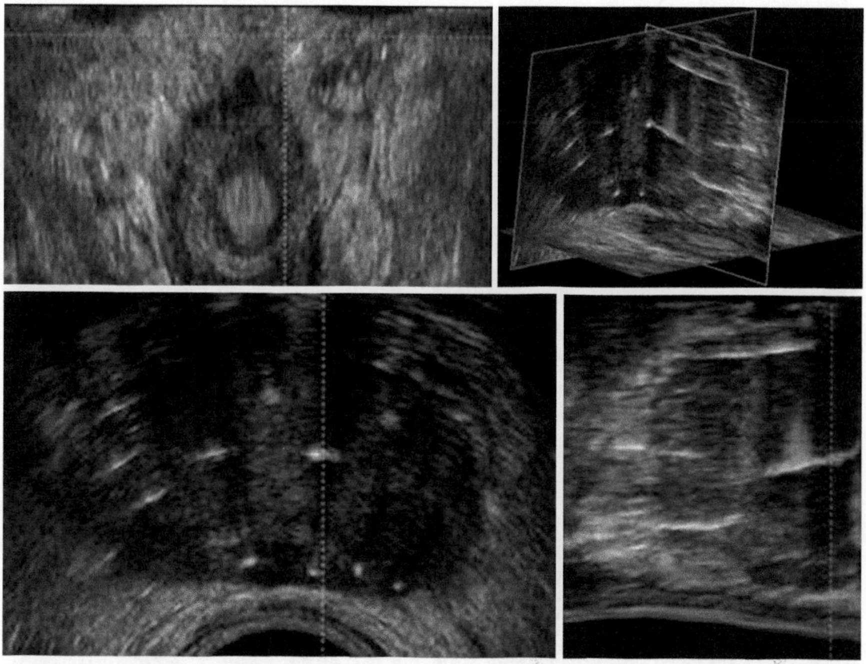

Fig. 9.11 3D TRUS post-implant

leaving some other sources unidentified, due to overlapping or suboptimal ultrasound imaging. Thus the coordinates of unidentified sources will not be available, resulting in some degree of dosimetric uncertainty (Paulo et al. 2010). This is an extremely important issue with clinical relevance and must be considered in any consideration of TRUS-based post-implant dosimetry.

9.6 Comparative Imaging and Fusion Imaging for Post-implant Dosimetry

Target delineation uncertainty will in general vary with patient, modality and observer (Remeijer et al. 1999) and the question is how to assimilate all the imaging data available. Different imaging of the prostate yields different contours (Smith et al. 2007) for the same prostate and it is clear therefore that these different imaging techniques (Fig. 9.12) can subsequently yield different measurements of implant quality for the same implant (Kalkner et al. 2006; Debois et al. 1999; Lee et al. 2002; Al-Qaisieh et al. 2002). Each technique has advantages and various image fusion methods have been used for post-implant imaging to attempt to combine these various advantages (Steggerda et al. 2006; Archer et al. 2010). The basic premise is to co-register the image data from a technique in which the sources are clearly visible, such as CT or fluoroscopy, with another technique that demonstrates the soft tissue margins more clearly such as MRI or TRUS. The advantages are clear; the major

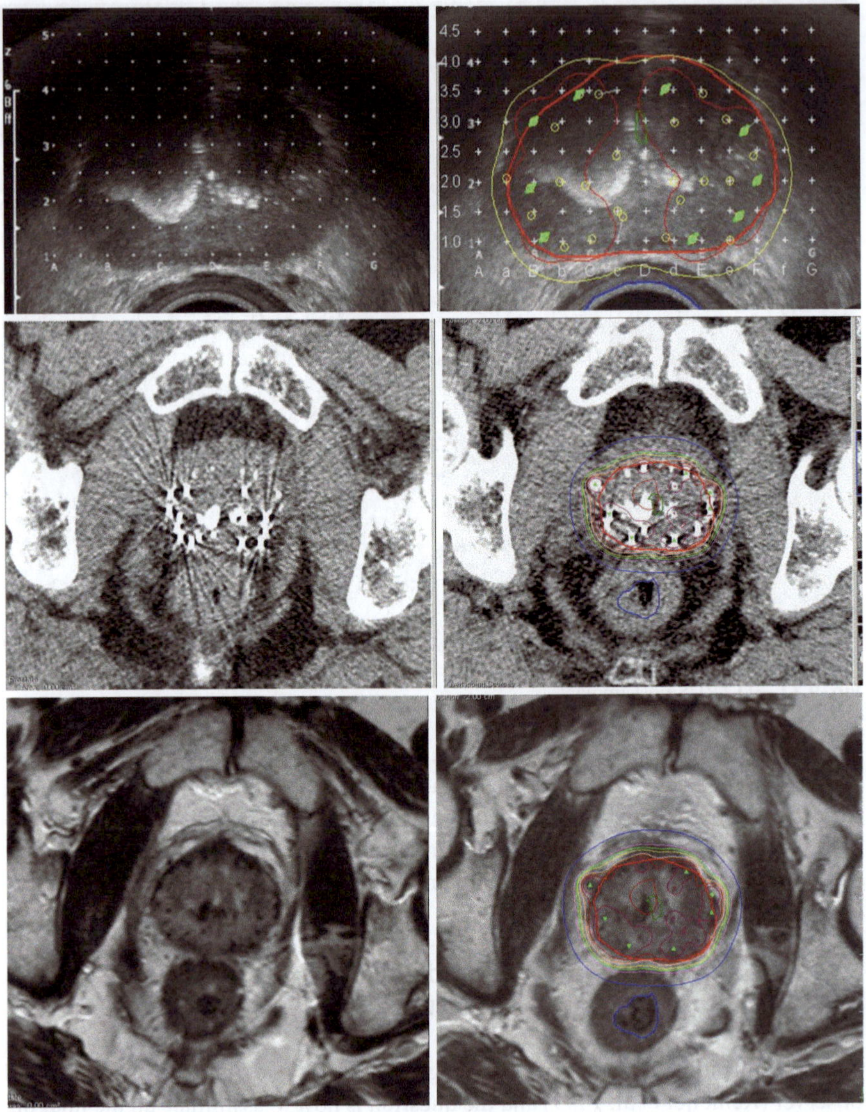

Fig. 9.12 Different imaging yields different dosimetry

drawbacks are cost and potential inconvenience to the patient who is unlikely, on an individual basis, to gain any clinical benefit from more complicated post-implant dosimetry. Image registration and fusion is also time-consuming and needs dedicated medical physics support. Furthermore, there are more fundamental issues that need to be considered when fusing pre-, intra- and postoperative image data sets. As already discussed, the intraoperative setting is different: the patient is in a lithotomy as opposed to supine position and registration of these different data sets can be very difficult and often a compromise. Interpretation of results must therefore be viewed with some caution. The most widely used imaging for post-implant dosimetry, CT scanning, overestimates the true prostate volume by about 30 % (Zhanrong et al. 2007). Both intra- and inter-observer variation can and should improve with experience, but the basic problem remains that CT-based dosimetry may not accurately reflect dosage to the prostate and structures at risk. Of all the potential fusion combinations suggested and investigated for improving our assessment of implant quality, CT-MRI fusion is the commonest and most widely published (Tanaka et al. 2006a, b; Crook et al. 2004; Polo et al. 2004). Different fusion software exists for combining the data sets and the fused images may then be used to calculate post-implant dosimetric indices. It has been advocated that CT-MRI fusion is a useful learning tool for prostate contouring and provides feedback to help improve performance and accuracy on CT delineation. This aspect of CT-MRI fusion for post-implant dosimetry alone is a valuable use of what is a time-consuming and often expensive process. If improvements in CT-based post-implant dosimetry in comparison to the CT-MRI fusion-based data can be made, then the CT-MRI fusion process might not need to be routinely performed in all patients. Selected implants could be chosen for CT-MRI fusion on a quality control basis at intervals. Acquiring and fusing two different image sets currently prohibits the use of image fusion on a routine basis, although recent software developments in CT-TRUS fusion may change this.

9.7 Effects of Timing on Post-implant Imaging

For low-dose rate brachytherapy, the dose is calculated by integrating the dose rate from the time of the implant to infinity, taking into account the natural decay of the sources. This assumes that the geometry of the implanted prostate and anatomy as determined at the time of the implant does not alter with time. However, the prostate swells during the implant procedure due to oedema and haemorrhage and this oedema will gradually resolve during the following weeks. Obviously, source locations and consequently the dose rate distribution will change during this time with changing volume and geometry, and there will be a dynamic distribution of measured dosimetry depending on the time as well as the technique of imaging. Various investigators have described the oedematous change in the prostate gland during and after the implant procedure and analysed the temporal post-implant resolution of this oedema (Reed et al. 2005; Dogan et al. 2002; Taussky et al. 2005; Moerland 1998; McNeely et al. 2004). Oedema itself does not have any specific appearances on imaging and generally only manifests as an increase in the overall observed volume of the gland and/or a change in the geometry of the gland. The timing of post-implant image acquisition affects dosimetry as the prostate initially swells and later regresses back to its preimplant volume (Steggerda et al. 2007). After prostate implantation, dosimetry is usually based on a single imaging session and has to assume no geometric change in the prostate during the time of dose accumulation. Dose cannot be assumed to have accumulated in a particular area of the prostate with certainty if the geometry of the prostate changes with respect to the planned location of those sources in the gland. In addition, it is also recognised that the changes in prostate geometry may not be symmetrical and may even be different for different preimplant prostate volumes. It is difficult therefore to be prescriptive in how to deal with the different data provided by different imaging at different time intervals after the implant. Craniocaudal shift of the implanted sources over time with respect to the prostate has also been recognised on imaging and is a further confounding factor.

Imaging performed immediately post-implant, also known as “day zero imaging,” results in lower measured dosimetric indices than if the dosimetry is based on imaging performed at any time thereafter. Day zero imaging is still performed at many centres, the advantages being immediate feedback and convenience for the patient. At institutions where a catheter is placed during the imaging procedure in order to localise the urethra, day zero imaging, with the urinary catheter still in situ, should aid visualisation of the urethra and the dose distribution in the organ can be more accurately determined.

Several authors have suggested that the most representative time to image the patient is 2–3 weeks following the procedure. Previous work has shown (Taussky et al. 2005) only small changes in these parameters at day 8 and day 30 compared to the day of implantation using MRI-CT instead of TRUS-CT image fused data. If the CT scan is postponed to about 1 month after implantation, there is a reasonable assumption that there will be little further change in prostate volume afterwards. Although this might be based on clinical experience and intuition, it does not guarantee that the dose cumulated over time in the prostate and adjacent anatomical structures will be accurately estimated from the dose rate distribution based on imaging at this time. Nevertheless, consistency of the timing of post-implant imaging is the key and it is recommended that each centre should adopt a uniform policy for when the imaging is to be performed.

9.8 Structures at Risk: Imaging for Post-implant Dosimetry

Analysis of implant-related morbidity is an important aspect of any quality assurance programme for prostate brachytherapy. Whatever imaging is used, the major structures at risk from prostate LDR brachytherapy should be considered in any evaluation of post-implant imaging. These are the urethra, the rectum, the penile bulb and the neurovascular bundles. Morbidity related to brachytherapy can be correlated with the calculated doses to these structures but efforts to standardise critical structure dosimetry and reporting, although improving as outcome data matures and results are published, have been less success than in standard prostate gland dosimetry (Nag et al. 2002).

Knowledge of the varied appearances of these critical structures on different imaging techniques is essential for accurate delineation. Correlation of urethral dosimetry with post-implant urethral obstructive and irritative symptoms may help clarify the aetiology of post-implant urinary toxicity and dose constraints to the urethra adjusted as necessary (Allen et al. 2005; Leclerc et al. 2006). The urethra has a structure that is best seen on MRI scans and usually impossible to visualise on a

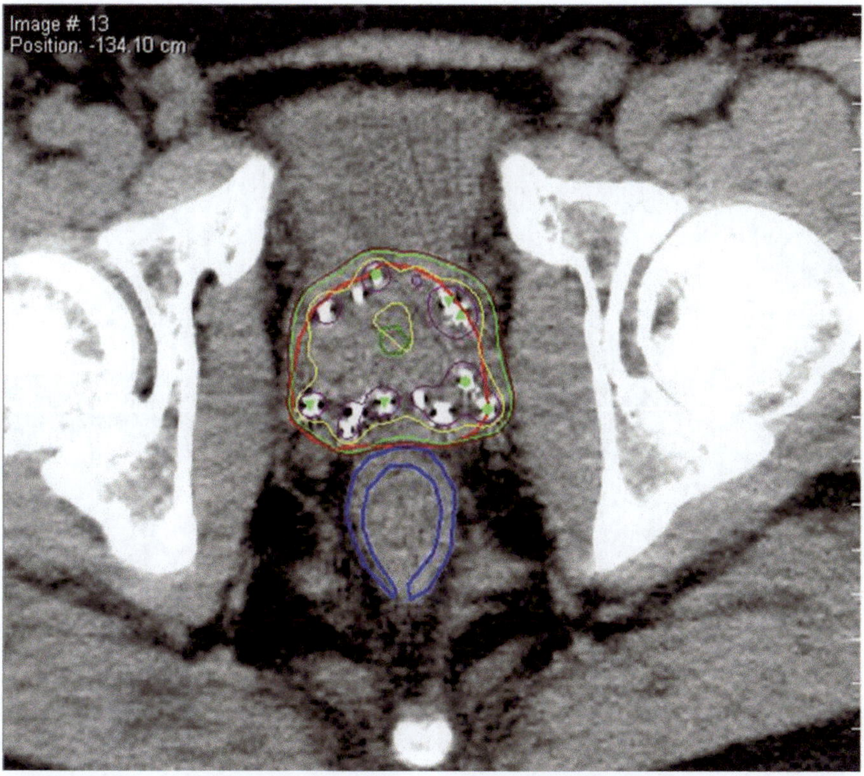

Fig. 9.13 Post-implant CT for dosimetry with arbitrary delineation of nonvisualised urethra

CT scan without a urinary catheter in position. It is adequately visualised on TRUS during the implant with placement of a urinary catheter or instillation of intraurethral gel. Traditionally, the urethra has arbitrarily been designated as a point source in the centre of the prostate (Fig. 9.13), but this can be far from representative. High calculated doses to the urethra would be expected to be associated with increased urinary toxicity. The estimated dose to the urethra increases with time as the prostate oedema settles and the sources contract towards the centre of the prostate where the urethra is located as was also reported by Waterman and Dicker (2000). Segmental urethral dosimetry and stratifying associated toxicity into irritative and obstructive cases may improve our understanding of the underlying mechanisms associated with urethral toxicity. MRI allows improved definition of segmental urethral anatomy including the bladder outlet, transition zone, lower sphincter and membranous and bulbar urethra. The urethra, for example, has several distinct regions definable on MRI and dose to these regions can vary widely. The urethra can also potentially be distorted by the TRUS probe and assessment of its exact position within the gland may therefore vary from the post-implant CT or MRI images.

The most common rectal complication associated with LDR prostate brachytherapy is proctitis, usually resulting in rectal bleeding. Many authors have successfully

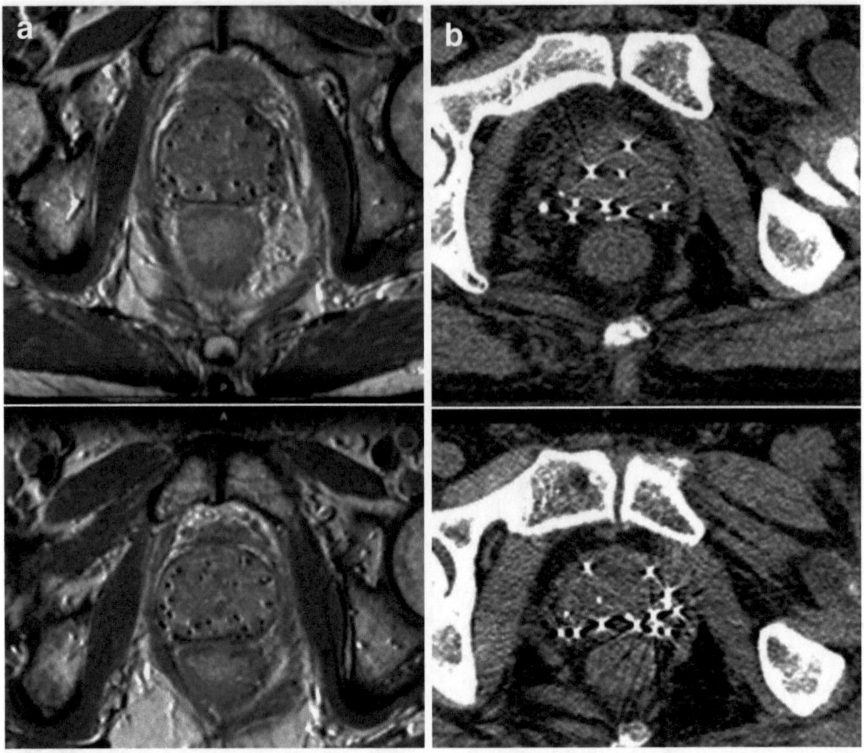

Fig. 9.14 Axial CT (**a**) and MRI (**b**) on same patient scanned consecutively showing different appearances of rectum

correlated rectal complications with the calculated post-implant doses to the rectum (Merrick et al. 1999; Han and Wallner 2001). However, differences in the definition of the rectal target volume have made valid comparison of results difficult. The rectum may often have a very different appearance on different imaging techniques: compare the intraoperative TRUS to the post-implant CT or MRI scan (Fig. 9.14). Varying degrees of rectal filling will clearly influence the delineated volume and so affect the measured dosimetry. Inherent measured dosimetry for the rectum therefore exists between pre-plan and post-implant analyses, due to the different imaging techniques, timings and body positions used. The inner wall of the rectum is generally defined on the post-implant imaging by the edge of the lumen, taking care to exclude any faeces (Snyder et al. 2001). If the lumen cannot be identified, the inner wall was approximated based on diameter of the outer rectal wall and thickness of the rectal wall on consecutive sections. The serial changes in rectal measured dosimetry are now well recognised (Pinkawa et al. 2009) when trying to interpret dosimetry results with regard to rectal toxicity.

The multifactorial nature of sexual dysfunction makes it difficult to precisely quantify the incidence of this aspect of brachytherapy-related toxicity following implant. Many authors have attempted to correlate doses to the erectile organs

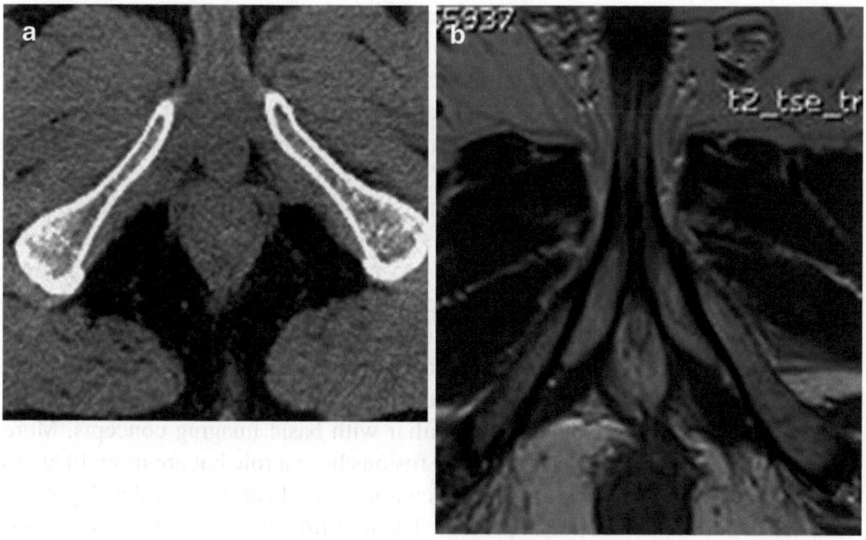

Fig. 9.15 Axial CT (**a**) and MRI (**b**) through penile bulb

(penile bulb, neurovascular bundles) with sexual morbidity following brachytherapy. However, again we have the issues of delineation on imaging when considering results. The neurovascular bundles are not well defined on CT scans and so often arbitrary point sources are chosen as surrogate markers in order to calculate dosimetry. As these locations are based on prostate margin delineation as seen on CT, a further magnitude of uncertainty is introduced into the calculations. The penile bulb may be the most significant organ in terms of sexual dysfunction post-brachytherapy and this structure is more readily defined on TRUS and MRI but again is often misinterpreted on CT scans (Fig. 9.15). The penile bulb lies 1–2 cm inferior to the apex of the prostate gland. Work has been performed recently using MRI to localise the neurovascular bundles for prostate brachytherapy (McLaughlin et al. 2005). Previous work has demonstrated a dose–volume effect for radiation dose to the penile bulb (Merrick et al. 2002), but more studies are needed to be able to reliably define dose constraints for these structures.

Conclusion

Post-implant dosimetry provides the foundation for our current knowledge regarding implant quality. Clinical outcomes have been directly correlated with implant quality (Potters et al. 2003). In general, multimodality imaging studies confirm that CT overestimates both the true volume of the prostate and the volume as compared with the measurements recorded on other imaging methods particularly TRUS imaging. Imaging technology continues to improve but fundamental limitations will always be present based on the inherent laws of imaging physics. Clinical bias, experience and the inherent limitations of the imaging contribute to a range of uncertainties about the accuracy of any measured dosimetric indices. Radiation oncologists may be more concerned about unintentional inclusion of rectal tissue than over-inclusion of bladder base and tissue beyond the prostate apex, and variations will always exist in the contouring of the prostate and structures at risk. Consistent results can, however, be achieved with training and experience (Al-Qaisieh et al. 2009). A well-structured programme for post-implant dosimetry should be established within all centres. Without post-implant dosimetry, systematic errors can be overlooked and, as with any treatment for early-stage prostate cancer, suboptimal results may take many years to manifest. At present, there is no evidence that intraoperative dosimetry can replace post-implant analysis and despite its acknowledged limitations, CT-based dosimetry alone has stood the test of time (Al-Qaisieh et al. 2009). Analysis of post-implant dosimetry data should be team-based and it is helpful if team members are familiar with basic imaging concepts. More complex imaging using various image fusions have a role but are more likely to be of relevance in the evaluation of technique modifications and developments rather than routine practice at present. The multifocality of prostate cancer and the influence of the dominant tumour focus may need more detailed consideration in the future, as new dosimetry paradigms seek to define sub-volumes of increased and reduced dose.

References

Al-Qaisieh B, Ash D, Bottomley D et al (2002) Impact of prostate volume evaluation by different observers on CT-based postimplantation dosimetry. Radiother Oncol 62:267–273

Al-Qaisieh B (2003) UK Prosate Brachytherapy Group. Pre- and post-implant dosimetry: an inter-comparison between UK prostate brachytherapy centres. Radiother Oncol 66: 181–183

Al-Qaisieh B, Smith D, Brearley E et al (2007) ComprehensiveI-125 multi seed comparison for prostate brachytherapy: dosimetry and visibility analysis. Radiother Oncol 84(2):140–147

Al-Qaisieh B, Witteveen T, Carey B et al (2009) Correlation between pre- and postimplant dosimetry for iodine-125 seed implants for localized prostate cancer. Int J Radiat Oncol Biol Phys 75(2):626–630

Allen Z, Merrick G, Butler W et al (2005) Detailed urethral dosimetry in the evaluation of prostate brachytherapy-related urinary morbidity. Int J Radiat Oncol Biol Phys 62:981–987

Archer P, Puttagunta S, Rhode K et al (2010) An analysis of intraoperative versus post-operative dosimetry with CT, CT/MRI fusion and XMR for the evaluation of permanent prostate brachytherapy implants. Radiother Oncol 96(2):166–171

Crook J, Milosevic M, Catton P et al (2002) Interobserver variation in postimplant computed tomography contouring affects quality of prostate brachytherapy. Brachytherapy 1:66–73

Crook J, Mclean M, Yeung I et al (2004) MRI-CT fusion to assess post brachytherapy prostate volume and the effects of prolonged oedema on dosimetry following transperineal interstitial permanent prostate brachytherapy. Brachytherapy 3:55–60

De Brabandere M, Kirisits C, Peeters R et al (2006) Accuracy of seed reconstruction in prostate postplanning studied with a CT- and MRI-compatible phantom. Radiother Oncol 79(2):190–197

Debois M, Oyen R, Maes F et al (1999) The contribution of magnetic resonance imaging to the three-dimensional treatment planning of localized prostate cancer. Int J Radiat Oncol Biol Phys 45:857–865

Dogan N, Mohideen N, Glasgow G et al (2002) Effect of prostatic oedema on CT-based postimplant dosimetry. Int J Radiat Oncol Biol Phys 53:483–489

Dubois D, Prestidge B, Hotchkiss L et al (1998) Intraobserver and interobserver variability of MR imaging- and CT-derived prostate volumes after transperineal interstitial permanent prostate brachytherapy. Radiology 207:785–789

Fuks Z, Leibel S, Wallner K et al (1991) The effect of local control on metastatic dissemination in carcinoma of the prostate: long-term results in patients treated with 125I implantation. Int J Radiat Oncol Biol Phys 21(3):537–547

Han B, Wallner K (2001) Dosimetric and radiographic correlates to prostate brachytherapy-related rectal complications. Int J Cancer 96(6):372–378

Holupka E, Meskell P, Burdette E et al (2004) An automatic seed finder for brachytherapy CT postplans based on the Hough transform. Med Phys 31(9):2672–2679

Jaffray D, Siewerdsen J, Edmundson et al (2002) Flat-panel cone-beam CT on a mobile isocentric C-arm for image-guided brachytherapy. Proc SPIE 4682:209–217

Kalkner K, Kubicek G, Nillson J et al (2006) Prostate volume determination: differential volume measurements comparing CT and TRUS. Radiother Oncol 81(2):179–183

Leclerc G, Lavallée M, Roy R et al (2006) Prostatic edema in ^{125}I permanent prostate implants: dynamic rectal dosimetry taking volume changes into account. Med Phys 33:574–583

Lee W, Roach M, Michalski J et al (2002) Interobserver variability leads to significant differences in quantifiers of prostate implant adequacy. Int J Radiat Oncol Biol Phys 54:457–461

McLaughlin P, Narayana V, Drake D et al (2002) Comparison of MRI pulse sequences in defining prostate volume after permanent implantation. Int J Radiat Oncol Biol Phys 54:703–711

McLaughlin P, Narayana V, Meirovitz A et al (2005) Vessel-sparing prostate radiotherapy dose limitation to critical erectile vascular structures (internal pudental artery and corpus cavernosum) defined by MRI. Int J Radiat Oncol Biol Phys 61:20–31

Merrick G, Butler W, Dorsey A et al (1999) Rectal dosimetric analysis following prostate brachytherapy. Int J Radiat Oncol Biol Phys 43:1021–1102

Merrick G, Butler W, Wallner K et al (2002) The importance of radiation doses to the penile bulb vs. crura in the development of postbrachytherapy erectile dysfunction. Int J Radiat Oncol Biol Phys 54:1055–1062

McNeely L, Stone N, Presser J et al (2004) Influence of prostate volume on dosimetry results in real-time 125I seed implantation. Int J Radiat Oncol Biol Phys 58:292–299

Moerland M (1998) The effect of oedema on post implant dosimetry of permanent iodine-125 prostate implants: A simulation study. J Brachyther Int 14:225–231

Nag S, Bice W, DeWyngaert K et al (2000) The American Brachytherapy Society recommendations for permanent prostate brachytherapy postimplant dosimetric analysis. Int J Radiat Oncol Biol Phys 46:221–230

Nag S, Ellis R, Merrick G et al (2002) American Brachytherapy Society recommendations for reporting morbidity after prostate brachytherapy. Int J Radiat Oncol Biol Phys 54(2):462–470

Paulo A, Salembwier C, Venselaar J et al (2010) Review of intraoperative imaging and planning techniques in permanent seed prostate brachytherapy. Radiotherapy Oncol 94(1):12–23

Pinkawa M, Asadpour B, Piroth M et al (2009) Rectal dosimetry following prostate brachytherapy with stranded seeds – comparison of transrectal ultrasound intra-operative planning (day 0) and computed tomography-postplanning (day 1 vs. day 30) with special focus on sources placed close to the rectal wall. Radiother Oncol 91(2):207–212

Polo A, Cattani F, Vavassori A et al (2004) MR and CT image fusion for postimplant analysis in permanent prostate seed implants. Int J Radiat Oncol Biol Phys 60:1572–1579

Potters L, Cao Y, Calugaru E et al (2001) A comprehensive review of CT-based dosimetry parameters and biochemical control in patients treated with permanent prostate brachytherapy. Int J Radiat Oncol Biol Phys 50:605–661

Potters L, Huang D, Calugaru E et al (2003) Importance of Implant dosimetry for patients undergoing prostate brachytherapy. Urol 62(6):1073–1077

Reed D, Wallner K, Ford E et al (2005) Effect of post-implant edema on prostate brachytherapy treatment margins. Int J Radiat Oncol Biol Phys 63:1469–1473

Remeijer P, Rasch C, Lebesque J et al (1999) A general methodology for three-dimensional analysis of variation in target volume delineation. Med Phys 26:931–994
Roy J, Wallner K, Harrington P et al (1993) CT-based evaluation method for permanent implants: application to prostate. Int J Radiat Oncol Biol Phys 26(1):163–169
Salembier C, Lavagnini P, Nickers P et al (2007) Tumour and target volumes in permanent prostate brachytherapy: a supplement to the ESTRO/EAU/EORTC recommendations on prostate brachytherapy. Radiother Oncol 83:3–10
Siewerdsen J, Moseley D, Burch S et al (2005) Volume CT with a flat-panel detector on a mobile, isocentric C-arm: pre-clinical investigation in guidance of minimally invasive surgery. Med Phys 32:241–254
Smith W, Lewis C, Bauman G et al (2007) Prostate volume contouring: a 3D analysis of segmentation using 3DTRUS, CT, and MR. Int J Radiat Oncol Biol Phys 67(4):1238–1247
Snyder K, Stock R, Hong S et al (2001) Defining the risk of developing Grade 2 proctitis following 125I prostate brachytherapy using a rectal dose–volume histogram analysis. Int J Radiat Oncol Biol Phys 50(2):335–341
Steggerda M, Schneider C, van Herk M et al (2005) The applicability of simultaneous TRUS-CT imaging for the evaluation of prostate seed implants. Med Phys 32:2262–2271
Steggerda M, Luc M, Moonen H et al (2007) The influence of geometrical changes on the dose distribution after I-125 seed implantation of the prostate. Radiother Oncol 83(1):11–17
Su Y, Davis B, Herman M et al (2004) Prostate brachytherapy seed localization by analysis of multiple projections: identifying and addressing the seed overlap problem. Med Phys 31:1277–1287
Su Y, Davis B, Herman M et al (2005) Examination of dosimetry accuracy as a function of seed detection rate in permanent prostate brachytherapy. Med Phys 32(9):3049–3056
Tanaka O, Hayashi S, Matsue M et al (2006a) Comparison of MRI-based and CT/MRI fusion–based postimplant dosimetric analysis of prostate brachytherapy. Int J Radiat Oncol Biol Phys 66:597–602
Tanaka O, Hayashi S, Sakurai K et al (2006b) Importance of the CT/MRI fusion method as a learning tool for CT-based postimplant dosimetry in prostate brachytherapy. Importance of the CT/MRI fusion method as a learning tool for CT-based postimplant dosimetry in prostate brachytherapy dosimetry. Radiother Oncol 81(3):303–308
Taussky D, Austen L, Toi A et al (2005) Sequential evaluation of prostate edema after permanent seed prostate brachytherapy using CT-MRI fusion. Int J Radiat Oncol Biol Phys 62:974–980
Thomas S, Wachowicz K, Fallone B (2009) MRI of prostate brachytherapy seeds at high field: a study in phantom. Med Phys 36(11):5228–5235
Todor D, Cohen G, Amols H et al (2002) Operator-free, film-based 3D seed reconstruction in brachytherapy. Phys Med Biol 47(12):2031–2048
Waterman F, Dicker A (2000) The impact of postimplant oedema on the urethral dose in prostate brachytherapy. Int J Radiat Oncol Biol Phys 47:661–664
Zhang M, Zaider M, Worman M et al (2004) On the question of 3D seed reconstruction in prostate brachytherapy: the determination of x-ray source and film locations. Phys Med Biol 49(19): 335–345
Zhanrong G, Wilkins D, Eapen L et al (2007) A study of prostate delineation referenced against a gold standard created from the visible human data. Radiother Oncol 85(2):239–246

Dosimetry Planning for Permanent Seeds

10

B. Al-Qaisieh

10.1 Introduction

During the last 10 years, planning procedures for permanent seed implantation have evolved. Technical developments, like biplane ultrasound probes, stepping devices with positioning feedback or treatment planning systems with the potential for using real-time techniques, have given the user the ability to perform treatment planning in the operating theatre (OR) using permanently updated patient image data to obtain optimised results. The operator acquires feedback from the live images and can adjust the treatment plan accordingly. The main approaches to planning permanent seed implants of the prostate are preplanning, intraoperative planning, interactive planning and dynamic dose calculation. These four methods will be described in this chapter following a brief overview of planning technique.

Prior to clinical implementation, the treatment planning software (TPS) and the hardware equipment must be commissioned and tested. This should include investigation of the ultrasound probe, stepping unit and template grid prior to their first use. During commissioning of the TPS, all required data relating to the used source type(s), such as the base data for the planning algorithm, must be entered and checked by the medical physicists. Dose calculation results of the TPS should also be carefully checked against manual computations of a collection of representative dose points. It is recommended that the TG-43 formalism for the dose calculation of permanent seed implants be used (Rivard et al. 2004). This formalism takes the orientation of the sources into account. However, in most cases, the orientation of the seeds in the patient is not exactly known; thus, the point-source approach can be used in these instances (Nath et al. 2009). Improved imaging is necessary to determine the seeds' orientation.

B. Al-Qaisieh
St James's Institute of Oncology, Medical Physics and Engeneering,
The Leeds Teaching Hospital NHS TRUST, West Yorkshire, UK
e-mail: bashar@medphysics.leeds.ac.uk

G. Kovács, P. Hoskin (eds.), *Interstitial Prostate Brachytherapy*,
DOI 10.1007/978-3-642-36499-0_10, © Springer-Verlag Berlin Heidelberg 2013

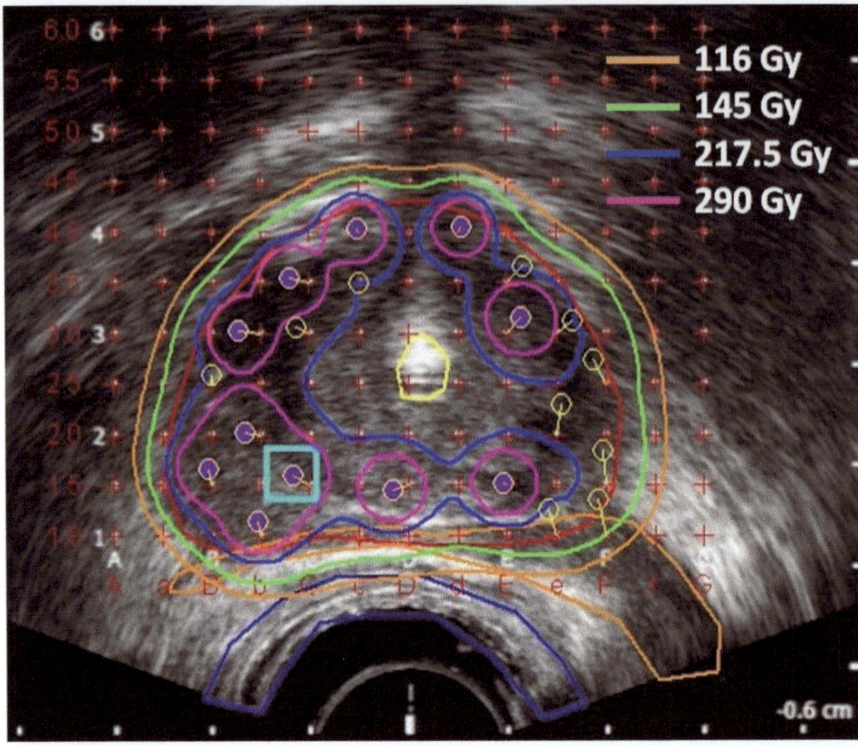

Fig. 10.1 Transrectal ultrasound in transversal view. Prescription dose is 145 Gy (*green isodose line*). Needle paths are indicated by *yellow circles*, source positions by *purple filled circles*. The short *yellow lines* at the needle paths indicate the deviation of the needle path from the template coordinate position

Due to the fact that the half-life of the seeds (either I-125, Pd-103 or even Cs-131) is quite short, the sources are ordered from the vendor to be delivered with the required activity for the implant day. To determine the required number of sources to be ordered, nomograms are used which require an estimate of the prostate volume. In most cases, a few 'extra' seeds can be helpful to account for situations in which the prostate volume may change between the day of the volume determination and the day of the implant.

The prescription dose for an I-125 permanent seed implant is recommended by the GEC-ESTRO to be 145 Gy (Salambier et al. 2007). The guidelines of the ABS allow a variation of between 140 and 160 Gy (Davis et al. 2012). The clinical target volume (CTV) is the prostate gland capsule with a small margin and is then equal to the planning target volume (PTV). For T1 and T2 tumours, GEC-ESTRO guidelines recommend that the expansion from the capsule is a three-dimensional margin of 3 mm to create the PTV (Salambier et al. 2007), but this might be constrained at the bladder neck and rectal wall. For treatment planning, two organs at risk (OAR) have to be considered: the rectum and the urethra; see Fig. 10.1. Although other organs

Table 10.1 Dose criteria for target volumes and organs at risk according to the GEC-ESTRO recommendations

Target volume	
CTV	D90 ≥ 145 Gy, V100 ≥ 95%, V150 ≤ 50%
Organs at risk	
Rectum	D_{2cc} ≤ 145 Gy, $D_{0.1cc}$ < 200 Gy
Urethra	D10 < 150% (217.5 Gy), D30 < 130% (188.5 Gy)

such as the penile bulb and the neurovascular bundle will be irradiated, not enough data exists to suggest what the maximum tolerated doses are. In Table 10.1, the dose criteria for the organs at risk and the target are listed according to the GEC-ESTRO recommendations. It should be noted that these parameters differ slightly from the ABS guidelines (Davis et al. 2012). When the whole prostate gland is treated, the V100 for the prostate should be ≈100 % (Nath et al. 2009).

For a given source strength, the geometry (i.e. the distribution) of the seeds in the target volume is the only parameter which effects the shape of the dose distribution. The positioning of the sources in the treatment planning process must therefore be performed carefully. There is no clear consensus as to the optimal single source strength in seed implants. For I-125, it might vary between 0.4 U and 0.8 U per seed (Davis et al. 2012); however, all seeds in an implant should have the same source strength, although techniques using cooler seeds centrally have also been described.

In principle the treatment planning procedure for single and stranded seeds is the same. When using stranded seeds with fixed distances between sources, these dimensions have to be configured in the TPS. If strands with varying distances are in use and are assembled in the OR, treatment planning can be performed in the same manner as single seeds. Some centres prefer a combination of stranded and single seed implants (Langley et al. 2012).

The treatment planning and dose calculation of the seed implant can be performed in different ways:

1. In *manual forward planning*, the seeds are placed manually in the TPS until the constraints are reached. This can be time consuming as more than 80 seeds are often required. While determining seed positions, it should be considered that the dose distribution around a source is three-dimensional and viewing in planes other than transaxial can be advantageous. As any limitations of the imaging will have a direct impact on both the complexity and the quality of the plan, it is important to consider the volume in 3D and, therefore, it is extremely helpful if the medical physicist is present in the OR while the TRUS probe is set up and the images for planning are acquired. This allows modifications to the setup to be made if required.
2. *Geometrical optimisation* uses geometrical information only. The seed positions are computed by the distances of the seeds. This form of optimisation is very fast but uses no dose information relating to the CTV or OARs. To save time, geometrical optimisation can be used as a first step in the planning process which must then be followed by refinement using manual forward planning.

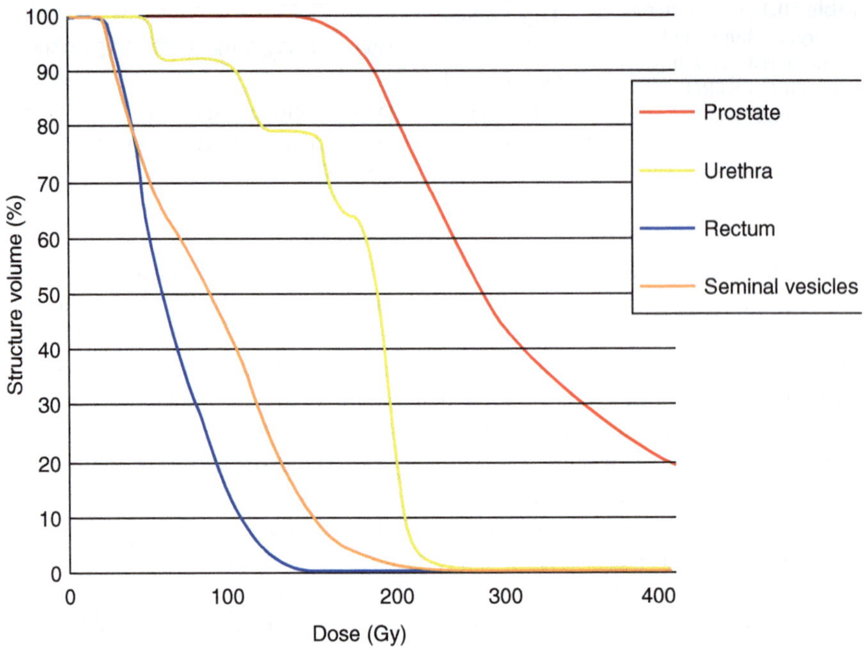

Fig. 10.2 Cumulative dose-volume histogram of a typical seed implant with 145 Gy prescription dose

3. *Inverse planning* uses a set of dose constraints linked with weightings. To design a set of dose constraints for the TPS, the dose values of the GEC-ESTRO guidelines might be helpful as a starting point. These constraints can then be adapted until a 'class solution' is found which produces a clinically acceptable plan. Once established, the constraints and associated weightings can be stored in the TPS as a template. The preparation of such templates can be time consuming, because often a significant number of adjustments are required from the initial dose constraints specified in the recommendations. It should also be noted that different TPSs use different optimisation approaches and thus templates of dose constraints will not be universally suitable across all TPSs without adaption.

For the evaluation of a treatment plan, dose-volume histograms (DVHs) are used with those that display cumulative dose being the most common, an example of which is shown in Fig. 10.2.

Once a clinically acceptable treatment plan has been produced, a report is generated which summarises the patient demographics, number of sources and needles, source activity, dose parameters and organ volumes in addition to a needle loading report. This important element of the report specifies the location of each individual seed within each needle including needle depths. An example of such a report is given in Fig. 10.3.

Needle number	Retraction (cm)	Hole location	Number seeds
1	0.50	c3.0	3
2	0.50	d3.0	3
3	0.50	b2.5	3
4	0.00	c2.5	5
5	0.00	d2.5	5
6	0.50	E2.5	4
7	0.00	b2.0	4
8	2.50	C2.0	2
9	0.00	e2.0	3
10	1.00	B1.5	2
11	0.50	b1.5	3
● *12*	*0.00*	*c1.5*	*4*
● *13*	*0.00*	*d1.5*	*4*
14	0.50	e1.5	3
15	0.00	b1.0	3
16	0.50	C1.0	4
17	0.50	D1.0	3
18	0.50	E1.0	4
19	0.00	e1.0	3

● = *Special loading*

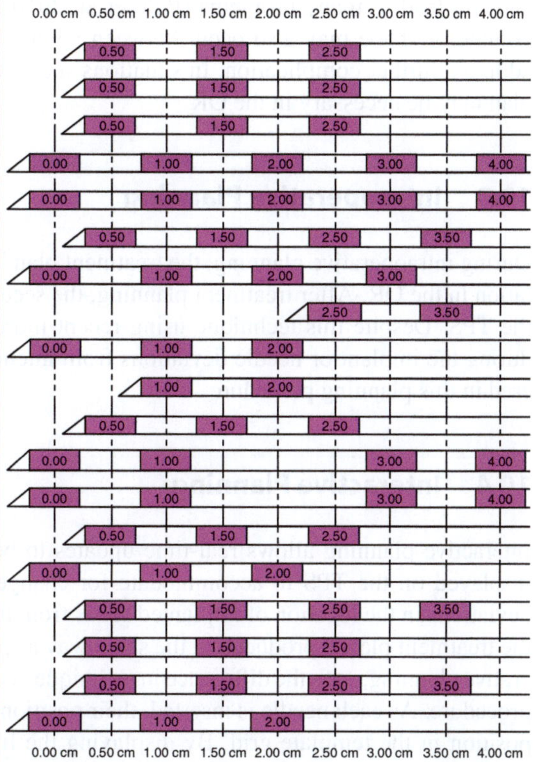

Fig. 10.3 In a needle loading report, the position of the sources within every implant needle and the needle depth is depicted. Moreover, it contains information about the template coordinate, number of sources per needle and if special loadings are existent. The latter are of importance when using stranded seeds with standardised seed-spacer sequences

The following paragraphs explain different planning approaches such as preplanning, intraoperative planning, interactive planning and dynamic dose calculation.

10.2 Preplanning

When following a preplanning protocol, a treatment plan is produced based on the imaging carried out during the volumetry analysis. Although in this case the prediction of the number of seeds required will be more accurate than using a nomogram, this technique has many inherent inaccuracies and should be used with care. One of the main issues is related to reproducibility. On the day of the implant, both the patient and the TRUS probe must be set up in exactly the same position as when the volume study was initially acquired. Trying to achieve this can be extremely difficult and can lead to many problems (Polo et al. 2010). Changes in either the prostate volume or shape may also occur between volume measurement and implantation adding a further complication. In situations such as these, changes to the treatment plan may be necessary in the OR.

10.3 Intraoperative Planning

During intraoperative planning, the treatment plan is created shortly before implantation in the OR. After treatment planning, the seeds are implanted as computed by the TPS. Despite this technique using recent morphological image data, changes during the implant or needle deviations from intended positions cannot be considered in this planning procedure.

10.4 Interactive Planning

Interactive planning allows real-time updates to be made to the dose distribution displayed on the TPS to accommodate for changes in the prostate and to reflect deviations in the location of implanted seeds from their intended positions. Initially, the treatment plan is produced in the same way as previously described for intraoperative planning, with the difference in technique occurring during the actual implant procedure. As each needle is inserted, their position might deviate from the planned position in the template grid. By displaying the live image data on the TPS, this deviation in needle position and seed placement can simply be adjusted on screen. By recalculating the dose following each update in needle position, a more accurate representation of the dose within the patient can be displayed. It is also possible using this technique to modify the loading pattern of the needles by reassessing the treatment plan following the insertion of several needles and before actually depositing the seeds within the patient. Final assessment of the delivered plan at the end of the implant may also highlight the need for additional needles.

10.5 Dynamic Dose Calculation

Dynamic dose calculation requires the position of every single seed to be known after implantation thus allowing the 3D dose matrix to be constantly updated (Todor et al. 2003). Although this technique provides the most accurate representation of the delivered dose, its complexity means that it is not straightforward to implement clinically. Recent studies relating to dynamic dose calculation have been reported involving several different imaging modalities. TRUS has been used in combination with fluoroscopy (Su et al. 2007). Sources were reconstructed by the fluoroscopy images and fused with matched seed images identified in the TRUS. Differences $<0.2\ \%$ in D90 of the prostate between CT post-planning and TRUS-fluoroscopic planning using this fusion technique were reported. Another approach has used a

C-arm cone-beam CT (Westendorp et al. 2007). After TRUS-based seed implant in the OR, a C-arm cone-beam CT was acquired and the reconstructed seed positions registered with the seed locations of the TRUS data set. Poor coverage of the target could be eliminated by implanting further sources in the 'cold' areas of the gland. At the end of the procedure, a further C-arm cone-beam CT was obtained. A mean number of four 'extra' were seeds implanted in their study of ten patients due to this cone-beam technique. Dose parameters of the prostate could be increased from 95 % to 100 % for D90 and from 83 % to 90 % for the V100.

References

Davis BJ, Horwitz EM, Lee WR et al (2012) American Brachytherapy Society consensus guidelines for transrectal ultrasound-guided permanent prostate brachytherapy. Brachytherapy 11:6–19

Langley SE, Laing RW (2012) 4D Brachytherapy, a novel real-time prostate brachytherapy technique using stranded and loose seeds. BJU Int 109(Suppl 1):1–6

Nath R, Bice WS, Butler WM et al (2009) AAPM recommendations on dose prescription and reporting methods for permanent interstitial brachytherapy for prostate cancer. Report of the AAPM Task Group 137. Med Phys 36:5310–5322

Polo A, Salembier C, Venselaar J, Hoskin P, PROBATE group of the GEC ESTRO (2010) Review of intraoperative imaging and planning techniques in permanent seed prostate brachytherapy. Radiother Oncol 94:12–23

Rivard MJ, Coursey BM, DeWerd L et al (2004) Update of AAPM TG 43: a revised AAPM protocol for brachytherapy dose calculations. Med Phys 31:633–674

Salambier C, Lavagnini P, Nickers P et al (2007) Tumour and target volumes in permanent prostate brachytherapy: a supplement to the ESTRO/EAU/EORTC recommendations on prostate brachytherapy. Radiother Oncol 83:3–10

Su Y, Davis BJ, Furutani KM, Herman MG, Robb RA (2007) Seed localization and TRUS fluoroscopy fusion for intraoperative prostate brachytherapy dosimetry. Comput Aided Surg 12:25–34

Todor DA, Zaider M, Cohen GN, Worman MF, Zelefsky MJ (2003) Intraoperative dynamic dosimetry for prostate implants. Phys Med Biol 48:1153–1171

Westendorp H, Hoekstra CJ, van't Riet A, Minken AW, Immerzeel JJ (2007) Intraoperative adaptive brachytherapy of iodine-125 prostate implants guided by C-arm cone-beam computed tomography-based dosimetry. Brachytherapy 6:231–237

11 HDR Planning

Frank-André Siebert

11.1 Introduction

In contrast to LDR techniques, HDR brachytherapy (BT) prostate techniques are less standardized, leading to a variety of planning techniques, dose prescriptions, and target volume concepts. The most common form of HDR BT for prostate is the use of HDR as a boost technique. In this case, the BT is delivered in combination with external beam in a dedicated, but not standardized, fractionation scheme (Kovács et al. 2005). HDR as monotherapy is also possible but is still subject of ongoing clinical research (Martin et al. 2004; Kovács et al. 2005). There are two common target volume concepts; the first involves using the prostate gland as a single CTV, with or without a margin, and the second uses two CTVs with CTV1 defined as the whole prostatic gland and CTV2 defined as the peripheral zone (Galalae et al. 2002; Aebersold et al. 2004). Prior to the start of treatment, maximum doses for organs at risk (urethra and rectum) must be defined. The GEC-ESTRO report recommends that urethral doses be kept below 10 Gy per fraction and rectal doses below 6 Gy per fraction (Kovács et al. 2005). Table 11.1 lists a variety of dose prescriptions and target volume definitions for HDR prostate boost techniques.

Although the scope of this chapter does not include a discussion of quality assurance procedures, it should be noted that both the software and hardware equipment must be commissioned and tested on a regular basis, ensuring that national legislation is considered and adhered to. A good overview of quality assurance in brachytherapy is given in the GEC-ESTRO handbook *A Practical Guide to Quality Control of Brachytherapy Equipment* edited by Venselaar and Calatayud (2004).

Similar to LDR treatment planning, several technical approaches exist for HDR prostate. With a preplanning method, a treatment plan is generated and implant

F.-A. Siebert
Clinic of Radiotherapy (Radiooncology), UKSH, Campus Kiel,
Arnold-Heller-Str. 3, Haus 23, D-24105 Kiel, Germany
e-mail: siebert@onco.uni-kiel.de

G. Kovács, P. Hoskin (eds.), *Interstitial Prostate Brachytherapy*,
DOI 10.1007/978-3-642-36499-0_11, © Springer-Verlag Berlin Heidelberg 2013

Table 11.1 Overview of target volume definitions, fractionation, and prescription doses for HDR prostate boost techniques

	Center	EBRT	# BT fx	Gy/fx	Target vol
Borghede et al.	Göteborg	50	2	10	Tumor volume
Dinges et al.	Berlin	45	2	10	Prostate capsule
Kovács et al.	Kiel	40/50	2	15	Peripheral zone
Martinez et al.	Royal Oak	45	3	5.5–10.5	Prostate capsule
Mate et al.	Seattle	50.4	4	3–4	Prostate capsule

Kovács et al. (2005)

needles are inserted "virtually" in the TPS. The 3D plan is generated using either computed tomography (CT)-based images or transrectal ultrasound (TRUS) data sets. An important issue that must be considered is the possibility of organ shift or rotation and organ deformation during the implant process. In particular changes of the prostate shape after insertion of implant needles should be adapted in the planning procedure. In addition to US- and CT-based treatment planning, MRI-based planning is also possible; a study has been reported in the literature for a cohort of ten patients using MRI data in conjunction with inverse treatment planning (Citrin et al. 2005).

The number of implant needles varies from patient to patient and is dependent on the technique employed; fewer needles are required, typically 8–12, for techniques where needles are implanted in the periphery of the prostate and more, typically 10–15, when the aim is to encompass the whole prostate with the prescription dose (Kovács et al. 2005). In contrast to LDR methods, there are two parameters available for optimization during the planning process: the dwell position and the dwell time, thus allowing a high flexibility in creating an optimal dose distribution. Despite this, the needle placement is still of great importance. If the applicators are not implanted in an appropriate geometry, this cannot be adjusted by varying the dwell times resulting in a suboptimal dose distribution. If on the other hand the needle positions are adequate, a very conformal dose distribution can be reached. For the preplanning technique, the patient must be set up on the day of the implant in the same way as when planned such that the planned needle geometry is reproduced as closely as possible. Alternatively, the treatment plan must be adapted to match the situation at the implant day.

11.2 Ultrasound Real-Time Planning

In ultrasound-based real-time planning, implant needles are implanted under ultrasound guidance (Kini et al.1999). Often the prostate is "fixed" by the first two needles placed on the right and left side of the gland. These needles are then followed by the anterior catheters. The reason for this is twofold: first, pubic arch interference can be detected earlier; second, the US image quality is not significantly reduced due to shadowing effects behind the already implanted needles. Using transverse and sagittal viewing planes, the needles can be inserted in the prostate gland. Transverse ultrasound images are acquired through the use of a stepping device which enables the ultrasound probe to be retracted from the rectum in equidistant steps (e.g., 5 or 2.5 mm). These images form the basis for the 3D treatment planning process. Moreover, the implant needles can be reconstructed in real-time mode. In a phantom study, accuracies better than 1 mm for the needle tip definition were found (Siebert et al. 2009). After delineating the prostate gland, rectal wall, and the urethra, the planning process can start. The impact of patient movement as well as anatomic alterations has been studied and an average needle displacement of 1 mm and of 0.57 mm of the prostate has been shown (Milickovic et al. 2011). These displacements were considered to result in acceptable dosimetric results.

In forward planning, the dwell times are manually adjusted until the result is satisfactory. Afterloading dwell positions are typically defined in 5 mm steps for each catheter. There are different software tools available within the TPS which allow the user to easily change the dwell times of the dwell positions. An alternative is the use of dose-shaping tools whereby isodose lines can be dragged to the desired positions using the computer mouse and the dwell times are automatically calculated accordingly. The dwell times should always be manually inspected following the use of dose-shaping tools to avoid unwanted hot or cold spots. Figure 11.1 shows

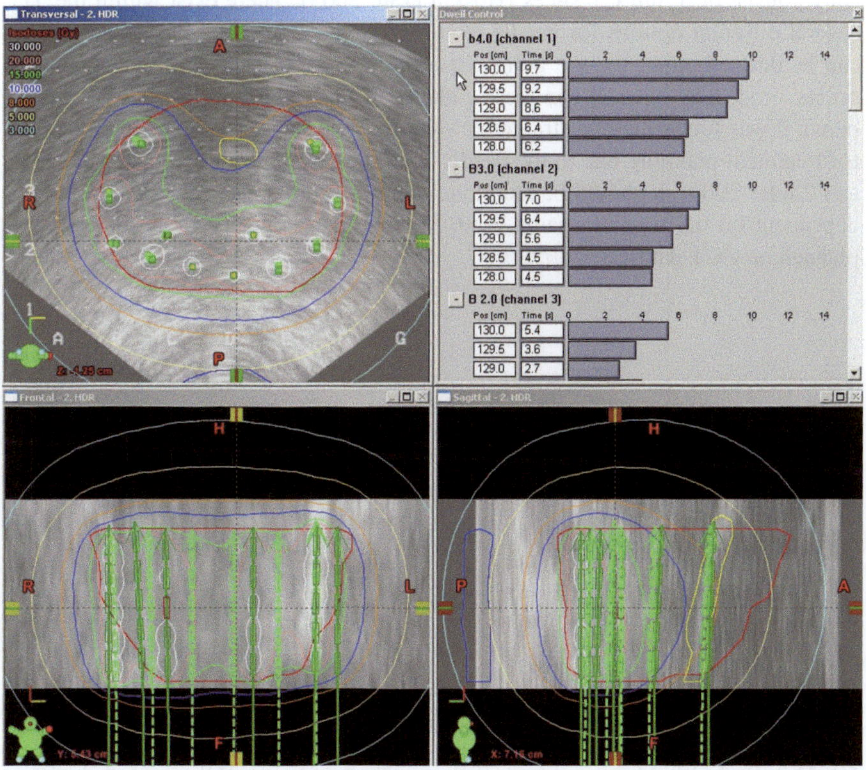

Fig. 11.1 Screenshot of an ultrasound-based HDR real-time plan. A dose of 15 Gy (*green isodose line*) is prescribed to the periphery of the prostate gland. Upper right dwell times for the first two channels are shown. The first dwell position is at 130 cm

an example of a typical isodose distribution for an HDR prostate treatment plan using the prostate gland as CTV1 and the peripheral zone as CTV2.

Geometrical optimization uses a least square algorithm to compute dwell times of the dwell positioning. Although this is a very fast technique, it is only based on the distances of the dwell positions and thus the contours of the organs are not taken into account. As a result, this method requires further manual refinement by the user after calculation.

Maximum doses for organs at risk can also be implemented within a template for inverse planning. Similar to inverse LDR treatment planning, dose constraints and weighting factors must be defined. Using this method, an adequate dose distribution can often be achieved; however, a manual inspection after the calculation process should be carried out as further manual optimization is often required.

11.3 CT-Based Treatment Planning

The treatment planning process is similar when using CT image sets. These typically consist of 5 mm CT slices (Hoskin et al. 2003). There exist within the TPS several different options for needle reconstruction. In the TPS, implanted needles can be defined either by one point, assuming a parallel needle geometry, by two points, or slice by slice. When these catheter reconstruction techniques were compared, it was found that the straight or slice-by-slice reconstruction should be used for treatment planning due to the needles being often implanted nonparallel (Fung and Zaider 2000). It should be mentioned that the effect of needle displacement is dependent on the type of needle used. Figure 11.2 shows an HDR prostate case planned on a CT dataset.

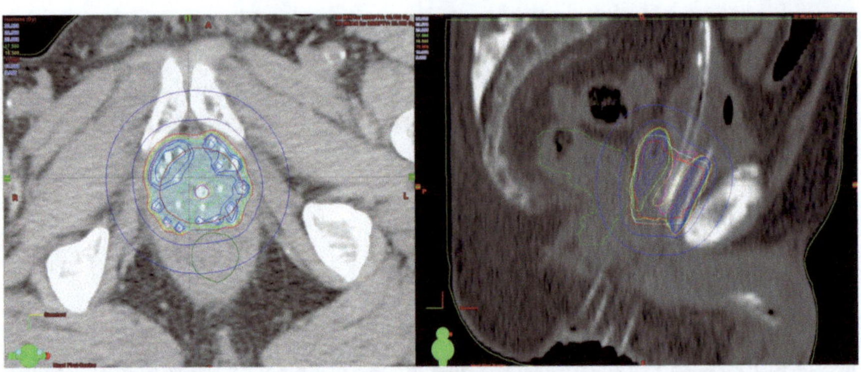

Fig. 11.2 On the left side, a transversal CT slice, on the right a sagittal CT view, both displaying implanted catheters. Fraction dose is 15 Gy as illustrated by the *red isodose line*

Catheter fixation is especially important for CT-based planning because the patient is moved from the CT suite to the treatment couch. Any movements of the needles must be avoided or minimized. Where changes do occur, it is important that these are taken into account. In one study, inter-fraction changes over 36 h were observed (Hoskin et al. 2003). Without adaption of the actual catheter position, the mean D90 of the prostate dropped from 99.5 % at the first fraction to 63.4 % at the second fraction. With correction, a mean D90 of 96.4 % could be reached. Pieters et al. (2006) presented results obtained using self-anchoring and self-expanding catheters that can be fixed in the prostate for use in pulsed-dose rate (PDR) brachytherapy of the prostate. A mean displacement of 1.2 mm was reported during a 3-day treatment.

MRI will give better definition of the prostate anatomy but catheter tracking may be less predice. In some centers, fused CT/MR images may be used for definition of the CTV taking advantage of the better soft tissue imaging on CT while still using CT for the catheter tracking and dosimetry calculations.

11.4 Dose Calculation

For both TRUS- and CT-based HDR treatment planning approaches, the state-of-the-art dose calculation method is the TG-43 formalism (Rivard et al. 2004). This method allows fast calculation of dose distributions and is thus well suited for real-time planning techniques. However, it is commonly known that the TG-43 algorithm does not take into account tissue inhomogeneities and attenuation effects of the applicators themselves, and there is a low sensitivity for HDR catheters (Rivard et al. 2009). This means that modern TPS algorithms which better model attenuation, shielding, and scattering, such as Monte-Carlo methods or model-based dose calculation algorithms, are not expected to lead to large changes in the absorbed dose for HDR brachytherapy of the prostate when compared to TG-43-based calculations.

As described previously, different planning approaches can be used to reach an adequate dose distribution. Different geometrical optimization methods with an inverse-planning simulated annealing (IPSA) algorithm have been compared (Lachange et al. 2002). The study involved 34 HDR prostate boost patients with a dose of 18 or 19.5 Gy delivered in three fractions. It was found that the mean prostate volume receiving 100 % of the prescription dose (V100) was 96.3 and 94.5 % (depending on the applied single dose) for the inverse algorithm (IPSA) and between 92.1 and 88.8 % for different methods of geometrical optimization.

In a similar study with a cohort of 20 patients, two fractions of 9.5 Gy HDR boost were administered (Jacob et al. 2008). The authors compared anatomy-based optimization (IPSA), geometric optimization, and dose point optimization. They found that anatomy-based optimization resulted in the best target coverage. The mean V100 of the prostate was 92.5 % for IPSA, 93.7 % for dose point optimization, and 84.2 % for geometric optimization. For the V125 of the urethra, they reported ranges of 0.01–0.62 cm^3 for IPSA, 0.13–1.86 cm^3 for dose point optimization, and 0.00–1.26 cm^3 for

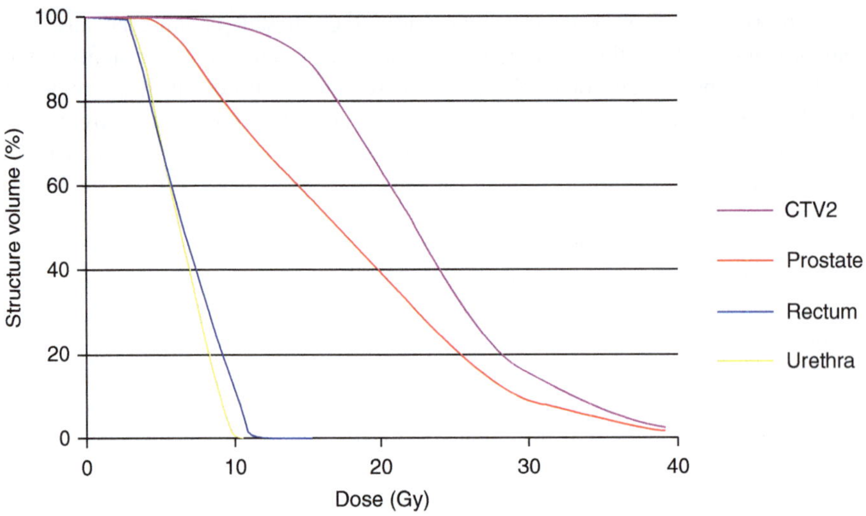

Fig. 11.3 DVHs of a typical HDR prostate treatment plan. Prescription dose is 15 Gy to the peripheral zone (*CTV2*). The prostatic gland should be encompassed with 8–9 Gy

geometric optimization. Figure 11.3 illustrates the DVHs of an inverse-planned HDR prostate treatment plan.

References

Aebersold DM, Isaak B, Thalmann G, Behrensmeier F, Kolotas C, Kranzbühler H, Mini R, Greiner RH (2004) Applicability and dosimetric impact of ultrasound-based preplanning an high-dose-rate brachytherapy of prostate cancer. Strahlenther Onkol 180:351–357

Citrin D, Ning H, Guion P, Li G, Susil RC, Miller RW, Lessard E, Pouliot J, Huchen X, Capala J, Coleman CN, Camphausen K, Ménard C (2005) Inverse treatment planning based on MRI for HDR prostate brachytherapy. Int J Radiat Oncol Biol Phys 61:1267–1275

Fung AYC, Zaider M (2000) Accuracy in catheter reconstruction in computed tomography planning of high dose rate prostate brachytherapy. Med Phys 27:2165–2167

Galalae R, Kovács G, Schultze J, Loch T, Rzehak P, Wilhelm R, Bertermann H, Buschbek B, Kohr P, Kimmig B (2002) Long-term outcome after elective irradiation of the pelvic lymphatics and local dose escalation using high-dose-rate brachytherapy for locally advanced prostate cancer. Int J Radiat Oncol Biol Phys 52:81–90

Hoskin PJ, Bownes PJ, Ostler P, Walker K, Bryant L (2003) High dose rate afterloading brachytherapy for prostate cancer: catheter and gland movement between fractions. Radiother Oncol 68:285–288

Jacob D, Raben A, Sarkar A, Grimm J, Simpson L (2008) Anatomy-based inverse planning simulated annealing optimization in high-dose-rate prostate brachytherapy: significant dosimetric advantage over other optimization techniques. Int J Radiat Oncol Biol Phys 72:820–827

Kini VR, Edmundson GK, Vicini FA, Jaffray DA, Gustaffson G, Martinez AA (1999) Use of three-dimensional radiation therapy planning tools and intraoperative ultrasound to evaluate high dose rate prostate brachytherapy implants. Int J Radiat Oncol Biol Phys 43:571–578

Kovács G, Pötter R, Loch T, Hammer J, Kolkman-Deurloo IK, de la Rosette JJ, Bertermann H (2005) GEC/ESTRO-EAU recommendations on temporary brachytherapy using stepping sources for localized prostate cancer. Radiother Oncol 74:137–148

Lachange B, Bélivau-Nadeau D, Lessard E, Chrétien M, Hsu CJ, Pouliot J, Beaulieu L, Vigneault E (2002) Early clinical experience with anatomy-based inverse planning dose optimization for high-dose-rate boost of the prostate. Int J Radiat Oncol Biol Phys 54:86–100

Martin T, Baltas D, Kurek R, Röddiger S, Kontova M, Anagnostopoulos G, Dannenberg T, Buhleier T, Skazikis G, Tunn U, Zamboglou N (2004) 3-D conformal HDR brachytherapy as monotherapy for localized prostate cancer. Strahlenther Onkol 180:225–232

Milickovic N, Mavroidis P, Tselis N, Nikolova I, Katsilieri Z, Kefala V, Zamboglou N, Baltas D (2011) 4D analysis of influence of patient movement and anatomy alteration on the quality of 3D U/S-based prostate HDR brachytherapy treatment delivery. Med Phys 38:4982–4993

Pieters BR, van der Grient JNB, Blank LE, Koedooder K, Hulshof MC, de Reijke TM (2006) Minimal displacement of novel self-anchoring catheters suitable for temporary prostate implants. Radiother Oncol 80:69–72

Rivard MJ, Coursey BM, DeWerd L et al (2004) Update of AAPM TG 43: a revised AAPM protocol for brachytherapy dose calculations. Med Phys 31:633–674

Rivard MJ, Venselaar JLM, Beaulieu L (2009) The evolution of brachytherapy treatment planning. Med Phys 36:2136–2153

Siebert FA, Hirt M, Niehoff P (2009) Imaging of implant needles for real-time HDR-brachytherapy prostate treatment using biplane ultrasound transducers. Med Phys 36:3406–3412

Venselaar JLM, Pérez-Calatayud J (eds) (2004) A practical guide to quality control of brachytherapy equipment, ESTRO booklet no. 8. ESTRO, Belgium. ISBN 90-804532-8

Postimplant Dosimetry

12

Marinus Adriaan Moerland

12.1 Introduction

Permanent prostate brachytherapy has become a well-established treatment option for localized prostate cancer (Stock et al. 1998; Beyer and Brachman 2000; Blasko et al. 2002; Battermann et al. 2004; Hinnen et al. 2009). The techniques are generally based on the transperineal implantation technique guided by transrectal ultrasound imaging (Holm et al. 1983). In the past decade, several innovations have been introduced like intraoperative planning, inverse planning, seeds in strand configurations, an afterloader for automatic seed delivery, and various types of iodine seeds and strands (Nag et al. 2001, 2008; Anagnostopoulos et al. 2002; Kaplan et al. 2004; Van Gellekom et al. 2004; Rivard et al. 2005, 2007; Wei et al. 2005; Lessard et al. 2006; McLaughlin et al. 2006; Lin et al. 2007; Al-Qaisieh et al. 2007; Radford Evans et al. 2007; Zelefsky et al. 2007). However, despite these innovations, authors describe deviations between the actual and planned dose distributions due to errors in needle localization, errors in seed delivery, prostate deformation between needle insertion and retraction, individual edema resolution dynamics, and seed migration (Nag et al. 2008; McLaughlin et al. 2006; Waterman et al. 1998; Taschereau et al. 2000; Van Gellekom et al. 2002; Yue et al. 2002; McNeely et al. 2004; Lagerburg et al. 2005; Wang et al. 2006; Reed et al. 2007; Chen et al. 2008). Therefore, the international organizations recommend postimplant dosimetry as a valuable tool in a permanent prostate brachytherapy program to control and assure the quality of the implants and to establish accurate dose–response relationships (Ash et al. 2000; Nag et al. 2001; Salembier et al. 2007). This chapter describes the various aspects of postimplant dosimetry.

M.A. Moerland
Department of Radiation Oncology, University Medical Center Utrecht,
Heinel Berglaan 100, 3584 CX,
Utrecht, The Netherlands
e-mail: m.a.moerland@umcutrecht.nl

G. Kovács, P. Hoskin (eds.), *Interstitial Prostate Brachytherapy*,
DOI 10.1007/978-3-642-36499-0_12,

12.2 Timing of Postimplant Dosimetry

Several authors have investigated postimplant prostate volume changes and how they influence the dose that is actually administered to the prostate. In the short term, the prostate swells due to edema caused by piercing the prostate with needles (Moerland et al. 1997; Moerland 1998; Waterman et al. 1998); a long-term effect may be radiation-induced shrinkage of the prostate (Dale et al. 1994). A study of serial CT scans on ten patients who received either I-125 or Pd-103 seed implants found that edema was present in all of the implanted prostates (Waterman et al. 1998). The magnitude of the edema, defined as the ratio of the post- to pre-implant volume, ranged from 1.33 to 1.96 (mean: 1.52). This resolved exponentially in time with an edema half-life which varied from 4 to 25 days (mean: 9.3 days). Willins and Wallner (1998) acquired pre- and postimplant CT scans from 11 patients and found that the prostate returned to the pre-implant size within 2 months of the procedure. Moerland et al. (1997) acquired serial MRI scans on 21 patients and found that postimplant prostate edema increased volume by 1.3 ± 0.2. Tanaka et al. (2007) performed TRUS-based preplanning and CT/MRI fusion-based postimplant dosimetry on 74 patients. Prostatic edema was largest on day 1 postimplant (mean prostate volume 36 % larger than preplan volume) and thereafter decreased over time with mean prostate volume 9 % larger on day 30 postimplant. Crook et al. (2004) performed TRUS-based preplanning and CT/MRI fusion-based postimplant dosimetry on 241 patients. In 12 % of patients, residual edema at 1 month was observed: mean pre-implant volume was 34.8 cc and 1-month volume was 46.1 cc. Steggerda et al. (2007) acquired combined TRUS-CT scans of 13 patients 1 day, 1 month, and 3½ months after seed implantation and observed 7 % prostate volume increase at day 1 compared to pre-implant prostate volume. Prostate volume returned to baseline value at 1 month postimplant. Sloboda et al. (2010) acquired MRI scans on days −1, 0, 14, and 28. The average relative volume was 1.18 ± 0.14 (one standard deviation) on day 0 and resolved linearly in time to 0 % at 1 month. Prostate dimension changes due to edema were anisotropic, about 10 % in each of the anteroposterior and superior-inferior directions and about 0 % in the left-right direction.

Accurate estimation of a patient's individual edema resolution dynamics would require multiple postimplant evaluations to calculate the cumulative dose to the prostate and the critical organs. For practical reasons, postimplant dosimetry is generally based on just one evaluation moment, and therefore, it is important to determine the optimal time of acquiring the images for postimplant dosimetry.

Before this time, when the implant is expanded due to swelling, an underestimation of the dose to the prostate will be calculated and, after the optimal time, when the implant has shrunk again, the dose will be overestimated. If edema is ignored, the calculated physical total dose is not a good representation of the real dose in the prostate (Yue et al. 1999a, b; Chen et al. 2000). Several groups have performed simulation studies to determine the optimal timing of postimplant dosimetry on the basis of edema resolution dynamics as described above. Moerland (1998) calculated the effect of edema on the cumulative prostate dose and the effect of timing on the outcome of the implant quality assessment. Prostate edema was modelled as increased prostate volume at day 0 that resolves exponentially over time postimplant.

For a typical prostate volume increase of 30 % with a half-life of 9 days, the cumulative dose was found to be 2 % smaller than the dose of an ideal implant without edema. Errors in quality assessment were larger (underestimation up to 20 % at day 0) depending upon the magnitude and half-life of prostate edema. The simulation study showed that the optimal timing is about 4 weeks after the implantation with errors generally smaller than 5 % for clinical ranges of prostate edema magnitudes and half-lives. Yue et al. (1999a, b) developed a dynamic biomathematical model for the estimation of the optimum timing of a CT scan for dose evaluation of a permanent implant. For different combinations of edema magnitude and half-life, which were based on the study of Waterman et al. (1998), optimum time windows for postimplant dosimetry were identified for I-125 and Pd-103 implants. Errors were generally less than 5 % if dosimetry was performed 4–10 weeks after the implantation for an I-125 implant and 2–4 weeks for a Pd-103 implant. Steggerda et al. (2007) acquired combined TRUS-CT scans of 13 patients 1 day, 1 month, and 3½ months after seed implantation and calculated geometry corrected dose distributions. Postimplant dosimetry based on the day 1 scan underestimated V150 of the prostate (18 ± 10 %) and V120 of the urethra (47 ± 32 %). The dose to the bladder wall (D2cc) was overestimated (31 ± 35 %). Dose parameters based on scans 1 month or later after the implantation were within 5 % of the geometry corrected values. Whereas most authors regarded the physical dose to find the optimal timing for postimplant dosimetry, Van Gellekom et al. (2002) calculated the biologically effective dose (BED) for a range of radiobiologic and edema parameters and found that the optimal timing of BED evaluation for I-125 seed implants is 25 days after implantation.

In conclusion, measurement of the patient's individual edema resolution dynamics would require multiple postimplant evaluations for an accurate calculation of the cumulative dose to the prostate and the critical organs. For practical reasons, postimplant dosimetry is generally based on just one evaluation. The optimal timing for I-125 implants is estimated to be around 1 month postimplant. This is also reflected in the ESTRO/EAU/EORTC recommendations on permanent seed implantation for localized prostate cancer (Ash et al. 2000; Salembier et al. 2007).

12.3 Imaging for Postimplant Dosimetry

First attempts to perform postimplant dosimetry originate from the 1980s and were based on radiograph images (Amols and Rosen 1981; Altschuler et al. 1983; Biggs and Kelley 1983; Siddon and Chin 1985). Although radiographs do not show the relationship with the prostate, these methods gave some idea of the implanted volume, the geometry of the implant, and the dose distribution. Seed localization relative to the prostate can be gained by US, CT, or MRI examinations (Moerland et al. 1997; Narayana et al. 1997; Willins and Wallner 1997; Vicini et al. 1999; Polo et al. 2004; Villeirs et al. 2005). On US the seeds cause artifacts, which hamper evaluation of the prostate and the seed distribution, especially in the ventral region of the prostate away from the US probe in the rectum. Since seed localization is cumbersome on transrectal ultrasound (TRUS), several authors developed methods of registration of TRUS anatomy images with other imaging modalities like fluoroscopy, cone-beam CT, or CT (Westendorp et al. 2007; Fallavollita et al. 2010). A CT-based method for postimplant dosimetry was described by Roy et al. (1993). Seed positions and prostate contours were digitized on axial CT slices with 3 mm thickness and 3 mm scan spacing. Since then, various authors improved the technique by using thinner CT slices and automated seed localization (Brinkmann and Kline 1998; Bice et al. 1999; Holupka et al. 2004). Although the soft tissue contrast of CT is poor, many postimplant dosimetry studies are based on CT as the imaging modality. The prostate at the base and the apex are especially difficult to delineate (Dubois et al. 1998; McLaughlin et al. 2010). In CT-based postimplant dosimetry studies, there is a risk that the presence of the seeds reduces prostate delineation to "circling the seeds" resulting in overestimation of the prostate dose especially at the base of the prostate, which is more difficult to cover with seeds (McLaughlin et al. 2006; Moerland et al. 2009). MRI has excellent soft tissue contrast and several authors indicate that MRI is superior in defining the prostate and the periprostatic area (Villeirs et al. 2005). Therefore, MRI should be the imaging modality of choice for accurate postimplant dosimetry (Salembier et al. 2007). MRI for postimplant dosimetry is mostly registered with CT for seed localization (Polo et al. 2004; Moerland et al. 2009), although methods to identify seeds on MRI also have been investigated (Dubois et al. 1997).

In the following, the postimplant dosimetry technique at the University Medical Center Utrecht is described. MRI/CT-based postimplant dosimetry was implemented as a routine part of our prostate brachytherapy program since the end of 2003. Postimplant dosimetry is performed 4 weeks after the implantation procedure to get on average the best estimate of cumulative dose in the prostate taking into account prostate edema resolution dynamics (Moerland 1998; Waterman et al. 1998; Yue et al. 1999a, b; Van Gellekom et al. 2002). CT images are used for accurate seed localization and MR images for accurate prostate delineation (see Fig. 12.1). The MR images are acquired with a 1.5 T or 3 T MR scanner (Philips Medical Systems, the Netherlands) and the protocol consists of T1-weighted axial spin-echo (SE) images ($1\times1\times4$ mm^3 resolution), T2-weighted axial turbo spin-echo (TSE) images ($1\times1\times4$ mm^3 resolution), and a balanced steady-state free precession (bSSFP) 3D acquisition ($0.7\times0.7\times1$ mm^3 resolution). Prostate and other structures are delineated on the T2-weighted images, whereas the 3D bSSFP images are used for registration with the CT images. The CT datasets are acquired with 120 kV and 80 mA, spiral scanning, matrix$=512^2$, field of view (FOV) up to 30 cm, slice thickness 1 mm (was 3 mm up to 2011), and no gap (Philips Medical Systems, the Netherlands). The postimplant number of seeds was counted on plain radiography images until 2011, and since then directly on the maximum intensity projection (MIP) images which are reconstructed from the CT images with 1 mm slice thickness. MR and CT datasets are imported in the Sonographic Planning of Oncology Treatment (SPOT) system (Nucletron, the Netherlands). Seed locations are automatically detected in the CT images using the automatic seed finder module of the SPOT system. Registration between CT-based seed localizations and MRI-based prostate delineation is done with the SPOT image module for manual fusion of the high-intensity seed signals on CT and the seed signal voids on MRI. The seeds are clearly visible

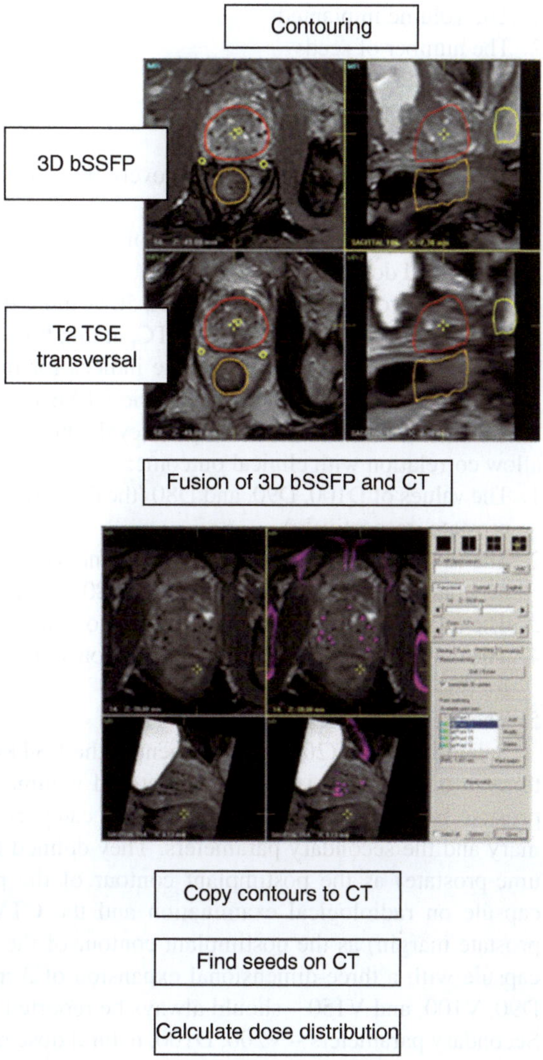

Fig. 12.1 Flow map of postimplant dosimetry at the University Medical Center Utrecht. Step 1: contouring on T2-weighted TSE and 3D bSSFP images; step 2: fusion of 3D bSSFP and CT images; step 3: copy the contours to CT; step 4: find the seeds on CT; step 5: calculate the dose distribution

as signal voids on the 3D bSSFP images and as small voids on the T2-weighted images, and therefore do not hamper delineation. The postimplant dose distributions are calculated according to equation 11 of the general 1D formalism of the AAPM TG 43 update report and taking into account strength and characteristics of the seed type (Rivard et al. 2004).

12.4 Reporting of Postimplant Dosimetry

Several organizations have published recommendations on reporting in permanent prostate brachytherapy (Ash et al. 2000; Nag et al. 2000; Salembier et al. 2007). In 2000, ESTRO/EAU/EORTC recommended the following indices for all patients:

1. The volume implanted
2. The number of seeds
3. The number of needles used
4. The total activity implanted
5. The prescribed dose
6. The D90, that is, the dose that covers 90 % of the prostate volume as defined from postimplant imaging
7. The V100, that is, the percentage of the prostate volume that has received the prescribed dose
8. V150, the volume that has received 50 % more than the prescribed dose

According to ESTRO/EAU/EORTC, at that time there was insufficient information to recommend dose and volume indices for rectum and urethra. In the same year, the American Brachytherapy Society (ABS) recommended that the following has to be reported to allow adequate evaluation of postimplant dosimetry and to allow correlation with clinical outcome:

1. The values of D100, D90, and D80 (the dose that covers 100, 90, and 80 % of the prostate, respectively)
2. The values of V200, V150, V100, V90, and V80 (the fractional volume of the prostate that receives 200, 150, 100, 90, and 80 % of the prescribed dose, respectively)
3. The total volume of the prostate (in cc) obtained from postimplant dosimetry
4. The number of days between implantation and the date of the imaging study used for dosimetric reconstruction
5. The urethral and rectal doses

Salembier et al. (2007) supplemented the ESTRO/EAU/EORTC recommendations to develop consistency in target and volume definition for permanent seed prostate brachytherapy and to define two categories of dose parameters, the primary and the secondary parameters. They defined the CTV-P (clinical target volume-prostate) as the postimplant contour of the prostatic gland defined by the capsule on radiological examination and the CTV-PM (clinical target volume-prostate margin) as the postimplant contour of the prostatic gland defined by the capsule with a three-dimensional expansion of 3 mm. The primary parameters – D90, V100, and V150 – should always be reported for both CTV-P and CTV-PM. Secondary parameters – V200, D100, natural dose rate (NDR), homogeneity index (HI), and conformal index (CI) – may also be reported although their value in relation to outcome is not proven and should be a focus for further research. Concerning organs at risk, D2cc for the rectum and D10 for the urethra are the primary parameters. Secondary parameters, D0.1cc and V100 for rectum and D0.1cc, D30, and D5 for urethra, may also be reported. No parameters can be given at present regarding penile bulb and neurovascular bundles. Further details regarding contouring of prostate, urethra, and rectum can be found in Salembier et al. (2007).

12.5 Outcomes of Postimplant Dosimetry

As mentioned in the introduction of this chapter, several authors have described deviations between the realized and planned dose distributions due to errors in needle localization, errors in seed delivery, prostate deformation between needle insertion and retraction, individual edema resolution dynamics, and seed migration (Waterman et al. 1998; Taschereau et al. 2000; Van Gellekom et al. 2002; Yue et al. 2002; McNeely et al. 2004; Lagerburg et al. 2005; McLaughlin et al. 2006; Wang et al. 2006; Reed et al. 2007; Chen et al. 2008; Nag et al. 2008). Therefore, the international organizations recommend postimplant dosimetry as a valuable tool in a permanent prostate brachytherapy program to control and assure implant quality and to establish accurate dose–response relationships (Ash et al. 2000; Nag et al. 2001; Salembier et al. 2007).

Al-Qaisieh et al. (2009) showed in a study on 445 patients a strong correlation between transrectal US-based pre-implant and CT-based postimplant dosimetry, but the planned D90 was not achieved. The mean (±standard deviation) planned D90 was 183.4 (±12.1) Gy, while the D90 that was achieved was 145.5 (±20.4) Gy. A postimplant D90 of 145 Gy seems sufficient for biochemical control (Stock et al. 1998; Potters et al. 2001; Hinnen et al. 2009) and could be achieved by aiming for a D90 of around 180 Gy in the preplan. However, they also caution for excessive toxicities if a D90 of 180 Gy is actually achieved.

Intraoperative planning is expected to be more predictive than preplanning. Stone et al. (2003) applied intraoperative TRUS-based planning and demonstrated a difference in D90 of only 3.4 % with postimplant CT-based dosimetry in a study on 77 patients. The mean dose to 30 % of the urethra was 120 % of prescription in the intraoperative plan and 138 % on CT-based dosimetry.

Nag et al. (2008) reported on 82 implants in which the dosimetry was updated intraoperatively and CT-based dosimetry was performed a few hours later. They found mean relative differences of 16 and 10 % between intraoperative and postimplant D90 and V100 values and concluded that intraoperative US-based planning provides valuable real-time information, but postimplant dosimetry should remain the standard of care until studies have determined whether intraoperative or postimplant dosimetry better predicts patient outcomes.

Literature on intraoperative planning is ambiguous regarding predictive value and ability to make postimplant dosimetry superfluous. Moerland et al. (2009) evaluated 389 patients implanted in the period 2005–2007 by analysis of intraoperative plans (based on US scan after needle insertion) and 4-week MRI-based postplans. Intraoperative prostate dose parameters (mean ± standard deviation) amounted to D90 = 183 ± 13 Gy, VT100 = 98 ± 2 %, VT150 = 71 ± 8 %, and VT200 = 30 ± 7 %. Postimplant prostate dose parameters amounted to D90 = 161 ± 30 Gy, VT100 = 93 ± 7 %, VT150 = 70 ± 12 %, and VT200 = 38 ± 12 %. Figure 12.2 shows that the D90 values as derived from CT/MRI-based postimplant dosimetry ranged from 51 to 241 Gy. The challenge of prostate brachytherapy is to achieve adequate dose coverage for each individual patient. The mean of realized D90 values should be as close to the planned mean D90, and the standard deviation of the distribution of postimplant D90 values should be as narrow as possible. Postimplant dosimetry is a tool to find error sources in seed placement and to explain discrepancies between postimplant and intraoperative prostate dose distributions. We found higher postimplant D90 values for loose seed implants compared to stranded seed implants. However, the literature comparing loose seed and stranded seed implants is ambiguous regarding postimplant dosimetry

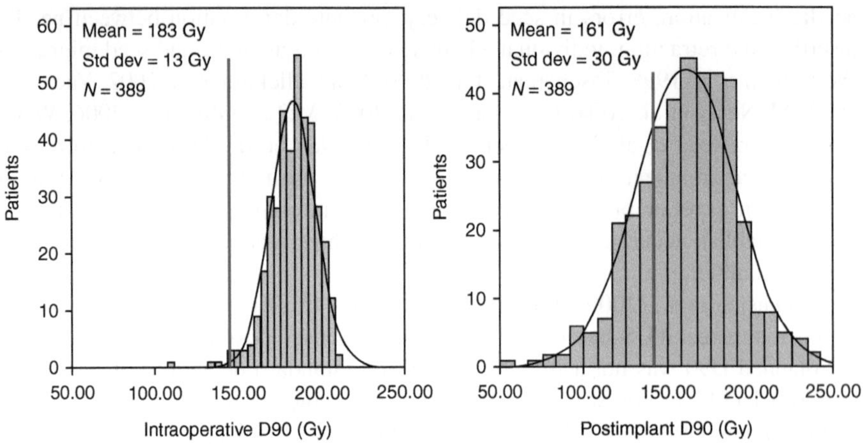

Fig. 12.2 Intraoperative and postimplant D90 values of 389 permanent prostate implants at the University Medical Center Utrecht in the period 2005–2007. The *red vertical lines* indicate the prescription dose of 145 Gy

results. Fagundes et al. (2004) concluded from a retrospective study of 473 patients that D90 values were superior in stranded seed patients compared to loose seed patients, and in a study of 80 patients subsequently treated with loose seeds and suture embedded seeds, Lin et al. (2007) reported that D90 improved from 143 to 155 Gy. Van Gellekom et al. (2004) analyzed nine patients who received a loose seed implant in the right half and a stranded seed implant in the left half of the prostate and found higher D90 values in the loose seeds part compared to the stranded seeds part (130 ± 21 vs. 110 ± 21 Gy).

Discussions on the use of loose seeds vs. stranded seeds are ongoing. It is hypothesized that strands if placed partially inferior to the prostate apex anchor in the surrounding tissue and are redrawn from the prostate after implantation, e.g., when the prostate is released from the locking needles or afterwards by muscle contractions (Moerland et al. 2009). This phenomenon is not likely to occur if loose seeds have been implanted. Similar findings were reported by McLaughlin et al. (2006) who observed in 11 of 28 patients that a shift of stranded seeds vs. prostate had a greater impact on final dosimetry than did prostate swelling. They adapted the technique to avoid placement of stranded seeds inferior to the prostate apex, which is a nice example how postimplant dosimetry is used to improve implant program quality.

Acher et al. (2010) analyzed the implants of 49 patients and found that the mean D90 difference between intraoperative and 4 weeks postoperative assessments ranged from 11 Gy to 20 Gy for 2 observers and depending on postoperative imaging (CT, CT/MR fusion, or combined X-ray and MR) with wide 95 % confidence limits. The worst agreement was for the CT estimations (95 % limits −34 to 63 Gy) of observer 2. The study by Acher et al. implied that the intraoperative dose assessment D90 of 170 Gy overestimated the postoperative D90 outcome in the order of 11–20 Gy and was only accurate enough to predict a postoperative D90 between 110 and 200 Gy. Again, the challenge of prostate brachytherapy is to achieve adequate dose coverage for each individual patient. Igidbashian et al. (2008) analyzed 127 patients who underwent prostate permanent seed brachytherapy using intraoperative planning. Implant quality was evaluated on CT at 28 days postimplant. 72.4 % of patients had a V100≥ and 74.8 % had a D90 ≥ 140 Gy. From analysis of area under the receiver operating characteristics for different models, they concluded that intraoperative TRUS-based dosimetry was not accurate enough to predict for postimplant CT-based dosimetry.

Results of comparisons between intraoperative and postimplant dosimetry should be interpreted with caution for a number of reasons. First, there may be errors in seed localization in both intraoperative US and postimplant CT or MRI (Dubois et al. 2001; Al-Qaisieh et al. 2007; Siebert et al. 2007; Ishiyama et al. 2008). Second, there may be errors in delineation of prostate and organs at risk on both intraoperative US and postimplant CT or MRI (Narayana et al. 1997; Dubois et al. 1998; Al-Qaisieh et al. 2002; Crook et al. 2002; Polo et al. 2004; Acher et al. 2008; Maletz et al. 2012). Third, the patient's individual edema resolution dynamics would require multiple postimplant evaluation moments for an accurate calculation of the cumulative dose to the prostate and the critical organs (see Sect. 13.2).

Nevertheless, postimplant dosimetry remains the standard of care in permanent prostate brachytherapy, preferably based on MRI acquired 4 weeks postimplant.

References

Acher P, Rhode K, Morris S et al (2008) Comparison of combined x-ray radiography and magnetic resonance (XMR) imaging-versus computed tomography-based dosimetry for the evaluation of permanent prostate brachytherapy implants. Int J Radiat Oncol Biol Phys 71:1518–1525

Acher P, Puttagunta S, Rhode K et al (2010) An analysis of intraoperative versus post-operative dosimetry with CT, CT-MRI fusion and XMR for the evaluation of permanent prostate brachytherapy implants. Radiother Oncol 96:166–171

Al-Qaisieh B, Ash D, Bottomley DM et al (2002) Impact of prostate volume evaluation by different observers on CT-based postimplant dosimetry. Radiother Oncol 62:267–273

Al-Qaisieh B, Smith DW, Brearley E et al (2007) Comprehensive I-125 multi-seed comparison for prostate brachytherapy: dosimetry and visibility analysis. Radiother Oncol 84:140–147

Al-Qaisieh B, Witteveen T, Carey B et al (2009) Correlation between pre- and postimplant dosimetry for iodine-125 seed implants for localized prostate cancer. Int J Radiat Oncol Biol Phys 75:626–630

Altschuler MD, Findlay PA, Epperson RD (1983) Rapid, accurate, three-dimensional location of multiple seeds in implant radiotherapy treatment planning. Phys Med Biol 28:1305–1318

Amols HI, Rosen II (1981) A three-film technique for reconstruction of radioactive seed implants. Med Phys 8:210–214

Anagnostopoulos G, Baltas D, Karaiskos P et al (2002) Thermoluminescent dosimetry of the selectseed 125I interstitial brachytherapy seed. Med Phys 29:709–716

Ash D, Flynn A, Battermann J, Urological Brachytherapy Group; EORTC Radiotherapy Group et al (2000) ESTRO/EAU/EORTC recommendations on permanent seed implantation for localized prostate cancer. Radiother Oncol 57:315–321

Battermann JJ, Boon TA, Moerland MA (2004) Results of permanent prostate brachytherapy, 13 years of experience at a single institution. Radiother Oncol 71:23–28

Beyer DC, Brachman DG (2000) Failure free survival following brachytherapy alone for prostate cancer: comparison with external beam radiotherapy. Radiother Oncol 57:263–267

Bice WS Jr, Dubois DF, Prete JJ et al (1999) Source localization from axial image sets by iterative relaxation of the nearest neighbor criterion. Med Phys 26:1919–1924
Biggs PJ, Kelley DM (1983) Geometric reconstruction of seed implants using a three-film technique. Med Phys 10:701–704
Blasko JC, Mate T, Sylvester JE et al (2002) Brachytherapy for carcinoma of the prostate: techniques, patient selection, and clinical outcomes. Semin Radiat Oncol 12:81–94
Brinkmann DH, Kline RW (1998) Automated seed localization from CT datasets of the prostate. Med Phys 25:1667–1672
Chen Z, Yue N, Wang X et al (2000) Dosimetric effects of edema in permanent prostate seed implants: a rigorous solution. Int J Radiat Oncol Biol Phys 47:1405–1419
Chen ZJ, Deng J, Roberts K et al (2008) On the need to compensate for edema-induced dose reductions in preplanned (131)Cs prostate brachytherapy. Int J Radiat Oncol Biol Phys 70:303–310
Crook J, Milosevic M, Catton P et al (2002) Interobserver variation in postimplant computed tomography contouring affects quality assessment of prostate brachytherapy. Brachytherapy 1:66–73
Crook J, McLean M, Yeung I et al (2004) MRI-CT fusion to assess postbrachytherapy prostate volume and the effects of prolonged edema on dosimetry following transperineal interstitial permanent prostate brachytherapy. Brachytherapy 3:55–60
Dale RG, Jones B, Coles IP (1994) Effect of tumour shrinkage on the biological effectiveness of permanent brachytherapy implants. Br J Radiol 67:639–645
Dubois DF, Prestidge BR, Hotchkiss LA et al (1997) Source localization following permanent transperineal prostate interstitial brachytherapy using magnetic resonance imaging. Int J Radiat Oncol Biol Phys 39:1037–1041
Dubois DF, Prestidge BR, Hotchkiss LA et al (1998) Intraobserver and interobserver variability of MR imaging- and CT-derived prostate volumes after transperineal interstitial permanent prostate brachytherapy. Radiology 207:785–789
Dubois DF, Bice WS Jr, Prestige BR (2001) CT and MRI derived source localization error in a custom prostate phantom using automated image coregistration. Med Phys 28:2280–2284
Fagundes H, Keys RJ, Wojcik MF et al (2004) Transperineal TRUS-guided prostate brachytherapy using loose seeds versus rapid-strand: a dosimetric analysis. Brachytherapy 13:136–140
Fallavollita P, Aghaloo ZK, Burdette EC et al (2010) Registration between ultrasound and fluoroscopy or CT in prostate brachytherapy. Med Phys 37:2749–2760
Hinnen KA, Battermann JJ, van Roermund JGH et al (2009) Long-term biochemical and survival outcome of 921 patients treated with I-125 permanent prostate brachytherapy. Int J Radiat Oncol Biol Phys 76:1433–1438
Holm HH, Juul N, Pedersen JF et al (1983) Transperineal 125 iodine seed implantation in prostatic cancer guided by transrectal ultrasonography. J Urol 130:283–286
Holupka EJ, Meskell PM, Burdette EC et al (2004) An automatic seed finder for brachytherapy CT postplans based on the Hough transform. Med Phys 31:2672–2679
Igidbashian L, Donath D, Carrier JF et al (2008) Poor predictive value of intraoperative real-time dosimetry for prostate seed brachytherapy. Int J Radiat Oncol Biol Phys 72:605–609
Ishiyama H, Nakamura R, Satoh T et al (2008) Intersoftware variability in post-implanted CT analysis for interstitial permanent brachytherapy for prostate cancer: differences in automatically detected seed location. Radiother Oncol 89:214–216
Kaplan ID, Meskell PM, Lieberfarb M et al (2004) A comparison of the precision of seeds deposited as loose seeds versus suture embedded seeds: a randomized trial. Brachytherapy 3:7–9
Lagerburg V, Moerland MA, Lagendijk JJ et al (2005) Measurement of prostate rotation during insertion of needles for brachytherapy. Radiother Oncol 77:318–323
Lessard E, Kwa SL, Pickett B et al (2006) Class solution for inversely planned permanent prostate implants to mimic an experienced dosimetrist. Med Phys 33:2773–2782
Lin K, Lee SP, Cho JS et al (2007) Improvements in prostate brachytherapy dosimetry due to seed stranding. Brachytherapy 6:44–48
Maletz KL, Ennis RD, Ostenson J et al (2012) Comparison of CT and MR-CT fusion for prostate post-implant dosimetry. Int J Radiat Oncol Biol Phys 82(5):1912–1917

McLaughlin P, Narayana V, Pan C et al (2006) Comparison of day 0 and day 14 dosimetry for permanent prostate implants using stranded seeds. Int J Radiat Oncol Biol Phys 64:144–150
McLaughlin PW, Evans C, Feng M et al (2010) Radiographic and anatomic basis for prostate contouring errors and methods to improve prostate contouring accuracy. Int J Radiat Oncol Biol Phys 76:369–378
McNeely LK, Stone NN, Presser J et al (2004) Influence of prostate volume on dosimetry results in real-time 125I seed implantation. Int J Radiat Oncol Biol Phys 58:292–299
Moerland MA (1998) The effect of edema on postimplant dosimetry of permanent iodine-125 prostate implants: a simulation study. J Brachytherapy Int 14:225–231
Moerland MA, Wijrdeman HK, Beersma R et al (1997) Evaluation of permanent I-125 prostate implants using radiography and magnetic resonance imaging. Int J Radiat Oncol Biol Phys 37:927–933
Moerland MA, van Deursen MJ, Elias SG et al (2009) Decline of dose coverage between intraoperative planning and post implant dosimetry for I-125 permanent prostate brachytherapy: comparison between loose and stranded seed implants. Radiother Oncol 91:202–206
Nag S, Bice W, DeWyngaert K et al (2000) The American Brachytherapy Society recommendations for permanent prostate brachytherapy postimplant dosimetric analysis. Int J Radiat Oncol Biol Phys 46:221–230
Nag S, Ciezki JP, Cormack R, Clinical Research Committee, American Brachytherapy Society et al (2001) Intraoperative planning and evaluation of permanent prostate brachytherapy: report of the American Brachytherapy Society. Int J Radiat Oncol Biol Phys 51:1422–1430
Nag S, Shi P, Liu B et al (2008) Comparison of real-time intraoperative ultrasound-based dosimetry with postoperative computed tomography-based dosimetry for prostate brachytherapy. Int J Radiat Oncol Biol Phys 70:311–317
Narayana V, Roberson PL, Pu AT et al (1997) Impact of differences in ultrasound and computed tomography volumes on treatment planning of permanent prostate implants. Int J Radiat Oncol Biol Phys 37:1181–1185
Polo A, Cattani F, Vavassori A et al (2004) MR and CT image fusion for postimplant analysis in permanent prostate seed implants. Int J Radiat Oncol Biol Phys 60:1572–1579
Potters L, Cao Y, Calugaru E et al (2001) A comprehensive review of CT-based dosimetry parameters and biochemical control in patients treated with permanent prostate brachytherapy. Int J Radiat Oncol Biol Phys 50:605–614
Radford Evans DA, Meyer T, Angyalfi S et al (2007) Enhanced efficiency and ergonomics of an intraoperative automated prostate brachytherapy delivery technique. Brachytherapy 6:254–257
Reed DR, Wallner KE, Merrick GS et al (2007) A prospective randomized comparison of stranded vs. loose 125I seeds for prostate brachytherapy. Brachytherapy 6:129–134
Rivard MJ, Coursey BM, DeWerd LA et al (2004) Update of AAPM Task Group No. 43 Report: a revised AAPM protocol for brachytherapy dose calculations. Med Phys 31:633–674
Rivard MJ, Evans DA, Kay I (2005) A technical evaluation of the Nucletron FIRST system: conformance of a remote afterloading brachytherapy seed implantation system to manufacturer specifications and AAPM Task Group report recommendations. J Appl Clin Med Phys 6:22–50
Rivard MJ, Butler WM, DeWerd LA, American Association of Physicists in Medicine et al (2007) Supplement to the 2004 update of the AAPM Task Group No. 43 Report. Med Phys 34:2187–2205
Roy JN, Wallner KE, Harrington PJ et al (1993) A CT-based evaluation method for permanent implants: application to prostate. Int J Radiat Oncol Biol Phys 26:163–169
Salembier C, Lavagnini P, Nickers P, GEC ESTRO PROBATE Group et al (2007) Tumour and target volumes in permanent prostate brachytherapy: a supplement to the ESTRO/EAU/EORTC recommendations on prostate brachytherapy. Radiother Oncol 83:3–10
Siddon RL, Chin LM (1985) Two-film brachytherapy reconstruction algorithm. Med Phys 12:77–83
Siebert FA, De Brabandere M, Kirisits C et al (2007) Phantom investigations on CT seed imaging for interstitial brachytherapy. Radiother Oncol 85:316–323
Sloboda RS, Usmani N, Pedersen J et al (2010) Time course of prostatic edema post permanent seed implant determined by magnetic resonance imaging. Brachytherapy 9:354–361
Steggerda MJ, Moonen LM, van der Poel HG et al (2007) The influence of geometrical changes on the dose distribution after I-125 seed implantation of the prostate. Radiother Oncol 83:11–17

Stock RG, Stone NN, Tabert A et al (1998) A dose–response study for I-125 prostate implants. Int J Radiat Oncol Biol Phys 41:101–108
Stone NN, Hong S, Lo YC et al (2003) Comparison of intraoperative dosimetric implant representation with postimplant dosimetry in patients receiving prostate brachytherapy. Brachytherapy 2:17–25
Tanaka O, Hayashi S, Matsuo M et al (2007) Effect of edema on postimplant dosimetry in prostate brachytherapy using CT/MRI fusion. Int J Radiat Oncol Biol Phys 69:614–618
Taschereau R, Pouliot J, Roy J et al (2000) Seed misplacement and stabilizing needles in transperineal permanent prostate implants. Radiother Oncol 55:59–63
Van Gellekom MP, Moerland MA, Kal HB et al (2002) Biologically effective dose for permanent prostate brachytherapy taking into account post implant edema. Int J Radiat Oncol Biol Phys 53:422–433
Van Gellekom MP, Moerland MA, Wijrdeman HK et al (2004) Quality of permanent prostate implants using automated delivery with seedSelectron versus manual insertion of RAPID Strands. Radiother Oncol 73:49–56
Vicini FA, Kini VR, Edmundson G et al (1999) A comprehensive review of prostate cancer brachytherapy: defining an optimal technique. Int J Radiat Oncol Biol Phys 44:483–491
Villeirs GM, L Verstraete K, De Neve WJ et al (2005) Magnetic resonance imaging anatomy of the prostate and periprostatic area: a guide for radiotherapists. Radiother Oncol 76:99–106
Wang JZ, Mayr NA, Nag S et al (2006) Effect of edema, relative biological effectiveness, and dose heterogeneity on prostate brachytherapy. Med Phys 33:1025–1032
Waterman FM, Yue N, Corn BW et al (1998) Edema associated with I-125 or Pd-103 prostate brachytherapy and its impact on post-implant dosimetry: an analysis based on serial CT acquisition. Int J Radiat Oncol Biol Phys 41:1069–1077
Wei Z, Ding M, Downey D et al (2005) Dynamic intraoperative prostate brachytherapy using 3D TRUS guidance with robot assistance. Conf Proc IEEE Eng Med Biol Soc 7:7429–7432
Westendorp H, Hoekstra CJ, van't Riet A et al (2007) Intraoperative adaptive brachytherapy of iodine-125 prostate implants guided by C-arm cone-beam computed tomography-based dosimetry. Brachytherapy 6:231–237
Willins J, Wallner K (1997) CT-based dosimetry for transperineal I-125 prostate brachytherapy. Int J Radiat Oncol Biol Phys 39:347–353
Willins J, Wallner K (1998) Time-dependent changes in CT-based dosimetry of I-125 prostate brachytherapy. Radiat Oncol Investig 6:157–160
Yue N, Dicker AP, Corn BW et al (1999a) A dynamic model for the estimation of optimum timing of computed tomography scan for dose evaluation of 125I or 103Pd seed implant of prostate. Int J Radiat Oncol Biol Phys 43:447–454
Yue N, Chen Z, Peschel R et al (1999b) Optimum timing for imagebased dose evaluation of I-125 and Pd-103 prostate seed implants. Int J Radiat Oncol Biol Phys 45:1063–1072
Yue N, Chen Z, Nath R (2002) Edema-induced increase in tumour cell survival for 125I and 103Pd prostate permanent seed implants – a bio-mathematical model. Phys Med Biol 47:1185–1204
Zelefsky MJ, Yamada Y, Cohen GN et al (2007) Intraoperative real-time planned conformal prostate brachytherapy: post implantation dosimetric outcome and clinical implications. Radiother Oncol 84:185–189

Dose–Response Relationship in Permanent Implant Brachytherapy for Prostate Cancer

13

Jean-Marc Cosset and Georges Wakil

13.1 Introduction

In radiation oncology, it is intuitive that higher doses of radiation could result in an increase in cell kill and consequently in higher relapse-free survival rates with higher overall survival rates. Unfortunately, it is not always the case: the doses necessary to locally eradicate some tumors are sometimes not compatible with the tolerance of some particularly radiosensitive surrounding tissues; in other cases, the prognosis is essentially linked to distant metastases, so that an improved local control will not actually improve overall survival.

When it comes to prostate cancer, recent data have however strongly suggested that there is a dose–response relationship with external beam radiation therapy (EBRT). The last update of the MD Anderson Cancer Center randomized trial of dose escalation documented an improvement in freedom from biochemical failure when increasing dose: 78 % for the 78 Gy arm versus only 59 % for the 70 Gy arm ($p=0.004$) (Kuban et al. 2008). In the last update of the Boston trial comparing a conventional dose of 70.2 Gy delivered with photons, and an escalated dose of 79.2 Gy delivered by a combination of photons and protons, men receiving high-dose radiation therapy were significantly less likely to have local failures, with a hazard ratio of 0.57. In this study, the 10-year American Society for Therapeutic Radiology and Oncology (ASTRO) biochemical failure

J.-M. Cosset (✉)
Department of Oncology/Radiotherapy, Institut Curie, Paris, France
e-mail: jean-marc.cosset@curie.net

G. Wakil
Department of Radiotherapy, Institut Curie, Paris, France

Hôpital Charles LeMoyne, Montréal, Canada

G. Kovács, P. Hoskin (eds.), *Interstitial Prostate Brachytherapy*,
DOI 10.1007/978-3-642-36499-0_13, © Springer-Verlag Berlin Heidelberg 2013

rates were 32.4 % in the conventional-dose arm and 16.7 % with high-dose radiation therapy ($p<0.0001$) (Zietman et al. 2010). The New York Memorial Sloan-Kettering Cancer Center (MSKCC) reported an improved 5-year relapse-free survival (RFS) with increasing radiation doses: 5-year RFS was 71, 66, 61, and 40 % for doses of 86.4, 81, 75.6, and <70.2 Gy, respectively (Zelefsky et al. 2008). In the British Medical Research Council (MRC) randomized trial RT01, patients were randomized between 74 Gy in 37 fractions and 64 Gy in 32 fractions: biochemical progression-free survival was 71 % (108 cumulative events) and 60 % (149 cumulative events) at 5 years in the escalated and the standard group, respectively (Dearnaley et al. 2007).

Furthermore, a multi-institutional study (pooling nine series) has shown that patients receiving doses greater than or equal to 72 Gy had a PSA disease-free survival of 69 % versus 63 % for doses less than 72 Gy (Kupelian et al. 2005). Finally Viani et al. (2009) published a meta-analysis of seven randomized trials with a total patient population of 2,812. Pooled results from these trials showed a significant reduction in the incidence of biochemical failure in prostate cancer patients treated with high-dose external irradiation (HDRT) ($p<0.0001$). In the subgroup analysis, patients classified as being at low ($p=0.007$), intermediate ($p<0.0001$), and high risk ($p<0.0001$) of biochemical failure all showed a benefit from HDRT. The meta-regression analysis also detected a linear correlation between the total dose of radiotherapy and biochemical failure ($BC=-67.3+[1.8\times$ radiotherapy total dose in Gy]; $p=0.04$). The authors conclude that their meta-analysis showed that HDRT is superior to conventional-dose radiotherapy in preventing biochemical failure in low-, intermediate-, and high-risk prostate cancer patients.

With such data available for EBRT, it is tempting to extrapolate to prostate cancer brachytherapy and to anticipate that there also should be a strong dose–response relationship with permanent implant prostate brachytherapy. However, it is not that simple, since dose specification is very different between EBRT and brachytherapy. In EBRT, dose can be specified either to the isocenter or, most often, to an isodose line that covers the prostate with a margin. This is significantly different from the way the dose is described in prostate brachytherapy, where the most commonly used dosimetric parameter is the D90, the dose delivered to 90 % of the gland from a dosimetric analysis of the postimplant computed tomography (CT) (Stock 2010; Kovács et al. 2005; Salembier et al. 2007).

13.2 Evidence for a Dose–Response Relationship in Permanent Implant Prostate Brachytherapy

When transperineal ultrasound-guided prostate brachytherapy began to be implemented in the late 1980s, very few data were available to guide physicians in defining the appropriate dose prescription with Iodine-125 seed implants (Stock 2010). Most authors used the empirically derived dose of 160 Gy used in the original paper by Hilaris from the New York MSKCC (Hilaris et al. 1974). This dose was felt to be equivalent to a 70 Gy EBRT delivered with conventional fractionation, based on the time-factor calculations accepted at that time. This equivalence is probably not correct if we base our calculations on the currently used BED (biologically equivalent dose) formula (Stock et al. 2006). Moreover, we now know that this traditional dose of 70 Gy leads to suboptimal results compared to present EBRT dose standards with a dose escalation up to about 78–80 Gy offering better results in terms of disease-free survival.

The New York Mount Sinaï group clearly showed that when patients were broken up into two groups, those with D90 values less than 140 Gy and those with a D90 greater than or equal to 140 Gy, there was a large difference in biochemical control rates (68 % versus 92 % at 4 years, $p < 0.02$) (Stock et al. 1998). This study is often considered a landmark study, with most authors trying since to achieve D90's greater than 140–145 Gy. A remarkable feature of this pioneer series is that, initially, a large proportion of patients had rather low D90 values. Stock et al. reported that in the 1990–1992 period, 69 % of their patients had a D90 below 120 Gy. Subsequently, this percentage rapidly decreased to reach 0 % after 1998. This means that the authors had a series of patients with large variations in D90 available. As can be intuitively expected, those patients with a very low D90 experienced more relapses, and dose clearly emerged as the most significant predictor of outcome in the multivariate analysis of this milestone series (Stock et al. 1998). Thereafter, the Mount Sinaï group published a long series of papers confirming the dose–response relationship found in their first study, and a number of other investigators seemed to subsequently validate their findings.

Potters (2001) reported that prostate implants with a D90 < 90 % of the prescribed dose had an 80.4 % 4-year PSA-RFS (relapse-free survival), while those with a D90 ≥ 90 % of the prescribed dose had a 92.4 % 4-year PSA-RFS ($p = 0.001$). Of note, no cutoff value was found for the V100 and D100. Wallner, in 2003 (Wallner et al. 2003), reported that the 3-year biochemical freedom-from-failure rate for patients with a D90 < 100 % of the prescription dose was 82 % versus 97 % for patients with a D90 ≥ 100 % of the prescription dose ($p = 0.01$). In 2007, Zelefsky published one of the rare multi-institutional studies on permanent implant prostate brachytherapy (Zelefsky et al. 2007). Eleven institutions combined their data on 2,693 patients treated with permanent interstitial brachytherapy as monotherapy for T1–T2 prostate cancers. The median follow-up was 63 months. In this series, among patients where the postimplantation I-125 dose to 90 % of the prostate (D90) was ≥130 Gy, the 8-year PSA relapse-free survival (PSA-RFS) was 93 %, compared with 76 % for those with lower D90 dose levels ($p < 0.001$). When restricted to patients with available postimplantation dosimetric information, D90 still emerged as a significant predictor of biochemical outcome ($p = 0.01$).

At that time, rather than using the conventional D90, some authors turned to BED (biologically effective dose). This was supposed to allow the comparison of different isotopes, different prescription doses, and protocols adding EBRT to brachytherapy. In 2006, the Mount Sinaï group analyzed their data using BED equations (Stock et al. 2006). In their series, the 10-year FFPF for BED < 100, >100–120, >120–140, >140–160, >160–180, >180–200, and >200 were 46, 68, 81, 85.5, 90, 90, and 92 %, respectively ($p < 0.0001$). BED and Gleason score had the greatest effect, with a p-value <0.0001 in multivariate analysis. The positive biopsy rates for BED≤100, >100–120, >120–140, >140–160, >160–180, >180–200, and >200 were 24 % (8/33), 15 % (3/20), 6 % (2/33), 6 % (3/52), 7 % (6/82), 1 % (1/72), and 3 % (4/131), respectively ($p<0.0001$). BED was the most significant predictor of biopsy outcome in multivariate analysis ($p=0.006$).

Other authors, using BED, confirmed those findings. In their series, Miles (2010) found that the BED, D90, and V100 were all highly correlated, and all were strongly correlated with biochemical relapse-free survival (bRFS). They identified the following cutoffs for predicting freedom from biochemical failure: D90≥110 Gy, V100≥74 %, and BED≥115 Gy. However, none of the covariates significantly predicted overall survival. Stone (2007) reported on the outcomes from a multi-institutional (six centers and 3,928 patients) dataset. Patients were divided in three dose groups: low dose (<140 BED), intermediate dose (140–200 BED), and high dose (>200 BED). Using the ASTRO definition of PSA failure, the 10-year PSA disease-free survival rates for low-risk patients were 69.8, 86, and 88.1 % in the low-, intermediate-, and high-dose groups, respectively ($p<0.0001$). For intermediate-risk patients, rates were 52.9, 74, and 94.3 % ($p<0.0001$); for high-risk patients, rates were 19.2, 61.8, and 90 %, respectively, for the three BED groups ($p<0.0001$). A subsequent paper by Taira found approximately the same results, with a BED cutoff at 116 Gy ($p<0.003$ for low-risk patients and <0.006 for intermediate-risk patients) (Taira et al. 2010). A Spanish group also found a cutoff D90 value of 147 Gy (obtained on postimplantation day 0) with a trend toward a significant correlation with bRFS when the standard ASTRO and nadir+2 definitions were used (Garrán et al. 2010).

With such a series of papers, most of which are summarized in the recent review by Stock (Stock 2010), one would think that the situation is clear and that the D90 (or BED) was unequivocally the most significant dosimetric parameter, with a very clear correlation between dose and clinical response. However, other recent data have shown that things are not quite as simple

13.3 Evidence Against a Dose–Response Relationship in Permanent Implant Prostate Brachytherapy

While a number of series have shown a seemingly unequivocal dose–response relationship in permanent implant prostate brachytherapy, some other recent studies reported different results.

One of the first reports in partial disagreement is in a paper published in 2006, and analyzing the data on 667 patients treated in Leeds between 1995 and 2001, Ash (2006) reported no significant correlation between D90 and outcome for the whole population, when comparing between patients who received greater versus less than 140 Gy ($p=0.43$); there was also no difference for those receiving more or less than 130 Gy ($p=0.14$). Subgroup analysis by risk group, however, showed that for *low-risk patients* there was a significant correlation between D90 and PSA control ($p<0.01$). No significant dose–response relationship was found between D90 and PSA in the intermediate- and high-risk population of patients.

In 2009, Morris (2009) reported on a series of 1,006 patients. In this series, with a median follow-up of 54 months, the actuarial 5-year rate of freedom from biochemical recurrence was 95.6 % +/– 1.6 %. Dosimetric values were not found to be predictive of biochemical recurrence on univariate or multivariate analysis. Analysis of dosimetric values by number of implants performed showed a statistically significant increase in all values with time (D90, V100, V150, and V200; $p<0.001$), but this did not translate into improved bNED (biochemical nonevidence of disease). The authors conclude: "In contrast to some previous studies, dosimetric outcomes did not correlate with biochemical recurrence in the first 1,006 patients treated with I-125 prostate brachytherapy at the British Columbia Cancer Agency."

Recently, Bittner (2010) used a different approach. Nineteen biochemically failed brachytherapy patients were matched to 74 dosimetric and clinically equivalent non-failures using a definition of biochemical progression-free survival (bPFS) as a prostate-specific antigen level of 0.40 ng/mL or less after nadir. A 5 mm annulus was constructed around the perimeter of each prostate. D90 and V100 at the anterior, posterior, superior, inferior, right lateral, and left lateral aspects of the annulus were evaluated for patients with biochemically controlled and failed disease. D90 and V100 parameters were compared between the controlled and failed groups using logistic regression. No statistically significant differences in prostate-specific antigen level, Gleason score, percent positive biopsies, or intraprostatic dosimetry were observed between the controlled and failed patients. Interestingly, the D90 and V100 at the anterior, posterior, superior, inferior, right lateral, and left lateral aspects of the annulus were not statistically different between biochemically controlled and failed groups. The authors concluded that in this study, there was no relationship observed between annular dosimetry and biochemical control.

Reporting on the results of the first 1,044 patients treated by the Paris group with a 6.7-year follow-up, Wakil (2010) showed no significant difference between patients receiving a D90 of more or less than 180 Gy ($p=0.47$).

13.4 Reasons for the Discrepancies

There are a number of reasons to explain why some authors found a significant dose–response relationship in permanent implant prostate brachytherapy and why others did not. Those reasons have been well explained in a few recent papers, particularly the one already mentioned by Ash et al. (2006), and the reviews by Morris and Stock (Morris et al. 2010; Stock 2010).

Ash et al. (2006) insisted on several points:

1. Oedema is constant after permanent implant prostate brachytherapy: such an oedema will temporarily—in the weeks following the implantation—increase the prostate volume. Therefore, if a dosimetric analysis is performed too soon, when oedema is still present, it will systematically find a D90 *lower* than the one obtained during the implantation, and also *lower* than the D90 calculated later on, at a time when the oedema has disappeared. Consequently, authors calculating their D90 at, for example, 3 weeks postimplant, are expected to obtain D90

values systematically lower than authors performing their postimplant dosimetry at 2 months. Such differences in the timing of postimplant dosimetry may therefore explain at least part of the discrepancies.

2. The possibility that the underdosed area could be situated where it is unlikely to have tumor. The "pure" D90 value alone does not give any insight about the prostate area that could be underdosed. A D90 of 145 Gy, with a V100 of 90 %, could be quite sufficient if the "well-irradiated" volume corresponds to the tumor area. In contrast, a plan with a D90 of 180 Gy with a V100 of 95 %, that leaves the tumor in the "5 %" underdosed volume, could lead to a local relapse. Not all prostate areas are equally at risk of relapse, and it has to be stressed that the D90 does not take this fact into account. It is well known that most tumors emerge in the prostatic peripheral zone and that the risk of finding a prostatic tumor is much smaller in the "adenoma" zone. The analysis of prostatectomy specimens has clearly demonstrated this usual distribution of tumor. As an additional example, a systematic resection of the median lobe in a recent series of 22 patients, before brachytherapy, did not show a single case of tumor involvement at that level (Cosset et al. 2011). These data therefore suggest that a limited underdosing located at the prostate base, in the adenoma zone (anterosuperior quadrant), is probably "acceptable" especially if biopsies were shown to be negative at that level, while the same underdosing—even limited—of the apex especially if biopsies were found to be positive at that level is definitively unacceptable, despite having the same D90.
3. The fact that biochemical control does not equate to local control because some patients fail outside the prostate, particularly in the intermediate- and high-risk group. According to Ash (2006), this is probably the reason why in their series D90 was found to be a good discriminator for low-risk patients only, where failure to achieve local control is likely to be the dominant cause of PSA failure.

Morris (2010) notes that the percentage of patients receiving androgen deprivation therapy (ADT) varies greatly from one center to another; he mentions, for example, that in their series (BCCA; British Columbia Cancer Agency), patients were more than twice as likely to receive ADT than those from the Mount Sinaï group (65 % versus 31 %). However, ADT was found in their series to contribute a less than 2 % improvement in the 5-year bNED, and therefore is unlikely to explain the apparent discrepancies. More relevant is probably the fact that an implant technique based exclusively on intraprostatic seed placement may require a higher D90 to deliver the same extraprostatic dose. This constitutes a significant difference between the Mount Sinai approach (intraprostatic loose seeds) and the BCCA technique, where many of the stranded seeds are deliberately placed outside the prostate: 20–80 % of the seeds in the BCCA experience (Morris et al. 2010). Morris (2010) also comments on the large uncertainties associated with the parameters used to calculate the BED, uncertainties in tumor repopulation, uncertainties due to the highly nonuniform dose distribution in brachytherapy (Ling et al. 1994), and uncertainties linked to the alpha/beta ratio (Lindsay et al. 2003), among others.

While clearly supporting the dose–response relationship in permanent prostate implant brachytherapy, Stock (2010) also recognizes that some other reasons could explain the apparent discrepancies in the literature. The first reason could be an inadequate follow-up: any study with a short follow-up may not have the power to discern differences in local control from dose escalation. The second reason is that most centers with experience are now reporting large series of patients, with sufficient follow-up and with a narrow range in delivered dose values; that is to say, fortunately most patients treated today are receiving "good quality" implantations, with very limited variations in D90. In such a situation, and considering all the uncertainties mentioned above, it is quite understandable that no role for D90 emerges in those modern series, while this role was obvious in the pioneer series where a number of patients received D90 of less than 120 Gy (see Chap. 2).

A last point, probably insufficiently emphasized, is the problem of the reproducibility of the prostate contours from one observer to another. Morris (2010) honestly reported that the D90 values of patients in a same institution have at least a 10 % variation when calculated from contours generated by multiple observers who are blinded to patient identity and the contours of other observers. It is well known that such large variations may occur (Crook et al. 2002; Xue et al. 2006).

Available data strongly suggest that there is a dose–response relationship in permanent implant brachytherapy for prostate cancer. However, one cannot expect this relationship to be a very close one, because a large number of reasons may blur, or even totally hide, the relationship between dose and clinical outcome. The main reasons, which may be responsible for such a blurring of the results, are the following:

- Variations in the prostate contours drawn by different radiation oncologists.
- Timing of the postimplant CT: too "early" CTs may be performed at a time when oedema is still present, thus leading to an underestimation of the D90.
- The underdosed area: if the underdosed area is located where no tumor is present (as is often the case for the anterosuperior zones and/or the median lobes), a "low" D90 will not translate into a lower RFS.
- Results are reported in terms of biochemical control which depends on local control (expected to be related to dose) and also on "distant" control (especially in high-risk patients), which has no relation to local dose.
- The percentage of patients receiving ADT (androgen deprivation therapy). Large variations from one series to another may introduce a bias in biochemical control in some instances.
- The specific implantation technique: techniques implanting exclusively intraprostatic seeds lead in most cases to a higher D90 than techniques that involve implanting a large percentage of seeds (usually stranded) extraprostatically.
- The uncertainties associated with most parameters used to calculate BED (biologically equivalent dose), when this approach is utilized.
- The follow-up, which may be inadequate in some series.
- The narrow range in D90 in most of the modern series.

In spite of these limitations, it is important to continue calculating D90. Such an evaluation of the dose appears to be particularly useful when starting the technique and remains useful thereafter to allow comparison between centers.

In conclusion, nowadays, D90 remains a valid parameter, but one should be aware of its limitations. It clearly still needs to be calculated, but its value should not be overestimated.

References

Ash D, Al-Qaisieh B, Bottomley D, Carey B, Joseph J (2006) The correlation between D90 and outcome for I-125 seed implant monotherapy for localised prostate cancer. Radiother Oncol 79(2):185–189

Bittner N, Merrick GS, Butler WM, Allen ZA, White B, Adamovich A, Wallner KE (2011) The correlation between annular treatment margins and biochemical failure in prostate brachytherapy patients with optimized intraprostatic dosimetry. Brachytherapy 10(5):409–415

Cosset JM, Barret E, Castro-Pena P, Cathelineau X, Galiano M, Rozet F, Pierrat N, Timbert M, Vallancien G (2011) One-step customized transurethral resection of the prostate and permanent implant brachytherapy for selected prostate cancer patients: technically feasible but too toxic. Brachytherapy 10(1):29–34

Crook J, Milosevic M, Catton P, Yeung I, Haycocks T, Tran T, Catton C, McLean M, Panzarella T, Haider MA (2002) Interobserver variation in postimplant computed tomography contouring affects quality assessment of prostate brachytherapy. Brachytherapy 1(2):66–73

Dearnaley DP, Sydes MR, Graham JD, Aird EG, Bottomley D, Cowan RA, Huddart RA, Jose CC, Matthews JH, Millar J, Moore AR, Morgan RC, Russell JM, Scrase CD, Stephens RJ, Syndikus I, Parmar MK, RT01 collaborators (2007) Escalated-dose versus standard-dose conformal radiotherapy in prostate cancer: first results from the MRC RT01 randomised controlled trial. Lancet Oncol 8(6):475–487

Garrán C, Ciérvide R, Cambeiro M, Moreno-Jiménez M, Ramos LI, Martínez-Monge R (2010) Relationship between day 0 dosimetric parameters and biochemical relapse-free survival in patients treated with transperineal permanent prostate interstitial brachytherapy with (125)I seeds. Brachytherapy 9(1):8–14

Hilaris BS, Whitmore WF Jr, Batata MA, Grabstald H (1974) Radiation therapy and pelvic node dissection in the management of cancer of the prostate. Am J Roentgenol Radium Ther Nucl Med 121(4):832–838

Kovács G, Pötter R, Loch T, Hammer J, Kolkman-Deurloo IK, de la Rosette JJ, Bertermann H (2005) GEC/ESTRO-EAU recommendations on temporary brachytherapy using stepping sources for localised prostate cancer. Radiother Oncol 74(2):137–148

Kuban DA, Tucker SL, Dong L, Starkschall G, Huang EH, Cheung MR, Lee AK, Pollack A (2008) Long-term results of the M. D. Anderson randomized dose-escalation trial for prostate cancer. Int J Radiat Oncol Biol Phys 70(1):67–74

Kupelian P, Kuban D, Thames H, Levy L, Horwitz E, Martinez A, Michalski J, Pisansky T, Sandler H, Shipley W, Zelefsky M, Zietman A (2005) Improved biochemical relapse-free survival with increased external radiation doses in patients with localized prostate cancer: the combined experience of nine institutions in patients treated in 1994 and 1995. Int J Radiat Oncol Biol Phys 61(2):415–419

Lindsay PE, Moiseenko VV, Van Dyk J, Battista JJ (2003) The influence of brachytherapy dose heterogeneity on estimates of alpha/beta for prostate cancer. Phys Med Biol 48(4):507–522

Ling CC, Roy J, Sahoo N, Wallner K, Anderson L (1994) Quantifying the effect of dose inhomogeneity in brachytherapy: application to permanent prostatic implant with 125I seeds. Int J Radiat Oncol Biol Phys 28(4):971–978

Miles EF, Nelson JW, Alkaissi AK, Das S, Clough RW, Broadwater G, Anscher MS, Chino JP, Oleson JR (2010) Biologically effective dose (BED) correlation with biochemical control after low-dose rate prostate brachytherapy for clinically low-risk prostate cancer. Int J Radiat Oncol Biol Phys 77(1):139–146

Morris WJ, Keyes M, Palma D, McKenzie M, Spadinger I, Agranovich A, Pickles T, Liu M, Kwan W, Wu J, Lapointe V, Berthelet E, Pai H, Harrison R, Kwa W, Bucci J, Racz V, Woods R (2009) Evaluation of dosimetric parameters and disease response after 125 iodine transperineal brachytherapy for low- and intermediate-risk prostate cancer. Int J Radiat Oncol Biol Phys 73(5):1432–1438

Morris WJ, Halperin R, Spadinger I (2010) Point: the relationship between postimplant dose metrics and biochemical no evidence of disease following low dose rate prostate brachytherapy: is there an elephant in the room? Brachytherapy 9(4):289–292; discussion 297–298

Potters L, Cao Y, Calugaru E, Torre T, Fearn P, Wang XH (2001) A comprehensive review of CT-based dosimetry parameters and biochemical control in patients treated with permanent prostate brachytherapy. Int J Radiat Oncol Biol Phys 50(3):605–614

Salembier C, Lavagnini P, Nickers P, Mangili P, Rijnders A, Polo A, Venselaar J, Hoskin P, GEC ESTRO PROBATE Group (2007) Tumour and target volumes in permanent prostate brachytherapy: a supplement to the ESTRO/EAU/EORTC recommendations on prostate brachytherapy. Radiother Oncol 83(1):3–10

Stock RG (2010) Counterpoint: there is a dose-response relationship in the low-dose rate brachytherapy management of prostate cancer. Brachytherapy 9(4):293–296

Stock RG, Stone NN, Tabert A, Iannuzzi C, DeWyngaert JK (1998) A dose-response study for I-125 prostate implants. Int J Radiat Oncol Biol Phys 41(1):101–108

Stock RG, Stone NN, Cesaretti JA, Rosenstein BS (2006) Biologically effective dose values for prostate brachytherapy: effects on PSA failure and posttreatment biopsy results. Int J Radiat Oncol Biol Phys 64(2):527–533

Stone NN, Potters L, Davis BJ, Ciezki JP, Zelefsky MJ, Roach M, Fearn PA, Kattan MW, Stock RG (2007) Customized dose prescription for permanent prostate brachytherapy: insights from a multicenter analysis of dosimetry outcomes. Int J Radiat Oncol Biol Phys 69(5):1472–1477

Taira AV, Merrick GS, Galbreath RW, Wallner KE, Butler WM (2010) Natural history of clinically staged low- and intermediate-risk prostate cancer treated with monotherapeutic permanent interstitial brachytherapy. Int J Radiat Oncol Biol Phys 76(2):349–354

Viani GA, Stefano EJ, Afonso SL (2009) Higher-than-conventional radiation doses in localized prostate cancer treatment: a meta-analysis of randomized, controlled trials. Int J Radiat Oncol Biol Phys 74(5):1405–1418

Wakil G, Gobaux V, Hajage D, Flam T, Thiounn N, Pontvert D, Pierrat N, Chauveinc L, Cosset JM (2011) Can intermediate-risk patients be safely treated with permanent implant prostate brachytherapy: long-term results of the first 1044 patients of the Paris Institut Curie/Cochin Hospital/Necker Hospital Group. Brachytherapy 10:S54. American Brachytherapy Society (ABS), San Diego. Abstract

Wallner K, Merrick G, True L, Sutlief S, Cavanagh W, Butler W (2003) 125I versus 103Pd for low-risk prostate cancer: preliminary PSA outcomes from a prospective randomized multicenter trial. Int J Radiat Oncol Biol Phys 57(5):1297–1303

Xue J, Waterman F, Handler J, Gressen E (2006) The effect of interobserver variability on transrectal ultrasonography-based postimplant dosimetry. Brachytherapy 5(3):174–182

Zelefsky MJ, Kuban DA, Levy LB, Potters L, Beyer DC, Blasko JC, Moran BJ, Ciezki JP, Zietman AL, Pisansky TM, Elshaikh M, Horwitz EM (2007) Multi-institutional analysis of long-term outcome for stages T1-T2 prostate cancer treated with permanent seed implantation. Int J Radiat Oncol Biol Phys 67(2):327–333

Zelefsky MJ, Yamada Y, Fuks Z, Zhang Z, Hunt M, Cahlon O, Park J, Shippy A (2008) Long-term results of conformal radiotherapy for prostate cancer: impact of dose escalation on biochemical tumor control and distant metastases-free survival outcomes. Int J Radiat Oncol Biol Phys 71(4):1028–1033

Zietman AL, Bae K, Slater JD, Shipley WU, Efstathiou JA, Coen JJ, Bush DA, Lunt M, Spiegel DY, Skowronski R, Jabola BR, Rossi CJ (2010) Randomized trial comparing conventional-dose with high-dose conformal radiation therapy in early-stage adenocarcinoma of the prostate: long-term results from Proton Radiation Oncology Group/American College of Radiology 95-09. J Clin Oncol 28(7):1106–1111

[illegible]

HDR Versus LDR Seeds

14

Peter Hoskin

14.1 Introduction

LDR seed brachytherapy is well established as an effective treatment for low-risk prostate cancer and increasingly recognised as effective in immediate-risk disease; biochemical relapse survival rates are equivalent to any other treatment modality in these settings, and the toxicity profile is well established with acute urinary disturbance and thereafter a low incidence of moderate or severe toxicity which will be predominately either urethral stricture or rectal bleeding. Erectile dysfunction likelihood is considered lower than other treatment modalities in patients who are potent at the time of implant.

HDR brachytherapy has been shown to be an efficient means of dose escalation when used as a boost with external beam radiotherapy. Biochemical relapse-free survival rates are equivalent to those achieved with high-dose intensity-modulated radiotherapy (IMRT) with the potential for lower rates of late toxicity. HDR used alone as monotherapy in the same way as LDR seeds is still relatively new with only a small number of groups reporting early results. Thus, true comparison between LDR and HDR is difficult since the comparison at a clinical level is currently between low-risk prostate patients treated with LDR monotherapy and intermediate- and high-risk patients treated with combined external beam and HDR. There is a smaller literature on the combination of LDR seeds with external beam radiotherapy and similarly on HDR monotherapy.

It is however possible to make theoretical comparison between the two modalities and draw some conclusions from the available clinical literature.

P. Hoskin
Mount Vernon Cancer Centre,
University College London, London, UK
e-mail: peterhoskin@nhs.net

G. Kovács, P. Hoskin (eds.), *Interstitial Prostate Brachytherapy*,
DOI 10.1007/978-3-642-36499-0_14, © Springer-Verlag Berlin Heidelberg 2013

14.2 Radiobiological Considerations

The radiobiology of low-dose-rate radiation delivery is fundamentally different to that of high-dose rate. Whilst there are models which attempt to compare LDR with HDR delivery using data from both external beam and brachytherapy, these are based on assumptions which may have some uncertainty.

14.2.1 Relative Biological Equivalence (RBE)

The RBE of LDR brachytherapy is considered by some to be higher than HDR with an RBE of 1.2 quoted. This is by no means certain and others have suggested there is no significant difference in RBE between LDR and HDR.

14.2.2 Alpha/Beta (α/β) Effect

Much has been made of the alpha/beta effect in prostate cancer. Most authors agree that it is lower than expected some suggesting values down to 1.2 or 1.5, whilst a more acceptable average would seem to be between 3 and 3.5 and a general consensus that with the exception of data from one paper that all estimates point to a value well below five. The implication of this is that there will be a significant fraction size effect when delivering radiation to a prostate cancer cell, with the most efficient cell kill being achieved by high doses per fraction as shown in Fig. 14.1. This immediately gives HDR brachytherapy a substantial advantage over any other form of radiation delivery. The geometric advantage of HDR brachytherapy based on the inverse square law means that very high fraction sizes of 10–20 Gy or more can be safely delivered within the limits of normal tissue tolerance and simple BED dose calculations show that doses well in excess of 100 Gy (2 Gy equivalent) can be achieved.

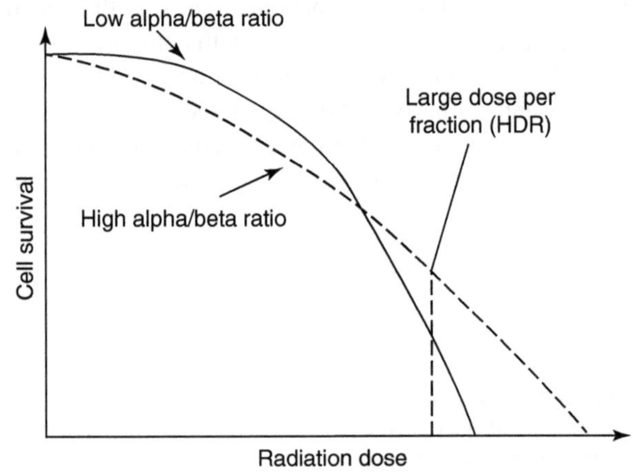

Fig. 14.1 Dose response curve for low and high alpha/beta

In contrast LDR brachytherapy is relatively inefficient even for a slowly proliferating tumour such as prostate where repopulation may not be a significant problem. Radiobiological data suggests a 2 Gy equivalent dose of 145 Gy LDR I-125 brachytherapy is 70–75 Gy. Thus, if the ability to deliver very high doses of radiation within the limits of normal tissue tolerance to the prostate is considered an advantage, then HDR brachytherapy is clearly ahead of LDR seeds. This may not be relevant for the management of low-risk prostate cancer, but where dose escalation is required for more bulky and high-grade tumours, then HDR has theoretical advantages.

14.3 Physical Implant Properties

LDR seeds are effective because they are retained within the prostate gland. Outside the prostate gland is a large venous plexus and above the prostate gland is the bladder. Seeds placed in these areas may not be retained, will not deliver reliable dosimetry and may embolise to distant sites. In general, therefore, an LDR implant is constrained to deliver dose to disease within the capsule of the prostate gland.

In contrast HDR catheters can be placed widely within the pelvic tissues around the prostate. Implantation of the periprostatic region can be readily achieved as can implantation of the seminal vesicles, shown in Fig. 14.2. HDR brachytherapy is

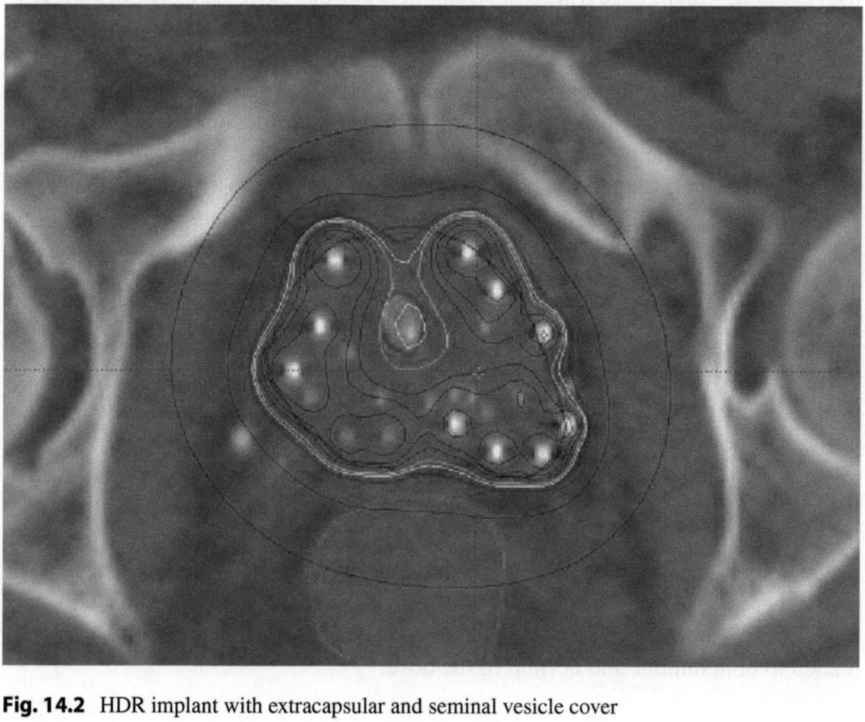

Fig. 14.2 HDR implant with extracapsular and seminal vesicle cover

therefore more likely to adequately treat intermediate- and high-risk disease where there is a significant risk of extracapsular extension and seminal vesicle involvement.

14.3.1 Dosimetry and Dose Delivery

The procedures for LDR and HDR brachytherapy, whilst both using transrectal ultrasound-guided transperineal implant techniques, are different in terms of their planning and verification procedures. The classical method of LDR seed implantation uses a two-stage technique in which a transrectal ultrasound volume study is performed from which a plan is derived of the seed distribution, and then as a second stage, seed implantation occurs with the hope and expectation that the setup used for the volume study can be reproduced for the implant procedure. It is recognised that this approach has many shortcomings, and therefore, most centres have moved to some form of interactive seed implantation in which the volume study and seed deposition are undertaken in the same procedure and seeds are tracked for their real position after delivery so that dosimetry can be updated and adjusted during the procedure. In the best centres, this provides very satisfactory implants, but there have been notorious examples of errors and incidents where seed deposition has been less than satisfaction. Clearly LDR seed brachytherapy is technically demanding requiring meticulous interaction between physicist and physician, notwithstanding which, once a seed or strand of seeds has been deposited in a particular position, the situation becomes irreversible and has to be accepted and included in the dose plan. There is therefore much less flexibility and room for manoeuvre. The only verification that can be undertaken other than real-time ultrasound plotting of the seed positions after deposition is post-plan dosimetry. This provides post hoc information on the quality of the implant, and there is a significant body of evidence, which suggests that dosimetric parameters in post-plan dosimetry can be related to biochemical relapse-free survival. What is not clear is how to address the problem of a bad implant once it has been completed and the patient discharged.

In contrast HDR brachytherapy incorporates careful dosimetric analysis of the implant before treatment delivery. All the available techniques whether ultrasound based or CT/MRI based reconstruct the implant once the applicators are in place, define dwell positions and can verify the applicator position both by measurement of the skin position and apposition of the template and applicators and by imaging to verify soft tissue relationships before treatment is delivered. Thus, a more accurate and tailored dose distribution can be achieved with greater reliability than with LDR techniques.

One study has compared the dosimetric parameters achieved using HDR brachytherapy and an LDR post-plan for the same patients. These results, shown in Table 14.1, demonstrated consistent advantages for the HDR plan for all parameters related to both tumour and normal tissue dose.

Table 14.1 Dosimetry parameters post-implant I-125 and HDR

	LDR post-plan	HDR
PTV D90	86.7 %	111.5 %
PTV V100	82.0 %	97.2 %
Urethral max	207.0 %	127.2 %
Rectal max	180.7 %	100.5 %
Conformity index	0.53	0.69

From Wang et al. 2006

14.4 Clinical Efficacy

14.4.1 Tumour Control

There is now a substantial literature demonstrating the efficacy of LDR brachytherapy for low-risk patients with 5- and 10-year biochemical relapse-free survivals of 85–90 %. There is greater controversy with regard to the role of this modality in intermediate- and high-risk patients although the results published suggest that it is no less effective in these groups than other treatments such as high-dose external beam radiotherapy.

The HDR literature is less mature and less extensive. However, when HDR brachytherapy is used as a boost with external beam radiotherapy typically after a dose of around 45 Gy, published data shows that it also achieves good results with low-, intermediate- and high-risk patients, which are again very similar to those achieved with LDR brachytherapy or high-dose IMRT. When the results of HDR brachytherapy with external beam are compared with those of LDR boosts after external beam, again no difference is clearly seen when comparing different series. There is of course no randomised data to evaluate this more rigorously.

HDR monotherapy is a newer approach with less mature data. The evidence available again suggests biochemical control rates at least as good as those achieved with LDR in low-risk patients and similar control rates to those with combined external beam and brachytherapy in intermediate- and high-risk patients (Table 14.2).

14.4.2 Toxicity

Given that efficacy rates seem very similar, it is important to consider toxicity profiles between the two modalities. When used in combination with external beam radiotherapy, no clear difference emerges. There is however a different profile of toxicity when LDR or HDR is used alone. The most striking toxicity with LDR brachytherapy is an acute prostatitis with associated dysuria, urgency and frequency. In addition 8–10 % of patients will require catheterisation in the immediate post-implant period which may be retained for some months. HDR brachytherapy produces a more acute disturbance which is seen within the first 2–4 weeks after

Table 14.2 Biochemical relapse-free survival for different modalities

	MSKCC risk category		
	Low (%)	Intermediate (%)	High (%)
8-year bRFS (ASTRO 1995 definition)			
IMRT 86 Gy[a]	98	85	70
LDR monotherapy[b]	88	77	58
HDR boost[c]	92	88	65
HDR monotherapy[d]	95	93	75

[a]Cahlon et al. 2007
[b]Henry et al. 2010
[c]Demanes et al. 2009
[d]Yoshioka et al. 2011

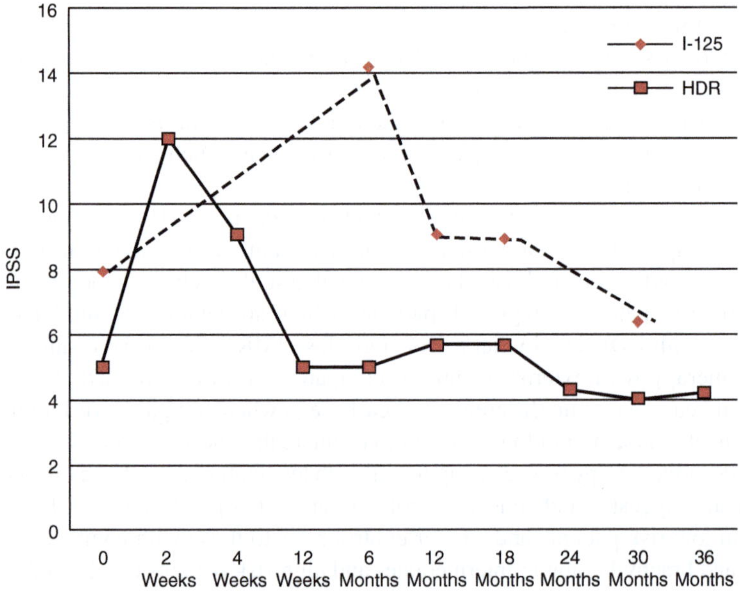

Fig. 14.3 IPSS scores after LDR and HDR prostate brachytherapy

treatment and settles in 4–6 weeks thereafter as shown in Fig. 14.3. A case-control study from one large centre has suggested that the acute toxicity of grade 3 or more severity which is seen with HDR brachytherapy in terms of dysuria, frequency, urgency and rectal pain is less than that seen with LDR brachytherapy, and similarly late toxicity in terms of dysuria, frequency, urgency and impotence is lower. Certainly a different profile of acute toxicity seems to be seen between the two modalities with LDR brachytherapy producing a more prolonged period of urinary dysfunction within the first few months post-implant. However, other late toxicities between the two modalities appear similar with a 5–10 % incidence of urethral stricture and rectal bleeding seen.

14.5 Economic Comparison

The equipment required for LDR brachytherapy and HDR brachytherapy is similar both employing the same technique of transrectal ultrasound-guided transperineal implantation and both requiring anaesthesia, either general or regional, for the implant procedure. Employing a similar technique, the environment, whether inpatient or day-case outpatient is also comparable. Similar levels of physician and physics support are also required. The principal difference then lies in the radiation source which for LDR brachytherapy is a permanent implant used on a single occasion for each patient, and therefore has a significant cost implication amounting on average to around €3,500 per patient. In contrast the HDR afterloader uses a single iridium source replaced every 3 months and which is used not only for prostate cancer but also for many other brachytherapy applications. There is however a capital cost for purchase and maintenance of the afterloader and the bunker within which it will have to sit. Based on an afterloader for which prostate accounts for 30 % of the workload, treating 50 patients per year, estimates suggest that the cost per patient is around €450 per patient.

Conclusions

LDR seed brachytherapy is a reliable and cost-effective alternative to radical prostatectomy or external beam radiotherapy in the management of early low-risk prostate cancer. It may also have an important role in intermediate- and higher-risk patients also. There is a large literature in which the experience of several thousand patients has now been reported from many centres across the world with remarkable consistency. It provides a very convenient treatment for the patient with a relatively low level of side effects, but in some circumstances, marked acute urinary dysfunction.

HDR brachytherapy has theoretical radiobiological advantages based on the low alpha/beta ratio of prostate cancer achieving significant dose escalation where this is considered important. It is also more flexible in enabling implantation of larger volumes particularly in the extracapsular region and seminal vesicles. Used in conjunction with external beam radiotherapy, it has been shown to be an effective means of dose escalation improving the therapeutic ratio with less toxicity. There is however less evidence to support the comparison between LDR seed brachytherapy alone and HDR brachytherapy alone.

HDR brachytherapy is more cost-effective in situations where there is sufficient volume of patients to justify the investment in an afterloader and bunker, together with the appropriate imaging equipment.

References

Wang Y, Sankreacha R, Al-Hebshi A, Loblaw A, Morton G (2006) Comparative study of dosimetry between high-dose-rate and permanent prostate implant brachytherapies in patients with prostate adenocarcinoma. Brachytherapy 5(4):251–255

Cahlon O, Zelefsky MJ, Shippy A, Chan H, Fuks Z, Yamada Y, Hunt M, Greenstein S, Amols H (2008) Ultra-high dose (86.4 Gy) IMRT for localized prostate cancer: toxicity and biochemical outcomes. Int J Radiat Oncol Biol Phys 71(2):330–337

Henry AM, Bashar A, Gould K et al (2010) Outcomes following iodine-125 monotherapy for localized prostate cancer: the results of Leeds 10-year single centre brachytherapy experience. Int J Radiat Oncol Biol Physics 76:50–56

Demanes DJ, Brandt D, Schour L, Hill DR (2009) Excellent results from high dose rate brachytherapy and external beam for prostate cancer are not improved by androgen deprivation. Am J Clin Oncol 32(4):342–347

Yoshioka Y, Komishi K, Somida Y et al (2011) Monotherapeutic high dose rate brachytherapy for prostate cancer: five year results of an extreme hypofractionation regimen with 54Gy in nine fractions. Int J Radiat Oncol Biol Phys 80(2):469–475, doi:10.1016/j.ijrobp.2010.02.013. Epub 2010 Jun 18

Results of Permanent Prostate Brachytherapy

15

Jan J. Battermann

15.1 Introduction

In the past 30 years, there has been a significant improvement in the assessment of patients with prostate cancer. After earlier dismal results implanting I-125 seeds in the prostate through an abdominal route (Fuks et al. 1991), Holm published the first experience with a perineal technique for permanent prostate brachytherapy (PPB) using ultrasonography guidance in the placing of needles (Holm et al. 1983). Promising results were published with this image-guided technique (Blasko et al. 1996; Battermann et al. 2004). Other improvements included staging by PSA, pathology using the Gleason score, imaging by ultrasonography and MRI and 3D planning systems. It should be stated that the Gleason score is based on whole section slices of prostatectomy material. For scores of 1–3, it is very hard for a pathologist to give an exact Gleason sum between 2 and 6, due to the small tissue samples. MRI is considered nowadays as the most accurate method of staging of localised prostate cancer. These improvements resulted in better dosimetry parameters and clinical outcome. Hence, the state-of-the-art PPB includes the use of modern imaging techniques for staging and for guidance of the needle placing into the prostate and 3D intraoperative planning. Because of screening programmes for early detection of prostate cancer in many western countries, there is an increase in number of prostate cancer patients but at the same time a decrease in mortality from prostate cancer. By now there are several papers with long-term experience of ten and more years (Zelefsky et al. 2007; Morris et al. 2009; Hinnen et al. 2010a, b; Henry et al. 2010; Taira et al. 2010). Since the natural history of prostate cancer is slow in the majority of patients, outcome is often stated as biochemical disease-free interval (bDFS). PSA decrease is slow after PPB and may take 5 years or more to reach PSA

J.J. Battermann
Head of the department of radiation oncology,
University Medical Center Utrecht, Correspondence, 3508 GA, Utrecht
e-mail: j.j.battermann@umcutrecht.nl

G. Kovács, P. Hoskin (eds.), *Interstitial Prostate Brachytherapy*,
DOI 10.1007/978-3-642-36499-0_15, © Springer-Verlag Berlin Heidelberg 2013

nadir values lower than 0.1 ng/ml (Grimm et al. 2001). Overall survival (OS) is not a good end point because many men will die of cardiovascular disease and not of prostate cancer (Bittner et al. 2008).

15.2 Result in Low-Risk Patients

According to a combination of PSA value, Gleason score and sum and T stage, patients can be categorised in three risk groups: low, intermediate and high risk. The GEC-ESTRO staging is depicted in Table 15.1. Different systems are in use in Europe and the USA. Especially in the intermediate group, some include T2b and T2c tumours, while others only include T2b. But also the definition of T2c is vague, where centres call T2c as positive biopsies in both lobes without palpable or visible tumours and others classify these patients as T1c (Table 15.2). In general only low- and intermediate-risk patients are considered candidates for PPB. Apart from the clinical relevance of risk categories, also the functional outcome needs attention as found in Table 15.3. In low-risk patients, defined as T1c–2c, PSA <10 ng/ml, Gleason sum ≤7, results are excellent from single institution or from combined data from several centres. Outcomes after more than 5 and 10 years show percentages of 82 % till 89 % for biochemical control (bNED) and approximately 95 % for disease-specific survival (see Table 15.4). Beyer and Brachman mention 85 % bNED in 128 low-risk patients with a median follow-up of 84 months (Beyer and Brachman 2000). Battermann presented data from 114 patients with bNED of 90 % at 4 years (Battermann et al. 2004). Hinnen from the same centre presented 72-month results from 232 patients with 88 % bNED (Hinnen et al. 2010a, b). Henry from Leeds reported 88 % bNED in 575 patients at median follow-up of 57 months. Crook described the 10-year experience in Toronto from 776 patients with an actuarial 7-year disease-free survival rate of 95 %. In 27 failures, 8 local relapses were confirmed (Crook et al. 2011). Several authors have presented data of PPB compared with surgery and/or EBRT, showing similar results of all three modalities (Kupelian et al. 2004; Potters et al. 2005; Sharkey et al. 2005; Tward et al. 2006;

Table 15.1 Factors influencing clinical outcome

Factor	Well	Fair	Poor
iPSA	<10	10–20	>20
Gleason sum	<7	=7	>7
Stage	T1c–2a	T2b,c	T3

Henry et al. (2010)

Table 15.2 Factors influencing functional outcome

Factor	Well	Fair	Poor
Volume	40	40–60	>60
IPSS	0–8	9–20	>20
TURP	No	No	Yes

Ash et al. (2000)

Table 15.3 Risk groups according to different definitions

Centre	Low	Intermediate	High
Seattle	PSA ≤10 and GS 2–6 and T1c–T2b	PSA >10 or GS ≥7 or T2b	2 or 3 factors
M. Sinai	PSA ≤10 and GS 2–6 and T1c–T2a	PSA 10–20 or GS =7 or T2b	2 or 3 factors, and/or PSA > 20, or GS 8–10, ≥T2c
Boston	PSA ≤10 and GS 2–6 and T1c–T2a	PSA 10–20 and/or GS =7 and/or T2b	2 or 3 factors, and/or PSA > 20, or GS 8–10, ≥T2c

Table 15.4 Results of low-risk patients

Author	Number of patients	Median follow-up (months)	% bNED
Beyer 2000	128	84	85
Grimm 2001	125	81	87
Battermann 2004	114	48	91
Sharkey 2005	528	72	87
Potters 2005	481	82	89
Zelefsky 2007		63	82
Hinnen 2010	232	72	88
Henry 2010	575	57	86
Taira 2011	575	148	98, 6

Colberg et al. 2007), although these data are never compared in randomised studies. Only in the Kupelian series, EBRT with insufficient dose <72 Gy ended with a significant lower bNED. Jabbari also included data from patients with a proton boost after EBRT and found at least comparable 5-year bNED results with a greater proportion of men achieving lower PSA nadirs with PPB compared with EBRT and proton boost (Jabbari et al. 2010).

15.3 Results in Intermediate-Risk Patients

In intermediate-risk patients (T1c–2c, Gleason 7, PSA 10–20 ng/ml), good results are also described by different authors (Datolli et al. 2007; Morris et al. 2009; Munro et al. 2010) (Table 15.5). Hinnen showed an improvement in outcome for intermediate-risk patients in the past decade compared with earlier experience (Hinnen et al. 2010a, b). However, for low-risk patients, he did not find an improvement, probably because the results already are very favourable. Definitions of intermediate-risk cases and selection criteria are different from series to series, and PPB may be combined with external beam radiotherapy (Datolli et al. 2007) and/or androgen deprivation therapy (ADT) (Morris et al. 2009; Munro et al. 2010) although neither the combination of EBRT plus seeds nor the use of seeds plus ADT has proven to be better than PPB alone. In four randomised trials regarding the role of ADT on local control in combination with EBRT, distant metastases, disease-free survival and overall

Table 15.5 Results of intermediate-risk patients

Author	Number of patients	ADT %	Median follow-up (months)	% bNED
Beyer 2000	345	0	84	66
Cosset 2008	276	68	43	94
Morris 2009	419	100	54	96
Taira 2010	144	0	74	96
Hinnen 2010	369	18	69	61
Henry 2010	430		57	86

survival, local control was better. For metastases one was positive (RTOG 8531), two were similar (RTOG 8610) and one (EORTC) was not conclusive. Disease-free survival was better in three (both RTOG studies and EORTC), but overall survival was only improved in the EORTC study. The American Brachytherapy Society (ABS) concluded: "ABS is currently unable to provide firm conclusions on the use of androgen deprivation in combination with PPB" (Potters 2000). Androgen deprivation had a negative effect on overall and cancer-specific survival, according to Beyer et al. (2005). Zelefsky found that high-dose EBRT (>72 Gy) has no advantage from adding androgen deprivation (Zelefsky et al. 1998; Kupelian et al. 2000). Also in combination with surgery, Bianco reported a similar cancer-specific survival with or without neoadjuvant hormonal therapy plus *prostatectomy* (Bianco 2005). There are conflicting data in the literature concerning the results in Gleason 3+4 and 4+3 for all treatment modalities. Wright looked at prostate cancer-specific mortality for Gleason 3+4 and 4+3 after surgery and radiotherapy and found an increased risk of recurrence or progression and specific mortality in those with Gleason 4+3 versus 3+4 (Wright et al. 2009). Merrick described a series of 530 patients with Gleason 3+4 (300) or 4+3 (230). At 10 years primary Gleason score did not impact survival, while death from cardiovascular disease or second malignancies was 9.6 times more common than death from prostate cancer (Merrick et al. 2007).

15.4 Results of Combined Therapy with External Beam and PPB

Combined treatment using EBRT plus PPB is given to widen the margin around the prostate, including the lymph nodes, and might be better for intermediate- and high-risk patients (Blasko et al. 2000). However, the EBRT dose is too low to kill even small tumour deposits. The majority of extracapsular extension is within 2 mm from the capsule (Teh et al. 2003). With seeds alone, a very high dose is given on both the prostate and a margin of 5 mm around the gland. Critz mentioned good results for combined treatment, but the results are not better than those of seeds alone (Critz and Levinson 2004). Sylvester et al. reported on 15-year data of combined treatment from the Seattle group. The authors found an 88 % bNED for low risk, 80 % for intermediate and 53 % for high-risk patients (Sylvester et al. 2007). However, several series show no difference in outcome with or without EBRT (Potters et al. 1999; Blasko et al. 2000; Kupelian et al. 2004). Combined therapy will be more expensive and may result in more side effects such as erectile dysfunction. Blasko found that combined therapy perhaps can more homogenise the total dose to the prostate, an advantage for centres with limited experience (Blasko et al. 2000).

15.5 Results and Dose Parameters

Dose escalation is well accepted in EBRT for prostate cancer. Zelefsky et al. reported on the dose escalation study at the MSKCC and found a significant improvement with 86.4 Gy versus standard 66 Gy in bNED and distant metastases-free survival in intermediate- and high-risk patients (Zelefsky et al. 2008). Also the studies from Kuban and Viani showed an improvement in clinical outcome with higher doses (Kuban et al. 2008; Viani et al. 2009). Viani found in his meta-analysis of randomised trials on higher-than-standard radiation doses a response curve after EBRT for all risk groups.

A relationship has also been found in PPB between D90 (dose that is received by 90 % of the target volume) and clinical outcome. Stock demonstrated a significant improvement in bNED with a D90 higher than 140 Gy versus lower D90 (92 % versus 68 % at 4 years). It should be mentioned that in the early years at the Mount Sinai Hospital, more than half of the patients received D90s less than 120 Gy that decreased to zero several years later (Stock et al. 1998). Zelefsky described results of 2,693 patients from a multicentre study and showed a significant 8-year bNED survival in patients treated with a D90 higher than 130 Gy (Zelefsky et al. 2007). Also Stone reported better clinical outcome with higher D90s (Stone et al. 2010). However, in more recent studies, the authors did not find a correlation between D90 and failure from relapse (Ash et al. 2006; Morris et al. 2009). Ash found similar results in patients with D90s higher or lower than 140 Gy in a cohort of 667 patients. Morris came to the same conclusion after evaluation of 1,006 patients. Piña found in 129 patients with D90s greater than 180 Gy not more urinary and gastrointestinal symptoms than in patients treated with D90s less than 180 Gy (Piña et al. 2010). It is advised to do post-implant dosimetry at 1 month. At that time oedema is resolved. Further volume reduction as the result of irradiation will give even higher D90s. The treatment technique is also of influence in dose parameters as was shown by Moerland as mentioned earlier (Moerland et al. 2009). When seeds are placed outside the prostate target area, D90s will be lower than in situations with all seeds within the prostate.

15.6 Results in Younger Patients

Data in the literature show that results after PPB in patients under 60 years of age are at least as good as in older patients (Merrick et al. 2006; Shapiro et al. 2009; Burri et al. 2010). Shapiro found that freedom from progression at 10 years after PPB, presenting with low, intermediate and high risk, was 91.3, 80.0 and 70.2 % compared to 91.8, 83.4 and 72.1 %, respectively, for men before 60 years versus men of 60 years or older. Interestingly, Merrick described high rates of cause-specific and biochemical progression-free survival after PPB in 145 consecutive men over 74 years of age. Overall survival and noncancer deaths were best predicted by tobacco status (Merrick et al. 2008).

15.7 Results in High-Risk Patients

A significant lower cure rate is found in high-risk (≥T3; Gleason >7; PSA>20 ng/ml) patients after all treatment options. This may be due to the fact that a substantial number of them will have microscopic metastases before prostate cancer is detected. In the treatment of these patients without traceable metastases, brachytherapy can be used, either alone or in combination with EBRT and/or ADT. Stone published good results in high-risk patients with D90s over 200 Gy (Stone 2010). However, patient selection must have played a role to achieve these good results. Many of these combined treatments are performed successfully using HDR brachytherapy, especially in high-risk patients (Martinez et al. 2003).

15.8 Second Primary Tumours

It is well known that radiation may induce cancer, but conflicting data are presented, either showing an increase in secondary primary cancers (SPCs) after EBRT of the prostate (Baxter et al. 2005) or not showing an increase (Bhojani et al. 2010). In Utrecht, we recently assessed the risk of SPCs after I-125 prostate brachytherapy compared to prostatectomy in a cohort of 1,888 patients treated with brachytherapy (63 %) or prostatectomy (37 %). Two hundred and twenty-three patients were diagnosed with a SPC, 136 (11.5 %) in the brachytherapy group and 87 (12.4 %) in the prostatectomy group. Patients ≤60 years had a significant increased risk of bladder cancer (Hinnen et al. 2011). This also was found earlier by Singh from the SEER registry in patients after surgery and/or EBRT (Singh et al. 2010). The Commission on Radiological Protection considers the risk of SPCs after PPB negligible (Cosset et al. 2004).

15.9 Discussion

Although the outcome of treatment is not substantially different between the established treatment options prostatectomy, EBRT and PPB, still more patients with low- and intermediate-risk prostate cancer are treated by surgery. Of course, results of EBRT or brachytherapy cannot be better than surgery, but in experienced hands toxicity after brachytherapy is much lower than after surgery. There is no indication that robot-assisted or laparoscopic prostatectomy will ensure a higher cure rate than open surgery nor will it lower the complication rate (Ficarra et al. 2009), although in an update of his article, better results with robot-assisted and laparoscopic prostatectomy were reported in terms of decreased blood loss, higher continence and potency rates (Coelho et al. 2010). Quality of life 6 years after PPB was similar to baseline, (Roeloffzen et al. 2010). Malcolm compared quality of life after open or robotic prostatectomy, cryoablation and brachytherapy. Brachytherapy was associated with higher urinary function, bother scores and sexual function compared to open and da Vinci prostatectomy (Malcolm et al. 2010). With PPB as in surgery, new developments have resulted in better outcomes, both in disease-free survival and in toxicity. The introduction of stranded seeds gave us more homogeneity in dose distribution and less seed loss, but comparative studies have not yet shown a difference between strands and loose seeds. However, with the use of an after loading technique with loose seeds, not only the dose distribution was better than with strands (Moerland et al. 2009) but also the biochemical outcome was better (Hinnen et al. 2010a, b). MRI gives the opportunity to better visualise tumour areas in the prostate, especially with dynamic contrast enhancement. This can be used to select patients for focal brachytherapy and hence further reduce side effects. For focal therapy, only low-risk patients with one defined tumour might be candidates. In a pilot study with patients after previous radiotherapy to the prostate, patients with focal brachytherapy had significant lower complications compared with patients who had been retreated on the whole gland (Moman et al. 2010). On the other hand, Isban studied 243 men with low-risk localised prostate cancer after RP. Despite unilateral stage at biopsy, bilateral or even nonorgan-confined cancer was reported in 64 % of all patients. The authors state that this alarming finding questions the safety and validity of hemi-ablative therapy (Isban et al. 2010). But if focal therapy is to be used, brachytherapy is the most appropriate technique in these situations.

Conclusions

It is clear that PPB is at least equal to surgery and therefore should be part of the armamentarium for low- and intermediate-risk prostate cancer patients. Excellent long-term biochemical disease-free survival for low- and intermediate-risk patients is mentioned in many series worldwide. Combined treatment with EBRT and brachytherapy boost does not result in a higher cure rate than seeds alone. Patients younger than 60 have equally good prospects for cure as older patients and should not be excluded from brachytherapy. Second primary tumours after seed implant are no reason for exclusion. Quality of life after PPB is, also after 6 years, not compromised.

References

Ash D et al (2000) ESTRO/EAU/EORTC recommendations on permanent seed implantation for localized prostate cancer. Radiother Oncol 57:315–321

Ash D, Al-Qaisieh B, Bottomley D et al (2006) The correlation between D90 and outcome for I-125 seed implant monotherapy for localised prostate cancer. Radiother Oncol 79:185–189

Battermann JJ, Boon TA, Moerland RA (2004) Results of permanent prostate brachytherapy, 13 years of experience of a single institution. Radiother Oncol 71:23–28

Baxter NN, Tepper JE, Durham SB et al (2005) Increased risk of rectal cancer after prostate radiation: a population-based study. Gastroenterology 128:819–824
Beyer DC, Brachman DG (2000) Failure free survival following brachytherapy alone for prostate cancer: comparison with external beam radiotherapy. Radiother Oncol 57:263–267
Beyer DC, McKeough T, Thomas T et al (2005) Impact of short course hormonal therapy on overall and cancer specific survival after permanent prostate brachytherapy. Int J Radiat Oncol Biol Phys 61:1299–1305
Bhojani N, Capitanio U, Suardi N et al (2010) The rate of secondary malignancies after radical prostatectomy versus external beam radiation therapy for localized prostate cancer: a population based study on 17,845 patients. Int J Radiat Oncol Biol Phys 76:342–348
Bianco FJ Jr, Scardino PT, Stephenson AJ et al (2005) Long-term oncologic results of salvage radical prostatectomy for locally recurrent prostate cancer after radiation therapy. Int J Radiat Oncol Biol Phys 62:448–453
Bittner N, Merrick GS, Galbreath R et al (2008) Primary causes of death after permanent prostate brachytherapy. Int J Radiat Oncol Biol Phys 72:433–440
Blasko JC, Ragde H, Luse RW et al (1996) Should brachytherapy be considered a therapeutic option in localized prostate cancer? Urol Clin North Am 23:633–650
Blasko JC, Grimm PD, Sylvester JE et al (2000) The role of external beam radiotherapy with I-125/Pd-103 brachytherapy for prostate carcinoma. Radiother Oncol 57:273–278
Burri RJ, Ho AY, Forsythe K et al (2010) Young men have equivalent biochemical outcomes compared with older men after treatment with brachytherapy for prostate cancer. Int J Radiat Oncol Biol Phys 77:1315–1321
Coelho RF, Rocco B, Patel MB et al (2010) Retropubic, laparoscopic, and robot-assisted radical prostatectomy: a critical review of outcomes reported by high-volume centers. J Endourol 24:2003–2015
Colberg JW, Decker RH, Khan AM et al (2007) Surgery versus implant for early prostate cancer: results from a single institution, 1992–2005. Cancer J 13:229–232
Cosset JM, Pinillos-Ashton L, Mettler F (2004) The recommendations of the International Commission on Radiological Protection (ICRP) for high-dose-rate brachytherapy and for permanent prostatic implants. Cancer 8(Suppl 1):S50–S55
Cosset JM, Flam T, Thiounn N et al (2008) Selecting patients for exclusive permanent implant at prostate brachytherapy: The experience of the Paris Institute Curie/Cochin Hospital/Necker Hospital group on 809 patients. Int J Radiat Oncol Biol Phys 71:1042–1048
Critz FA, Levinson K (2004) 10-Year disease-free survival rates after simultaneous irradiation for prostate cancer with a focus on calculation methodology. J Urol 172:2232–2238
Crook J, Borg J, Evans A et al (2011) 10-Year experience with I-125 prostate brachytherapy at the princess Margaret hospital: results for 1,000 patients. Int J Radiat Oncol Biol Phys 80: 1323–1329
Datolli M, Walner K, True L et al (2007) Long-term outcomes after treatment with brachytherapy and supplemental conformal radiation for prostate cancer patients having intermediate and high-risk features. Cancer 110:551–555
Ficarra V, Novara G, Artibani W et al (2009) Retropubic, laparoscopic, and robot-assisted radical prostatectomy: a systematic review and cumulative analysis of comparative studies. Eur Urol 55:1037–1063
Fuks Z, Leibel SA, Wallner KE et al (1991) The effect of local control on metastatic dissemination in carcinoma of the prostate: long term results in patients treated with I-125 implantation. Int J Radiat Oncol Biol Phys 21:537–547
Grimm PD, Blasko JC, Sylvester JE et al (2001) 10-Year biochemical (prostate specific antigen) control of prostate cancer with 125-I brachytherapy. Int J Radiat Oncol Biol Phys 51:31–40
Henry AM, Al-Qaisieh B, Gould K et al (2010) Outcomes following iodine-125 monotherapy for localized prostate cancer: the results of leeds 10-year single-center brachytherapy experience. Int J Radiat Oncol Biol Phys 76:50–56

Hinnen KA, Battermann JJ, van Roermond JGH et al (2010a) Long term biochemical and survival outcome of 921 patients treated with I-125 permanent prostate brachytherapy. Int J Radiat Oncol Biol Phys 76:1433–1438

Hinnen KA, Moerland MA, Battermann JJ et al (2010b) Loose seeds versus stranded seeds in I-125 prostate brachytherapy: differences in clinical outcome. Radiother Oncol 96:30–33

Hinnen KA, Schaapveld M, van Vulpen M et al (2011) Prostate brachytherapy and second primary cancer risk: a competitive risk analysis. J Clin Oncol 29:4510–4515

Holm HH, Juul N, Pedersen JF et al (1983) Transperineal 125-Iodine seed implantation in prostatic cancer guided by transrectal ultra-sonography. J Urol 130:283–286

Isban I, Karakiewitcz PI, Vogel S et al (2010) Unilateral prostate cancer cannot be accurately predicted in low-risk patients. Int J Radiat Oncol Biol Phys 77:784–787

Jabbari S, Weinberg VK, Shinohara K et al (2010) Equivalent biochemical control and improved prostate-specific antigen nadir after permanent prostate seed implant brachytherapy versus high-dose three-dimensional conformal radiotherapy and high-dose conformal proton beam radiotherapy boost. Int J Radiat Oncol Biol Phys 76:36–42

Kuban DA, Levy L, Tucker S et al (2008) Long term failure patterns and survival in a randomized dose-escalation trial for prostate cancer. Who dies of disease. Int J Radiat Oncol Biol Phys 72:S93

Kupelian PA, Mohan DS, Lyons J et al (2000) Higher than standard doses (> or = 72 Gy) with or without androgen deprivation in the treatment of localized prostate cancer. Int J Radiat Oncol Biol Phys 46:567–574

Kupelian PA, Potters L, Khuntia D et al (2004) Radical prostatectomy, external beam radiotherapy <72 Gy, external beam radiotherapy ≥72 Gy, permanent seed implantation, or combined seeds/external beam radiotherapy for stage T1-T2 prostate cancer. Int J Radiat Oncol Biol Phys 58:25–33

Malcolm JB, Fabrizio MD, Barone BB et al (2010) Quality of life after open or robotic prostatectomy, cryoablation or brachytherapy for localized prostate cancer. J Urol 183: 1822–1829

Martinez A, Gonzalez J, Spencer W et al (2003) Conformal high dose rate brachytherapy improves biochemical control and cause specific survival in patients with prostate cancer and poor prognostic factors. J Urol 169:974–979

Merrick GS, Wallner KE, Butler WM et al (2006) Brachytherapy in men aged < or = 54 years with clinically localized prostate cancer. BJU Int 98:324–328

Merrick GS, Galbreath RW, Butler WM et al (2007) Primary Gleason pattern does not impact survival after permanent interstitial brachytherapy for Gleason score 7 prostate cancer. Cancer 110:289–296

Merrick GS, Wallner KE, Galbreath RW et al (2008) Prostate brachytherapy in men ≥75 years of age. Int J Radiat Oncol Biol Phys 72:415–420

Moerland MA, van Deursen MJ, Elias SG et al (2009) Decline of dose coverage between intraoperative planning and post implant dosimetry for I-125 permanent prostate brachytherapy: comparison between loose seeds and stranded implants. Radiother Oncol 91:202–206

Moman MR, van den Berg CAT, Boeken Kruger AE et al (2010) Focal salvage guided by T2-weighted and dynamic contrast-enhanced magnetic resonance imaging for prostate cancer recurrences. Int J Radiat Oncol Biol Phys 76:741–746

Morris WJ, Keyes M, Palma D et al (2009) Evaluation of dosimetric parameters and disease response after ^{125}Iodine transperineal brachytherapy for low- and intermediate-risk prostate cancer. Int J Radiat Oncol Biol Phys 73:1432–1438

Munro NP, Al-Qaisieh B, Bownes P et al (2010) Outcomes from Gleason 7, intermediate risk, localized prostate cancer treated with iodine-125 monotherapy over 10 years. Radiother Oncol 96:34–37

Piña AG-I, Crook J, Borg J, Ma C (2010) Biochemical disease-free rate and toxicity for men treated with iodine-125 prostate brachytherapy with D90 ≥ 180 Gy. Int J Radiat Oncol Biol Phys 78:422–427

Potters L, Cha C, Oshinsky G et al (1999) Risk profiles to predict PSA relapse-free survival for patients undergoing permanent prostate brachytherapy. Cancer Sci Am 5:301–306
Potters L, Torre T, Ashley R et al (2000) Examing the role of neoadjuvant androgen deprivation in patients undergoing prostate brachytherapy. J Clin Oncol 18(6):1187–1192
Potters L, Morgenstern C, Calugaru E et al (2005) 12-year outcomes following permanent prostate brachytherapy in patients with clinically localized prostate cancer. J Urol 173:1562–1566
Roeloffzen EMA, Lips IM, van Gellekom MPR et al (2010) Health-related quality of life up to 6 years after I-125 brachytherapy for early-stage prostate cancer. Int J Radiat Oncol Biol Phys 76:1054–1060
Shapiro EY, Rais-Bahrami S, Morgenstern C et al (2009) Long-term outcomes in younger men following permanent prostate brachytherapy. J Urol 181:1665–1671
Sharkey J, Canto A, Solc Z et al (2005) Pd^{103} brachytherapy versus radical prostatectomy in patients with clinically localized prostate cancer: A 12-year experience from a single group practice. Brachytherapy 4:33–44
Singh AK, Mashtare TL, McCloskey SA et al (2010) Increasing age and treatment modality are predictors for subsequent diagnosis of bladder cancer following prostate cancer diagnosis. Int J Radiat Oncol Biol Phys 78:1086–1094
Stock RG, Stone NN, Tabert A et al (1998) A dose–response study for I-125 prostate implants. Int J Radiat Oncol Biol Phys 41:101–108
Stone NN, Stock RG, Cesaretti JA et al (2010) Local control following permanent prostate brachytherapy: effect of high biologically effective dose on biopsy results and oncologic outcomes. Int J Radiat Oncol Biol Phys 76:355–360
Sylvester JE, Grimm PD, Blasko JC et al (2007) 15-Year biochemical relapse free survival in clinical stage T_1-T_3 prostate cancer following combined external beam radiotherapy and brachytherapy; Seattle experience. Int J Radiat Oncol Biol Phys 67:57–64
Taira AV, Merrick GS, Galbreath RW et al (2010) Natural history of clinically staged low- and intermediate-risk prostate cancer treated with monotherapeutic permanent interstitial brachytherapy. Int J Radiat Oncol Biol Phys 76:349–354
Taira AV, Merrick GS, Butler W et al (2011) Long-term outcome for clinically localized prostate cancer treated with permanent interstitial brachytherapy. Int J Radiat Oncol Biol Phys 79:1336–1342
Teh BS, Bastasch MD, Wheeler TM et al (2003) IMRT for prostate cancer: defining target volume based on correlated pathologic volume of disease. Int J Radiat Oncol Biol Phys 56:184–191
Tward JD, Lee CM, Pappas LM et al (2006) Survival of men with clinically localized prostate cancer treated with prostatectomy, brachytherapy, or no definitive treatment: impact of age at diagnosis. Cancer 107:2392–2400
Viani GA, Stefano EJ, Alfonso SL (2009) Higher-than-conventional radiation doses in localized prostate cancer treatment: a meta-analysis of randomized, controlled trials. Int J Rad Oncol Biol Phys 74:1405–1418
Wright JL, Salinas CA, Kolb DW et al (2009) Prostate cancer specific mortality and Gleason 7 disease differences in prostate cancer outcomes between cases with Gleason 4+3 and 3+4 tumors in a population based cohort. J Urol 182:2702–2707
Zelefsky MJ, Leibel SA, Gaudin PB et al (1998) Dose escalation with three-dimensional conformal radiation therapy affects the outcome in prostate cancer. Int J Radiat Oncol Biol Phys 41: 491–500
Zelefsky MJ, Kuban DA, Levy B et al (2007) Multi-institutional analysis of long-term outcome for stages T1-T2 prostate cancer treated with permanent seed implantation. Int J Radiat Oncol Biol Phys 67:327–333
Zelefsky MJ, Yamada Y, Fuks Z et al (2008) Long-term results of conformal radiotherapy for prostate cancer: impact of dose escalation on biochemical tumour control and distant metastases-free survival outcomes. Int J Radiat Oncol Biol Phys 70:67–74

Results of HDR Prostate Brachytherapy Treatments

16

György Kovács

There are only a few phase III studies in the literature reporting on prostate HDR brachytherapy; most of the publications deal with retrospective monoinstitutional results or with the statistical analysis of multi-institutional data pools. Although results of prospective nonrandomized studies or retrospective data analysis may be influenced by potential selection bias and inconsistent data analysis, careful presentation can help to understand basic and meaningful trends within the specialty. Due to the fact that initial PSA level, Gleason score, and tumor stage each can affect results independently, the influence of these factors can change reported results if they are not equally distributed in the compared cohorts. Vicini et al. (2000) investigated the influence of predefined prognostic categories in the outcome following various forms of treatment from multiple institutions and stated that 5-year PSA results were similar for patients in low-risk and intermediate-risk groups, regardless of the form of therapy. However, different rates of 5-year bNED (biochemical no evidence of disease) were observed for the same form of the treatment at the same institution, depending on the number of prognostic factors to define specific prognostic groups. Also, significant influence of reported follow-up intervals on the outcome data was observed. Additionally, Lu demonstrated erroneous (15–30 %) interpretations of bNED data related to differences in follow-up (Lu 2000). Furthermore, definition of risk categories is different in different institutions (Table 16.1). It is also important to notice that postimplant, the course of PSA can decrease over a long time period – Deger et al. observed up to 5 years reaching values under 1.0 ng/ml (Deger et al. 2005).

Furthermore, there are also different PSA failure definitions presented in the literature. As prostate cancer treatment failure definition, PSA nadir plus 2 ng/ml or twice PSA raise >0.5 ng/ml has the best correlation with long-term clinical outcome

G. Kovács
Interdisciplinary Brachytherapy Unit, University of Lübeck,
Ratzeburger Allee 160, D-23562 Lübeck, Germany
e-mail: kovacsluebeck@gmail.com

G. Kovács, P. Hoskin (eds.), *Interstitial Prostate Brachytherapy*,
DOI 10.1007/978-3-642-36499-0_16, © Springer-Verlag Berlin Heidelberg 2013

Table 16.1 Different risk group specifications in different institutions

Institution	Low risk	Intermediate risk	High risk
Seattle	PSA ≤ 10	PSA > 10	2 *or* 3 factors
	and GS 2–6	*or* GS ≥ 7	
	and T1c–T2b	*or* ≥ T2b	
Mount Sinai	PSA ≤ 10	PSA 10–20	2 or 3 factors
	and GS 2–6	*or* GS = 7	*and/or* PSA > 20
	and T1a–T2a	*or* T2b	*and/or* GS 8–10
			and/or ≥ T2c
Boston	Idem	Idem, *and/or*	Idem

(Demanes et al. 2005). Last but not least, the involvement of brachytherapy in the treatment schedule of localized prostate cancer seems to lead to a lower risk of second malignancies following prostate radiotherapy (Huang et al. 2011).

High-dose-rate brachytherapy can be used for prostate cancer as:

1. Local dose escalation (boost) complementary to external beam radiation
2. HDR monotherapy
3. Salvage treatment

16.1 HDR Brachytherapy as Local Dose Escalation (Boost) Complementary to External Beam Radiation

When brachytherapy is administered complementary to external beam radiation, local dose escalation can be performed with less normal tissue involved in low-dose areas, and this is of interest in modern radiotherapy schedules (Tubiana 2005). Additionally, if brachytherapy is incorporated in the time schedule of external beam radiotherapy (Kovacs et al. 1999), overall treatment time of dose-escalated treatments can be shortened. Although some authors have suggested no strong clinical relevance of total treatment time in prostate cancer radiotherapy outcome (Lai et al. 1991), others have stated that overall treatment time and dose are significant determinants of the outcome of radiotherapy in low- and intermediate-risk patients treated to 70 Gy or higher (Thames et al. 2010). The authors also suggest that meaningful improvements in outcome may be reached by modest increases of total dose and decreases in overall treatment time.

HDR brachytherapy boost complementary to external beam was introduced in clinical practice at the end of the 1980s – shortly after the successful use of transrectal ultrasound (TRUS) in prostate seed implants (Brindle et al. 1989; Bertermann and Brix 1990; Kovacs et al. 1995). The advantage of the use of HDR brachytherapy boost compared to external beam local dose escalation is proven in a phase III trial (Corner et al. 2008), and many other groups have found favorable outcomes in retrospective investigations of large cohorts (Stromberg et al. 1997; Galalae et al. 2004; Pellizzon et al. 2003; Aström et al. 2005; Deger et al. 2005; Demanes et al. 2005).

There is evidence in the literature indicating no relevant influence of <6 months anti-androgen treatment (ADT) combined with radiotherapy when an HDR boost was used (Martinez et al. 2005) and no effect on the dose dependence of outcome in different risk groups (Galalae et al. 2006).

The question whether additional pelvic lymph node radiation in patients who have a >15 % chance for positive lymph nodes improves the outcome is still controversial (Kovacs and Galalae 2003; Hong et al. 2006).

From the clinical results, HDR brachytherapy combined with external beam radiation offers the best possible outcome for localized high-risk prostate cancer patients; however, there is no investigation in the literature comparing HDR boost with planned surgical resection and complementary adjuvant external beam radiotherapy. Patients in the low- or intermediate-risk groups may not be disadvantaged by brachytherapy as monotherapy; however, careful selection in the intermediate-risk group is advisory (Grimm et al. 2012). Experienced teams report 5-year biochemical relapse-free rates (bNED) about 70, 80, and 90 % in the high-, intermediate-, and low-risk groups (Kovacs et al. 1995; Martinez et al. 2001; Pellizzon et al. 2008; Demanes et al. 2011; Deger et al. 2005; Zwahlen et al. 2010).

Common significant acute toxicities (G2+) are small bowel (15–20 %), bladder reactions (40–50 %), and rectal discharge (35–40 %). Transient hematuria can be observed in 10–17 %.

Most frequent late severe toxicities are urethral stricture (4–8 %), gastrointestinal (2–7 %), and erectile dysfunction (20–60 % at 2 years); however, because erectile function is only one component of sexual function, it is necessary to assess sexual desire, satisfaction, intercourse frequency, and other factors when evaluating sexual function (Incrocci et al. 2002).

A relationship between irradiated volume and urethral toxicity has been demonstrated (Akimoto et al. 2005), and there is an increase in risk with increasing follow-up time (Aström et al. 2005). Transurethral prostate resection (TURP) should not be performed following brachytherapy to avoid frequent incontinence.

A significant difference in the overall quality of life in favor of the brachytherapy arm is seen at 12 weeks after treatment in randomized study comparing external beam with brachytherapy boost (Hoskin et al. 2007).

16.2 HDR Monotherapy

HDR monotherapy was first reported by the group from Osaka (Yoshioka et al. 2000). In recent years, different clinical trials were performed to prove the feasibility and outcome of HDR monotherapy in low-risk prostate cancer (Martinez et al. 2001; Martin et al. 2004; Corner et al. 2008). Some authors suggest potential advantages for HDR monotherapy over seed implants (Wang et al. 2006).

Hypofractionation in the treatment of prostate carcinoma is now established. HDR brachytherapy offers the most elegant and economical local irradiation method for this entity; however, no consensus exists in the literature regarding dose, fractionation, technique, and normal tissue constrains. Common schedules include

31.5–54 Gy total dose delivered in three to nine fractions of 8.5–10.5 Gy (one or two implantations) where total treatment time varies between 2 and 5 days (Martinez et al. 2001; Martin et al. 2004; Corner et al. 2008; Yoshioka et al. 2010; Demanes et al. 2011).

Five-year PSA failure rates are presented in the literature as 79–93 % with grade 2 or higher toxicity rates of 10–14 %.

The only matched-pair comparison of seed versus HDR monotherapy suggested an advantage for HDR in terms of both acute and late toxicity (Grills et al. 2004) which has subsequently been challenged (Stock 2006; Sylvester 2006). There is a relative lack of comparative and prospective clinical investigations – therefore, longer follow-up and prospective studies are needed to have more consistent data on this field.

16.3 HDR Brachytherapy as a Salvage Treatment

Published data for PSA failure after salvage radiotherapy report bNED survival rates of 10–77 % at up to 5-year follow-up. Analyzed patient cohorts are usually very inhomogeneous, and there is usually a lack of differentiation between microscopic and macroscopic recurrent disease. There is less experience with HDR brachytherapy as a salvage method in the literature and a lack of prospective investigations. In general, HDR technology is ideal for a high-quality treatment of local recurrences since the steep dose falloff of the source is the best prevention of normal tissue toxicity.

There are different salvage situations in the clinical practice; however, the majority deals only with case descriptions or small cohort feasibility series:

(a) *Local recurrences following previous radiotherapy with external beam or brachytherapy:*
Early publications state the feasibility of salvage HDR brachytherapy. Since recent papers deal with fewer than ten patients, it is hard to judge the results (Lee et al. 2007; Tarp et al. 2008). Ongoing clinical trials like the NCT00604526 trial will offer more information in the near future.

(b) *Local recurrences following previous radical prostatectomy:*
Macroscopic local recurrences require higher doses compared with microscopic disease. Therefore, local recurrences following radical prostatectomy need local dose escalation if macroscopic tumor is detectable. There is early experience in the literature showing the feasibility and favorable outcome of fractionated salvage HDR brachytherapy alone or complementary to external beam radiation in this setting (Niehoff et al. 2005).

References

Akimoto T, Ito K, Saitoh J-I et al (2005) Acute genitourinary toxicity after high-dose-rate (HDR) brachytherapy combined with hypofractionated external-beam radiation therapy for localized prostate cancer: correlation between the urethral dose in HDR brachytherapy and the severity of acute genitourinary toxicity. Int J Radiat Oncol Biol Phys 62(2):463–471

Aström L, Pedersen D, Mercke C et al (2005) Long term outcome of high dose rate brachytherapy in radiotherapy of localized prostate cancer. Radiother Oncol 74:157–161

Bertermann H, Brix F (1990) Ultrasonically guided interstitial high dose brachytherapy with iridium-192: technique and preliminary results in locally confined prostate cancer. In: Martinez AA, Orton CG, Mould RF (eds) Brachytherapy HDR & LDR. Proceedings brachytherapy meeting remote afterloading: state of the art, Nucletron USA, Columbia. pp 281–301

Brindle JS, Martinez A, Schray M et al (1989) Pelvic lymphadenopathy and transperineal interstitial implantation of Ir192 combined with external beam radiotherapy for bulky stage C prostate carcinoma. Int J Radiat Oncol Biol Phys 17:1063–1066

Corner C, Rojas AM, Bryant L et al (2008) A phase II study of high-dose-rate afterloading brachytherapy as monotherapy for the treatment of localized prostate cancer. Int J Radiat Oncol Biol Phys 72:441–446

Deger S, Boehmer D, Roigas J et al (2005) High-dose-rate brachytherapy with conformal radiation therapy of localized prostate cancer. Eur Urol 47:441–448

Demanes DJ, Rodriguez RR, Schour L et al (2005) High-dose-rate intensity modulated brachytherapy with external beam radiotherapy for prostate cancer: California endocurietherapy's 10-year results. Int J Radiat Oncol Biol Phys 61:1306–1316

Demanes DJ, Martinez AA, Ghilezan M et al (2011) High-dose-rate monotherapy: safe and effective brachytherapy for patients with localized prostate cancer. Int J Radiat Oncol Biol Phys 81(5):1286–1292

Galalae RM, Martinez AA, Mate T et al (2004) Long-term outcome by risk factors using conformal high-dose-rate brachytherapy (HDR-BT) boost with or without neoadjuvant androgen suppression for localized prostate cancer. Int J Radiat Oncol Biol Phys 58:1048–1055

Galalae RM, Martinez A, Nürnberg N et al (2006) Hypofractionated conformal HDR brachytherapy in hormone naive men with localized prostate cancer. Is escalation to very high biologically equivalent dose beneficial in all prognostic risk groups? Strahlenther Onkol 182:135–141

Grills IS, Martinez AA, Hollander M et al (2004) High dose rate brachytherapy as prostate cancer monotherapy reduces toxicity compared to low dose rate palladium seeds. J Urol 171: 1098–1104

Grimm P, Billiet I, Bostwick D et al (2012) Comparative analysis of prostate specific antigen free survival outcomes for patients with low-, intermediate and high risk prostate cancer by radical therapy. Results from the Prostate Cancer Results Study Group. BJU Int 109(Suppl 1):21–29

Hong TS, Tome WA, Jaradat H et al (2006) Pelvic nodal dose escalation with prostate hypofractionation using conformal avoidance defined (H-CAD) intensity modulated radiation therapy. Acta Oncol 45:717–727

Hoskin P, Motohashi K, Bownes P et al (2007) High dose rate brachytherapy in combination with external beam radiotherapy in the radical treatment of prostate cancer: initial results of a randomized phase three trial. Radiother Oncol 84:114–120

Huang J, Kestin LL, Ye H et al (2011) Analysis of second malignancies after modern radiotherapy versus prostatectomy for localized prostate cancer. Radiother Oncol 98:81–86

Incrocci L, Slob AK, Levendag P (2002) Sexual (dys)function after radiotherapy for prostate cancer: a review. Int J Radiat Oncol Biol Phys 52(3):681–693

Kovacs G, Galalae RM (2003) Fractionated perineal high-dose-rate temporary brachytherapy combined with external beam radiation in the treatment of localized prostate cancer: is lymph node sampling necessary? Cancer Radiother 7:100–106

Kovacs G, Galalae R, With B et al (1995) Optimierung der interstitiellen Brachytherapie bei lokal begrenzten Prostatakarzinom mit einer neuen Implantationstechnik. Strahlenther Onkol 171(12):685–688

Kovacs G, Galalae R, Loch T et al (1999) Prostate preservation by combined external beam and HDR brachytherapy by nodal negative prostate cancer. Strahlenther Onkol 175(Suppl II):87–89

Lai PP, Pilepich MV, Krall JM et al (1991) The effect of overall treatment time on the outcome of definitive radiotherapy for localized prostate carcinoma: the radiation therapy oncology group 75–06 and 77–06 experiences. Int J Radiat Oncol Biol Phys 21:925–933

Lee B, Shinohara K, Weinberg V et al (2007) Feasibility of high-dose-rate brachytherapy salvage for local prostate cancer recurrence after radiotherapy: the University of California-San Francisco experience. Int J Radiat Oncol Biol Phys 67(4):1106–1112

Lu J (2000) Statistical aspects of evaluating treatment and prognostic factors for clinically localized prostate cancer. Semin Urol Oncol 18:83–92

Martin T, Baltas D, Kurek R et al (2004) 3-D conformal HDR brachytherapy as monotherapy for localized prostate cancer: a pilot study. Strahlenther Onkol 180:225–232

Martinez A, Pataki I, Edmundson G et al (2001) Phase II prospective study of the use of conformal high-dose-rate brachytherapy as monotherapy for the treatment of favorable stage prostate cancer: a feasibility report. Int J Radiat Oncol Biol Phys 49:61–69

Martinez AA, Demanes JD, Galalae RM et al (2005) Lack of benefit from a short course of androgen deprivation for unfavorable prostate cancer patients treated with an accelerated hypofractionated regime. Int J Radiat Oncol Biol Phys 62(5):1322–1331. doi:10.1016/j.ijropb.2004.12.053

Niehoff P, Loch T, Nürnberg N et al (2005) Feasibility and preliminary outcome of salvage combined HDR brachytherapy and External Beam Radiotherapy (EBRT) for local recurrences after radical prostatectomy. Brachytherapy 4:141–145

Pellizzon AC, Nadalin W, Salvajoli JV et al (2003) Results of high-dose-rate afterloading brachytherapy boost to conventional external beam radiation therapy for initial and locally advanced prostate cancer. Radiother Oncol 66:167–172

Pellizzon ACA, Salvajoli J, Novaes J et al (2008) The relationship between the biochemical control outcomes and quality of planning of high-dose-rate brachytherapy as boost to external beam radiotherapy for locally and locally advanced prostate cancer using the RTOG-ASTRO phoenix definition. Int J Med Sci 5(3):113–120

Stock RG (2006) High-dose-rate versus low-dose-rate monotherapy in the treatment of localized prostate cancer: the case of low-dose-rate monotherapy. Brachytherapy 5:5–6

Stromberg JS, Martinez AA, Horwitz EM et al (1997) Conformal high-dose-rate iridium-192 boost brachytherapy in locally advanced prostate cancer: superior prostate specific antigen response compared with external beam treatment. Cancer J Sci Am 3(36):346–352

Sylvester JE (2006) High-dose-rate versus low-dose-rate monotherapy in the treatment of localized prostate cancer. The case for high-dose-rate monotherapy: an up and coming treatment option for low-risk prostate cancer. Brachytherapy 5:1–4

Tarp M, Hardcare M, Bennet R et al (2008) Prostate high-dose-rate brachytherapy as a salvage treatment of local failure after previous external or permanent seed irradiation for prostate cancer. Brachytherapy 7:231–236

Thames HD, Kuban D, Lavy LB et al (2010) The role of overall treatment time in the outcome of radiotherapy of prostate cancer: an analysis of biochemical failure in 4839 men treated between 1987 and 1995. Radiother Oncol 96(1):6–12

Tubiana M (2005) Dose-effect relationship and estimation of the carcinogenic effects of low doses of ionizing radiation: the joint report of the Academie des Sciences(Paris) and of the Academie Nationale de Medicine. Int J Radiat Oncol Biol Phys 63(2):317–319

Vicini FA, Martinez A, Hanks G et al (2002) An interinstitutional and interspeciality comparison of treatment outcome data for patients with prostate carcinoma based on predefined prognostic categories and minimum follow-up. Cancer 95:2126–2135

Wang Y, Sankreacha R, Al-Hebshi A et al (2006) Comparative study of dosimetry between high-dose-rate and permanent prostate implant brachytherapies in patients with prostate adenocarcinoma. Brachytherapy 5:251–255

Yoshioka Y, Nose T, Yoshida K et al (2000) High-dose-rate interstitial brachytherapy as a monotherapy for localized prostate cancer: treatment description and preliminary results of a phase I/II clinical trial. Int J Radiat Oncol Biol Phys 48:675–681

Yoshioka Y, Konishi K, Sumida I et al (2011) Monotherapeutic high-dose-rate brachytherapy for prostate cancer: five-year results of an extreme hypofractionation regimen with 54 Gy in nine fractions. Int J Radiat Oncol Biol Phys 80(2):469–475. doi:10.1016/j.ijrobp.2010.02.013

Zwahlen DR, Andrianopoulos N, Matheson B et al (2010) High-dose-rate brachytherapy in combination with conformal external beam radiotherapy in the treatment of prostate cancer. Brachytherapy 9:27–35

Results of PDR Treatments: The AMC PDR Experience

17

B.R. Pieters and E.D. Geijsen

17.1 Introduction

Pulsed-dose-rate (PDR) brachytherapy has been introduced for the treatment of intermediate-/high-risk prostate cancer. A dose of 28–35 Gy PDR brachytherapy is given as a boost to external beam radiotherapy of 46–50 Gy. The reported 5-year biochemical-free survival is 85.6–89.5 %. The incidence of grade 2–3 rectal complications is limited (12–15 %). The incidence of grade 2–3 urinary toxicity is 15–26.9 %. The figures on biochemical control and late toxicity are similar to HDR experience. PDR for prostate brachytherapy can be considered as an alternative to HDR brachytherapy.

17.2 Indication

The AMC has a long tradition of I-125 implants for low-risk prostate cancer (Blank et al. 2000). In accordance with the GEC/ESTRO-EAU recommendations, I-125 implants are not considered a treatment option for high-risk disease, e.g., T3 or poorly differentiated tumors (Kovács et al. 2005). The reason that these patients do relatively poorly with an I-125 implant only is because there is an increased risk of extracapsular invasion and (micro)metastases. To cover the presumed extracapsular invasion for this category of patients in the Academic Medical Center (AMC) in Amsterdam, a protocol was started in 2002 of brachytherapy in addition to external beam radiotherapy (EBRT). For brachytherapy a single-source modality was chosen for optimization of the dose distribution, and because of the excellent results obtained with low-dose rate I-125 implants and continued debate around the α/β ratio for prostate cancer, PDR was considered the optimal solution.

B.R. Pieters (✉) • E.D. Geijsen
Radiation Oncology, Academic Medical Center/University of Amsterdam,
Meibergdreef 9, 1105 AZ Amsterdam, Netherlands
e-mail: b.r.pieters@amc.uva.nl

G. Kovács, P. Hoskin (eds.), *Interstitial Prostate Brachytherapy*,
DOI 10.1007/978-3-642-36499-0_17, © Springer-Verlag Berlin Heidelberg 2013

17.3 Target Volume and Dose

The target volume for external beam radiotherapy was considered in the lower pelvis with the cranial border at the lower end of the sacroiliac joints. In 2009 the elective external beam portals were adapted to treat only the prostate and seminal vesicles with a 1 cm margin. This modification was introduced because of the lack of evidence from the literature that pelvic radiotherapy was of benefit in improving overall survival in prostate cancer treatment.

In the period of 2002–2007, 106 patients with prostate cancer were treated with the combination of EBRT and PDR brachytherapy boost. The majority of the patients were classified as intermediate- ($N=38$) or high-risk ($N=66$) disease in accordance to the National Cancer Comprehensive Network criteria. Nineteen patients of this cohort have used hormonal therapy. However, only ten of them have used hormonal therapy for a longer period than 6 months.

The EBRT dose was 46 Gy in 2 Gy daily fractions. At the start of the protocol, a PDR boost dose of 24 pulses of 1.04 Gy with 2.2 h interval was prescribed. The prescription was at the periphery of the prostate resulting in a clinical target volume (CTV) V100 of 95 %. Because at that time there was no experience with PDR brachytherapy for prostate cancer, the chosen dose was rather low. The EQD2 for an α/β ratio of 3 Gy or 1.5 Gy was 71.5 and 71.8 Gy, respectively. The brachytherapy dose was later increased to 24 times 1.1 Gy with a 2 h interval and finally to 1.2 Gy per pulse when more experience was gained with this treatment modality, and the dose-escalation treatment results of EBRT trials became available. In the latter schedule, the EQD2 for an α/β ratio of 3 and 1.5 Gy was 78.5 and 80.0 Gy, respectively.

17.4 Outcome

At a median follow-up time of 34.6 months, the 3- and 5-year biochemical relapse-free survival was 92.8 and 89.5 %, respectively (Fig. 17.1) (Pieters et al. 2011). Two patients died because of a second malignancy resulting in an overall survival at 3 and 5 years of 99 and 96 %, respectively (Fig. 17.2). These results are similar to patients with intermediate- and high-risk disease treated with EBRT combined with high-dose rate (HDR) brachytherapy (Åström et al. 2005; Kälkner et al. 2007; Phan et al. 2007; Vargas et al. 2006).

Acute toxicity was well tolerated. The International Prostate Symptom Score (IPSS) increased from a mean baseline value of 7 to a mean of 18 at 3 weeks after brachytherapy. By 6–12 weeks after treatment, the IPSS returned to baseline values.

PDR prostate brachytherapy resulted in a low incidence of late gastrointestinal (GI) toxicity in the treated cohort (Pieters et al. 2011). No late GI grade 3 or more events were observed. Late GI grade 2 toxicity at 3 and 5 years was 5.3 and 12.0 %, respectively. The most frequent GI complaints were fecal incontinence and rectal bleeding. Late genitourinary (GU) toxicity was more common. The 3- and 5-year incidence of late GU grade 2 or higher toxicity was 18.7 and 26.9 %, respectively.

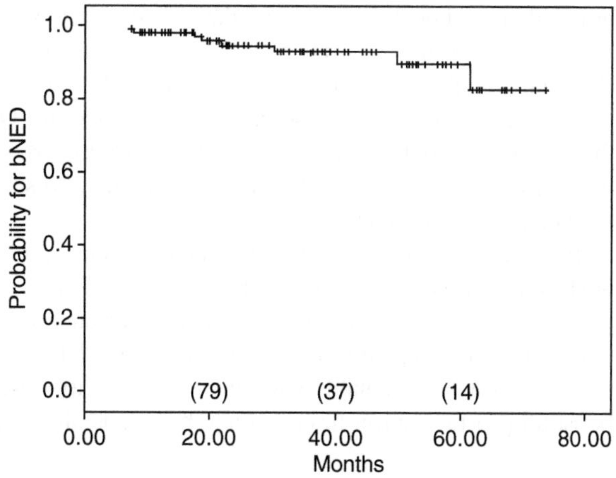

Fig. 17.1 Survival probability for biochemical non-evidence for disease (Reprinted from Pieters et al. (2011) with permission from Elsevier)

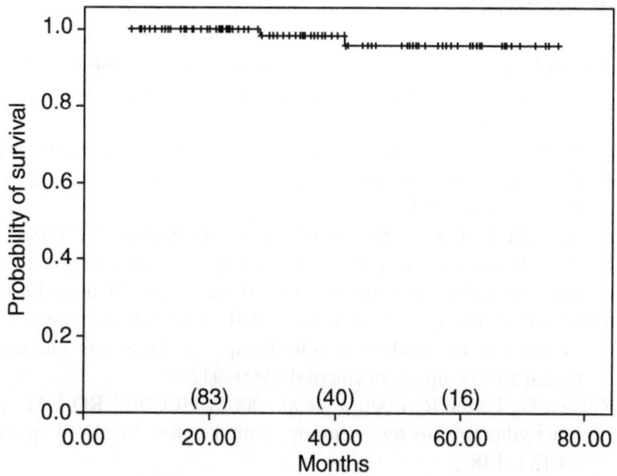

Fig. 17.2 Survival probability for overall survival (Reprinted from Pieters et al. (2011) with permission from Elsevier)

The most frequent GU toxicity was increased voiding frequency, although urethral stricture was also observed in two patients. These results are comparable to the experience with HDR boost (Åström et al. 2005; Phan et al. 2007; Vargas et al. 2006). Erectile function was preserved in 83 % of men that have not used hormonal therapy as part of their initial therapy.

With longer observation, it is obvious that toxicity can resolve with time, either spontaneously or after medical intervention. After 36 months of follow-up, no grade 3 GU toxicity was observed anymore (Pieters et al. 2010). The most severe sequelae were urethra strictures which were successfully treated with bladder neck incision or transurethral resection of the prostate. Erectile function can be improved with the use of phosphodiesterase type 5 inhibitors, and in the long-term a trend for an increase in impotency rate was observed.

At the university hospitals of Erlangen, 130 patients have been treated by an external beam radiotherapy (50.4 Gy) and a PDR brachytherapy boost. The PDR brachytherapy boost dose was escalated from 25 to 35 Gy. The 5-year biochemical nonevidence of disease was 85.6 %. Only one patient developed a grade 3 proctitis (Lettmaier et al. 2012). There were no cases of grade 3 urinary toxicity.

At Erlangen they have also gained experience in PDR brachytherapy as monotherapy for low-risk prostate cancer. The dose used was 65–70 Gy in 0.65–0.7 Gy/pulse. In a short median follow-up of 8 months, no severe grade 3–4 toxicity was seen (Geiger et al. 2008).

From the experience in Amsterdam and Erlangen, it has been found that an external beam radiotherapy and a PDR brachytherapy boost can result in a high dose to the prostate gland. The results for biochemical recurrence and overall survival were shown to be good and comparable to the results found in HDR boost studies. As expected the toxicity rate was acceptable with more GU toxicity observed than GI toxicity. PDR boost for prostate cancer has proven to be a good alternative to HDR boost.

References

Åström L, Pedersen D, Mercke C, Holmang S, Johansson KA (2005) Long-term outcome of high dose rate brachytherapy in radiotherapy of localised prostate cancer. Radiother Oncol 74:157–161

Blank LE, Gonzalez Gonzalez D, de Reijke TM, Dabhoiwala NF, Koedooder K (2000) Brachytherapy with transperineal (125)Iodine seeds for localized prostate cancer. Radiother Oncol 57:307–313

Geiger MH, Lotter M, Seeger AR, Sauer R, Strnad V (2008) Die alleinige interstitielle Pulse-Dose-Rate Brachytherapie in der Therapie des lokal begrenzten Prostatakarzinoms. Erfahungen hinsichtlich Durchführbarkeit und Toxizität. Strahlenther Onkol 184(Suppl 1):172

Kälkner KM, Wahlgren T, Ryberg M et al (2007) Clinical outcome in patients with prostate cancer treated with external beam radiotherapy and high dose-rate iridium 192 brachytherapy boost: a 6-year follow-up. Acta Oncol 46:909–917

Kovács G, Pötter R, Loch T et al (2005) GEC/ESTRO-EAU recommendations on temporary brachytherapy using stepping sources for localised prostate cancer. Radiother Oncol 74:137–148

Lettmaier S, Lotter M, Kreppner S, Strnad A, Fietkau R, Strnad V (2012) Long term results of a prospective dose escalation phase-II trial: interstitial pulsed-dose-rate brachytherapy as a boost for intermediate- and high-risk prostate cancer. Radiother Oncol 104:181–186

Phan TP, Syed AMN, Puthawala A, Sharma A, Khan F (2007) High dose rate brachytherapy as a boost for the treatment of localized prostate cancer. J Urol 177:123–127

Pieters BR, Rezaie E, Geijsen ED et al (2011) Development of late toxicity and IPSS resolution after external beam radiotherapy combined with PDR brachytherapy for prostate cancer. Int J Radiat Oncol Biol Phys 81:758–764

Pieters BR, Geijsen ED, Koedooder K et al (2011) Treatment results of PDR brachytherapy combined with external beam radiotherapy in 106 patients with intermediate- to high-risk prostate cancer. Int J Radiat Oncol Biol Phys 79:1037–1042

Vargas CE, Martinez AA, Boike TP et al (2006) High-dose irradiation for prostate cancer via a high-dose-rate brachytherapy boost: results of a phase I to II study. Int J Radiat Oncol Biol Phys 66:416–423

18 Incidence and Prognostic Factors for Complications After Permanent Interstitial Brachytherapy

Stefan Machtens

18.1 Introduction

Recent improvements in radioactive seed implantation techniques have established prostate brachytherapy as a highly effective treatment modality for localized prostate cancer, with long-term local and biochemical control similar to outcomes observed after radical prostatectomy and external beam radiation therapy (Grimm et al. 2012).

Nevertheless the patient undergoing this procedure and the provider offering this therapy have to be aware of the incidence and the prognostic factors for these complications.

Complications can be divided into acute and late events. These can occur in the urinary or gastrointestinal tract. Several prognostic factors have been identified in the recent past so that the knowledge about these factors may help to avoid these side effects.

The advantages of prostate brachytherapy compared with radical prostatectomy and external beam radiation therapy include lower rates of incontinence and sexual dysfunction and both early and late radiation proctitis (Machtens et al. 2006a, b).

18.2 Acute Rectal Side Effects

Acute radiation proctitis typically occurs in the first 6 months after implantation during which the majority of the radiation dose is delivered by the seeds.

It is characterized by intermittent rectal bleeding, diarrhea, mucous discharge, abdominal pain, and constipation.

S. Machtens
Department Urology and Paediatric Urology,
Marienkrankenhaus gGmbH, Dr. Robert-Koch-Str.16,
51465 Bergisch Gladbach, Germany
e-mail: stefan.machtens@mkh-bgl.de

G. Kovács, P. Hoskin (eds.), *Interstitial Prostate Brachytherapy*,
DOI 10.1007/978-3-642-36499-0_18,

Table 18.1 Modified Radiation Therapy Oncology Group rectal toxicity scale

	Characterization	Symptoms
Grade 1	Mild and self-limiting	Minimal, infrequent bleeding or clear mucous discharge, rectal discomfort not requiring analgesics, loose stools not requiring medications
Grade 2	Managed conservatively, lifestyle not affected	Intermittent rectal bleeding not requiring regular use of pads, erythema of rectal lining on proctoscopy, diarrhea requiring medications
Grade 3	Severe, alters patient lifestyle	Rectal bleeding requiring regular use of pads and minor surgical intervention, rectal pain requiring narcotics, rectal ulceration
Grade 4	Life-threatening and disabling	Bowel obstruction, fistula formation, bleeding requiring hospitalization, surgical intervention required

The most common symptom is diarrhea which can affect up to 75 % of patients but normally resolves shortly after onset. The physiological changes seen on sigmoidoscopy include inflammation, edema, and vulnerable rectal mucosa.

Conservative measures to improve these symptoms are described in a later chapter of the book.

18.3 Chronic Rectal Side Effects

In contrast to acute radiation proctitis, chronic rectal side effects may take up to 2 years to develop and are not associated with the occurrence of acute proctitis.

The typical time for manifestation of late or chronic radiation proctitis is between 6 months and 2 years. The common symptoms of chronic radiation proctitis are rectal bleeding, rectal urgency, rectal incontinence, pain, strictures, and mucous discharge. Severe complications in form of rectal fistulas and perforations are rare.

Chronic radiation proctitis is distinguished from acute radiation proctitis by changes including telangiectasias and small vasculopathy in the submucosa. These changes reduce bowel vascularization resulting in increased risk of mucosal ulceration and submucosal fibrosis. Biopsies taken from this altered rectal area can induce the development of rectal fistulas and have to be avoided during colonoscopy which are often undertaken to investigate rectal bleeding.

Late rectal adverse events can either be classified by the scoring criteria of the Radiation Therapy Oncology Group (RTOG) or by the National Cancer Institute Common Terminology Criteria for Adverse Events (version 3.0) (CTCAE).

Most commonly the RTOG scoring criteria are applied, as shown in Table 18.1.

18.4 Incidence of Rectal Side Effects

In general rectal outcomes have improved over the last years as experience with brachytherapy has grown worldwide. The use of ultrasound and intraoperative dynamic planning systems has improved seed placement and dosimetry. There is sufficient evidence that the probability of developing radiation proctitis increases with higher radiation dose to the rectal wall. The goal of improving tumor control must therefore be balanced against the increased risk of complications. The knowledge of non-brachytherapists in managing rectal side effects has improved.

In early experiences, the late rectal morbidity was relatively high and the rate of rectal ulceration was described as up to 10 %. In subsequent series, this side effect is seen in approximately 1 % of all patients.

The incidence of rectal fistulas in early series varied from 1 to 7 %, whereas in more contemporary series, the rates have been between 0 and 1 %.

The incidence of fistula or ulcer formation is higher in patients undergoing a rectal biopsy or a rectal coagulation in the diagnostic or therapeutic work-up of rectal bleeding. Therefore these procedures have to be regarded as contraindicated after a permanent interstitial brachytherapy (Gelblum et al. 2000; Theodorescu et al. 2000).

Improvements in brachytherapy techniques have also lowered the incidence of less severe side effects. Whereas rectal side effects that could be managed with conservative therapy (grade 2) ranged from 6 to 21 % in early reports, these rates have dropped to an average of 3.7–18 % in contemporary series (Beyer et al. 1997; Zeitlin et al. 1998).

Rectal bleeding accounted for the majority of late gastrointestinal complications, and approximately 80 % of all episodes occurred in the first 2 years after implantation.

The incidence of chronic radiation proctitis requiring medical management ranges between 3.7 and 10.4 %, and the incidence of chronic radiation proctitis requiring endoscopic treatment is <1 %.

18.5 Prognostic Factors for Radiation Proctitis

Several studies have identified prognostic factors for the development of radiation proctitis. There is sufficient evidence for a correlation between implant technique and the incidence of radiation proctitis. According to this evidence, the new ABS guidelines recommend to limit the intraoperative dose applied to <1 cc of the rectum to 100 % of the prescription dose ($RV_{100} < 1$ cc). On postimplant dosimetry typically performed 30 days after the implant, this dose should not cover more than 1.3 cc of the rectum ($RV_{100} < 1.3$ cc) (Davis et al. 2012).

Recent publications have demonstrated that prolonged catheterization and larger prostate size were associated with a higher rate of any acute rectal toxicity. Late rectal toxicity RTOG ≥ 2 was associated with higher rectal dose, acute rectal toxicity, and effects of the learning curve. Severe late rectal toxicity (RTOG ≥ 3) was rare (0.9–0.2 %) (Keyes et al. 2012).

There are also findings which implicate the possibility of intrinsic radiosensitivity, where genetic predisposition and different pathways of DNA repair mechanisms may contribute to enhanced toxicity in some patients. The mutation of the ataxia telangiectasia gene (ATM) was associated with a higher incidence of rectal bleeding. Presence of the ATM gene was the only independent predictor for rectal bleeding.

As the development of a rectal fistula is the most severe complication described, it is of special importance to place the most posterior row of the seeds far enough away from the rectum to keep the VR_{100} as low as possible.

The correlation between the distances of the most posterior seeds in relation to the edge of the rectum has been well described, demonstrating a 25 % rise in the maximum dose applied to the anterior rectal wall if the dose margin for seeds being placed 3 mm away from the rectum is decreased to 1–3 mm, raising the incidence of late rectal toxicity from 1 to 3 % (Waterman et al. 2003).

Other rectal dosimetric quantifiers were described as predictive for late rectal toxicities like %V25>25 % (25 % of the volume of the rectal wall receives more than 25 % of the prescription dose) and %V10>40 % (10 % volume of the rectal wall receives more than 40 % of the prescription dose) (Shah and Ennis 2006).

The influence of hormonal treatment on the development of late rectal toxicity after a permanent implant is controversial. Whereas some authors do not describe beneficial effects of hormonal deprivation on the avoidance of late rectal complications, others do (Gelblum et al. 2000; Shah and Ennis 2006).

18.6 Acute Urinary Morbidity

The acute urinary morbidity is either caused by the radiation itself or by the trauma due to needle insertion. Both can result in swelling of the prostate gland, or the needle insertion might induce a perineal hematoma or in most severe cases a bladder tamponade. Complete urinary retention is described with incidences between 5 and 22 % after LDR monotherapy and 5–14, 5 % after combined modality approaches (Benoit et al. 2000; Terk et al. 1998).

The probability of complete urinary retention correlates with the size of the prostate and the pretreatment International Prostate Symptom Score (IPSS). It develops in the first few days after the primary implant in most cases (Gelblum et al. 1999; Merrick et al. 2000).

The majority of patients demonstrate acute urinary morbidity in the form of dysuria, increased frequency, urge symptoms, and a reduced urinary flow (Arterbery et al. 1993; Kleinberg et al. 1994).

These symptoms typically return to baseline in 90 % of the treated men in the first year after treatment (Gelblum et al. 1999; Merrick et al. 2000).

The need for a transurethral resection (TUR-P) in cases of prolonged obstructive symptoms is described in between 0 and 8.7 % (Storey et al. 1999; Wallner et al. 1996).

The incidence of de novo urinary incontinence after an implant is described in between 0 and 19 % and increases in combination with a TUR-P up to 22 % (Bottomley et al. 2007; Nag et al. 1995; Wallner et al. 1997).

18.7 Chronic Urinary Morbidity

Chronic urinary morbidity can occur in the form of prolonged irritative or obstructive symptoms. Denovo urinary incontinence is rarely observed. The prevalent form appears to be an urge incontinence. Grade III urinary symptoms classified by RTOG criteria are reported in 1–3 % after permanent implants and were usually caused by excessive irradiation of the bladder neck or by prolonged irradiation-induced inflammation of the prostatic urethra (Brown et al. 2000; Zelefsky et al. 1999).

Operative intervention such as TUR-P before or after permanent implants can predispose to urethral strictures which are described in 12 % of treated patients (Ragde et al. 1997).

18.8 Erectile Dysfunction

Erectile function is the reason for many men to undergo a permanent interstitial brachytherapy for the treatment of prostate cancer as the chances to preserve a physiological erection in the time span between 1 and 6 years after an implant are between 49 and 94 % (Merrick et al. 2002a; Stember et al. 2012; Stock et al. 2001; Zelefsky et al. 1999).

The rate of erectile function preservation correlates most significantly with the preimplant erectile status.

Whereas 70 % of men who demonstrated an unaltered erectile function before the implant still showed a physiological erectile status 6 years after the implant, this rate dropped to 34 % in patients with already abnormal preimplant erectile function (Kao et al. 2000).

Recent data shows a correlation between the radiation dose delivered to the penile bulb and postimplant erectile function. Limiting the dose to the crus penis was able to protect the erectile function in one study but other authors could not confirm these findings (Merrick et al. 2002b).

The combination of permanent interstitial brachytherapy with external beam radiation therapy and/or hormonal deprivation reduces the probability of erectile function preservation (Potters et al. 2001).

The response rate to oral PDE inhibitors appears to be significantly higher than after alternative therapies (Merrick et al. 1999).

Conclusion

There is little doubt that modern prostate brachytherapy is convenient in relation to other alternatives with a short hospital stay or even the opportunity to perform the implant as a day-case procedure compared with 6–8 weeks of daily radiotherapy and 3–10 days in hospital plus several weeks of convalescence after surgery.

More than 20 years experience with modern transperineal prostate brachytherapy has proven the efficacy of this approach.

The biochemical relapse-free survival for similar-stage-risk patients is at least as good for brachytherapy as the alternatives, and the reported side effects and complications confirm the low risk of acute and long-term morbidity.

References

Arterbery VE, Wallner K, Roy J, Fuks Z (1993) Short – term morbidity from CT-planned transperineal 125J prostate implants. Int J Radiat Oncol Biol Phys 25:661–667

Benoit RM, Naslund M, Cohen JL (2000) Complications after prostate brachytherapy in the Medicare population. Urology 55:91–96

Beyer DC, Priestly JB Jr (1997) Biochemical disease-free survival following I-125 prostate implantation. Int J Radiat Oncol Biol Phys 37:559–563

Bottomley D, Ash D, Al-Quaisieh B et al (2007) Side effects of permanent I125 prostate seed implants in 667 patients treated in Leeds. Radiother Oncol 82:46–49

Brown D, Colonias A, Miller R, Benoit R, Cohen J, Arshoun Y (2000) Urinary morbidity with a modified peripheral loading technique of transperineal 125J implantation. Int J Radiat Oncol Biol Phys 47:353–361

Davis JD, Horwitz EM, Lee RW et al (2012) American Brachytherapy Society consensus guidelines for transrectal ultrasound-guided permanent prostate brachytherapy. Brachytherapy 11:6–19

Gelblum DY, Potters L (2000) Rectal complications associated with transperineal interstitial brachytherapy for prostate cancer. Int J Radiat Oncol Biol Phys 48:119–124

Gelblum DY, Potters L, Ashley R, Waldbaum R, Wang XH, Leibel S (1999) Urinary morbidity following ultrasound-guided transperineal prostate seed implantation. Int J Radiat Oncol Biol Phys 45:59–65

Grimm P, Billiet I, Bostwick D, Dicker A, Frank S, Immerzeel J, Keyes M, Kupelian P, Lee WR, Machtens S et al (2012) Comparative analysis of prostate-specific antigen free survival outcomes for patients with low, intermediate and high risk prostate cancer treatment by radical therapy. Results from the Prostate Cancer Results Study Group. BJU Int 109(Suppl 1): 22–29

Kao J, Stock RG, Stone NN (2000) Long-term erectile function following real-time ultrasound-guided brachytherapy for prostate cancer. J Urol 163:1276 (abstract)

Keyes M, Spadinger I, Liu M et al (2012) Rectal toxicity and rectal dosimetry in low-dose-rate iodine-125 permanent prostate implants: a long-term study in 1006 patients. Brachytherapy 11(3):199–208

Kleinberg L, Wallner K, Roy J, Zelefsky M, Artebery VE (1994) Treatment-related symptoms during the first year following transperineal 125J prostate implantation. Int J Radiat Oncol Biol Phys 28:985–991

Machtens S, Baumann R, Hagemann J et al (2006a) Long-term results of interstitial brachytherapy (LDR-brachytherapy) in the treatment of patients with prostate cancer. World J Urol 24(3):289–295

Machtens S, Karstens JH, Baumann R, Jonas U (2006b) Interstitial brachytherapy (LDR-brachytherapy) in the treatment of patients with prostate cancer. Eur Urol Suppl 5:514–521

Merrick GS, Butler WM, Lief JH, Stipetich RL, Abel LJ, Dorsey AT (1999) Efficacy of sildenafil citrate in prostate brachytherapy patients with erectile dysfunction. Urology 53:1112–1116

Merrick GS, Butler WM, Lief JH, Dorsey LT (2000) Temporal resolution of urinary morbidity following prostate brachytherapy. Int J Radiat Oncol Biol Phys 47:121–126

Merrick GS, Butler WM, Galbreath RW, Stipetich RL, Abel LJ, Lief JH (2002a) Erectile function after permanent prostate brachytherapy. Int J Radiat Oncol Biol Phys 52:893–902

Merrick GS, Butler WM, Wallner KE (2002b) The importance of radiation doses to the penile bulb vs. crura in the development of postbrachytherapy erectile dysfunction. Int J Radiat Oncol Biol Phys 54:1055–1062

Nag S, Scaperoth DD, Badalament R, Hall SA, Burgers J (1995) Transperineal palladium-103 prostate brachytherapy: analysis of morbidity and seed migration. Urology 45:87–92

Potters L, Torre T, Fearn PA, Leibel SA, Kattan MW (2001) Potency after permanent prostate brachytherapy for localized prostate cancer. Int J Radiat Oncol Biol Phys 50:1235–1242

Ragde H, Blasko JC, Grimm PD, Kenny GM, Sylvester JE (1997) Interstitial iodine-125 radiation without adjuvant therapy in the treatment of clinically localized prostate carcinoma. Cancer 80:442–448

Shah JN, Ennis RD (2006) Rectal toxicity profile after transperineal interstitial permanent prostate brachytherapy: use of a comprehensive toxicity scoring system and identification of rectal dosimetric toxicity predictors. Int J Radiat Oncol Biol Phys 64:817–824

Stember DS, Mulhall JP (2012) The concept of erectile function preservation (penile rehabilitation) in the patient after brachytherapy for prostate cancer. Brachytherapy 11:87–96

Stock RG, Kao J, Stone NN (2001) Penile erectile function after permanent radioactive seed implantation for treatment of prostate cancer. J Urol 165(2001):436–439

Storey MR, Landgren RC, Cottone R (1999) Transperineal iodine-125 implantation for treatment of clinically localized prostate cancer: 5-year tumor control and morbidity. Int J Radiat Oncol Biol Phys 43:565–571

Terk MD, Stock RG, Stone NN (1998) Identification of patients at increased risk for prolonged urinary retention following radioactive seed implantation of the prostate. J Urol 160:1379–1382

Theodorescu D, Gillenwater JY, Schneider BF, Koutrouvelis PG (2000) Prostatourethral-rectal fistula following prostate brachytherapy: incidence and risk factors. J Urol 163:1294 (abstract)

Wallner K, Roy J, Harrison L (1996) Tumor control and morbidity following transperineal iodine-125 implantation for stage T1/T2 prostatic carcinoma. J Clin Oncol 14:449–453

Wallner K, Lee H, Wassermann S, Dattoli M (1997) Low risk of urinary incontinence following prostate brachytherapy in patients with prior transurethral resection. Int J Radiat Oncol Biol Phys 37:565–569

Waterman FM, Dicker AP (2003) Probability of late rectal morbidity in 125 I prostate brachytherapy. Int J Radiat Oncol Biol Phys 55:342–353

Zeitlin SI, Sherman J, Raboy A, Lederman G, Albert P (1998) High dose combination radiotherapy for the treatment of localized prostate cancer. J Urol 160:91–95

Zelefsky MJ, Wallner KE, Ling CC, Raben A, Hollister T, Wolfe T (1999) Comparison of 5-year outcome and morbidity of three-dimensional conformal radiotherapy versus transperineal permanent iodine-125 implantation for early-stage prostatic cancer. J Clin Oncol 17:517–523

19 Salvage Treatment for Recurrent Prostate Cancer Following Brachytherapy: For Whom, When and Which?

Roos E. Stuurman-Wieringa, Hiren S. Sodha, Stavros Gravas, Jean J.M.H.C. de la Rosette, and Theo M. de Reijke

19.1 Introduction

Although patient selection and radiation delivery have improved over the past years, treatment may not always be successful because of inadequate coverage, poor treatment planning, low dose delivery and non-localized disease. Better staging modalities, improved biopsy protocols and strict follow-up can attribute to earlier and better detection of recurrences. Local recurrence after external beam radiotherapy or brachytherapy occurs in approximately 30 % of patients treated for localized prostate cancer (Baumert 2010). Twelve-year ASTRO-Kattan biochemical freedom from recurrence using risk stratification following permanent prostate brachytherapy in 1,449 patients with clinically localized prostate cancer was 89, 78 and 63 % in patients at low, intermediate and high risk, respectively (Potters et al. 2005).

Causes for local failure after brachytherapy are speculative but might include difficulties in achieving a geometrically appropriate distribution of seeds to achieve a satisfactory dose distribution within the prostate or inappropriate patient selection for the procedure in a monotherapy management plan (Kollmeier et al. 2003).

R.E. Stuurman-Wieringa, MD
Department of Urology, Reinier de Graaf Gasthuis,
P.O. box 5011, Delft, GA 2600,
The Netherlands

H.S. Sodha, MD • J.J.M.H.C. de la Rosette, MD, PhD • T.M. de Reijke, MD, PhD, FEBU (✉)
Department of Urology, Academic Medical Center,
P.O. box 22660, 1100 DD Amsterdam,
The Netherlands
e-mail: t.m.dereyke@amc.uva.nl

S. Gravas, MD, PhD
Department of Urology, University of Thessaly,
6-8, Feidiou Street, 41221 Larissa, Greece

G. Kovács, P. Hoskin (eds.), *Interstitial Prostate Brachytherapy*,
DOI 10.1007/978-3-642-36499-0_19, © Springer-Verlag Berlin Heidelberg 2013

19.1.1 PSA Evaluation and Relapse

Biochemical failure rate following brachytherapy is difficult to define. In contrast to undetectable PSA after radical prostatectomy, in patients who have radiation therapy, there is still PSA production by the remaining normal glandular prostatic tissue and the PSA drop is gradual. Following radical retropubic prostatectomy, two consecutive values of 0.2 ng/mL or greater appear to represent an international consensus definition for recurrent cancer (Boccon-Gibod et al. 2004; Moul 2000). After radiotherapy, the PSA level falls slowly and the optimal cut-off value for a favourable PSA nadir after radiotherapy is somewhat controversial. Achieving a PSA nadir of less than 0.5 ng/mL seems to be associated with a favourable outcome (Verhagen et al. 2006). The interval before reaching the nadir PSA may be very long and can sometimes take up to 3 years or more.

Multiple clinical characteristics have been described to define failure after radiotherapy: ASTRO (American Society for Radiation Oncology) criteria, Phoenix/Houston criteria, PSA doubling time and serum PSA only. Brachytherapy relapses are not described separately. The ASTRO designed a definition in 1996 for a biochemical failure after EBRT based on three consecutive PSA rises following a nadir with the date of failure to the point midway between the posttreatment PSA nadir and the first increase. This concept is known as backdating.

A new definition of radiation failure was established with the main aim to establish a better correlation between the definition and clinical outcome, which is known as the Phoenix/Houston criteria. The revised criteria in 2005 are the most frequently used nowadays and are considered the standard definition for biochemical failure after radiotherapy. It defines a biochemical failure as a PSA rise by 2 ng/mL or more above the nadir regardless of whether or not a patient received androgen deprivation therapy.

Any continuously rising PSA following a nadir after radiation therapy is an indicator of local recurrence, systemic metastatic spread or a combination of both. So-called PSA bounces (rise and decline), unrelated to recurrence, can occur after brachytherapy and high-dose external beam radiation therapy and can be disconcerting to both clinician and patient (Fig. 19.1). Approximately 30–50 % of patients treated with prostate brachytherapy experience a PSA bounce (Patel et al. 2004). The PSA bounce may mimic biochemical failure and may lead to the unnecessary administration of salvage therapy. Due to this phenomenon, the biochemical failure definition of a PSA rise by 2 ng/mL or more above the nadir is used. After radiotherapy, a late and slowly rising PSA is a sign of local failure. The timing and mode of treatment of PSA-only recurrence after primary therapy remains controversial. It has been shown that biochemical failure precedes clinical disease by 6–48 months (Lange et al. 1989).

PSA doubling time (PSADT) is a valuable prognostic factor in assessing the need for second- or even third-line treatment in men with biochemical progression of prostate cancer. PSADT has been significantly shorter in patients who developed metastases than in those who did not develop metastatic disease. Patients with local recurrence had a PSA doubling time of 13 months compared to 3 months for those with distant failure (Hancock et al. 1995). It should be based on at least three values

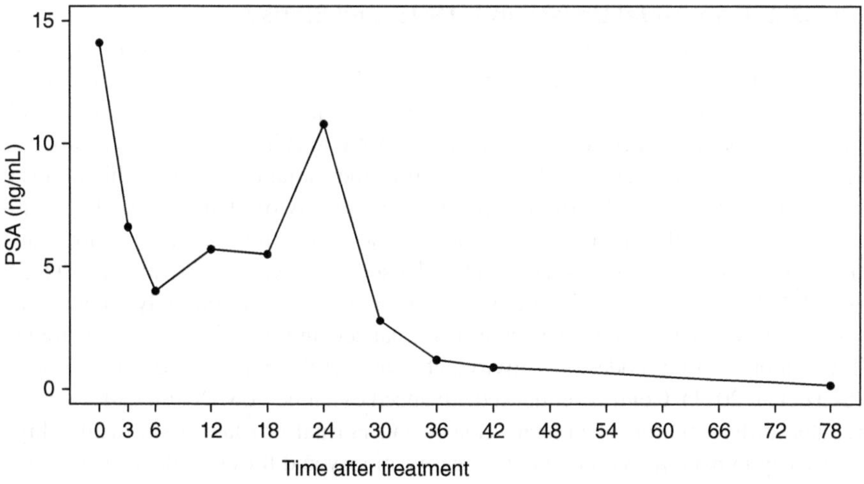

Fig. 19.1 PSA bounce following brachytherapy

separated by at least 3 months each and is best calculated with a mathematical log-slope method (www.normograms.mskcc.org/Prostate/PsaDoublingTime.aspx).

Patients should be followed up closely during the first years after primary curative treatment when the risk of failure is highest. PSA measurements are recommended at the following intervals: 3, 6 and 12 months posttreatment; every 6 months thereafter until 3years; and then annually until 10 years. The Phoenix/Houston biochemical relapse definition is considered the standard definition for biochemical failure after all radiotherapy modalities. Local failure after radiotherapy is documented by a positive prostatic biopsy and negative imaging studies. Prostate biopsy after radiotherapy is necessary only if local subsequent procedures (e.g. salvage prostatectomy) are indicated in the individual patient. Prior to extensive diagnostic workup in patients with PSA relapse following local treatment, men must be stratified into patients who are candidates for salvage therapy and those who are not. Patients must then be further stratified into candidates for local salvage treatment and those who might need systemic therapy. All diagnostic procedures should only be performed if these are likely to have therapeutic consequences.

19.1.2 Investigation for PSA Recurrence

19.1.2.1 Physical Examination

In case of a PSA-only relapse, a physical examination, and especially a digital rectal examination (DRE), is usually not helpful in determining the site of relapse. Due to the radiation, the prostate has undergone changes and it can be difficult to interpret the findings. A newly detectable nodule should raise the suspicion of local disease recurrence. Only in the case of a high-risk patient and a very early PSA relapse or in the case of local symptoms will a local progression can be identified by DRE.

19.1.2.2 Transrectal Ultrasound (TRUS) and Biopsy

TRUS cannot stand alone as a diagnostic tool in case of biochemical recurrence, but must usually be combined with biopsies to establish the presence of local disease recurrence. TRUS-guided biopsy is indicated for histological confirmation with proven biochemical recurrence if a salvage procedure is being considered. Biopsies should be performed at least 18 months after the radiation since the histological regression of tumour cells after RT may be prolonged (Cox et al. 1999). TRUS cannot reliably identify the areas in the prostate with recurrent/resistant tumour, and therefore, new developments such as Doppler sonography, contrast-enhanced TRUS (CE-TRUS) and elastography are being explored in order to possibly enhance the value of TRUS. CE-TRUS detected prostate cancer significantly better compared to TRUS alone with a modest sensitivity and a high PPV in a selected patient cohort (Seitz et al. 2011). Contrast-enhanced advanced dynamic flow Doppler allows more reliable differentiation of prostate cancer and normal prostate tissue with a high sensitivity in patients with previous negative biopsy and fewer artefacts than power Doppler images, thus providing a good basis for targeted prostate biopsy instead of systematic biopsy (Taymoorian et al. 2007). However, the role of these new imaging modalities in relapsing patients following radiation therapy remains to be established.

19.1.2.3 Magnetic Resonance Imaging (MRI)

Another local imaging technique in detecting local recurrence is the T2-weighted MRI. The reported sensitivities (26–44 %) and specificities (64–86 %) in the detection of tumour recurrence have been rather low due to the morphologic changes after irradiation including inflammation, glandular atrophy, fibrosis and prostate shrinkage. These changes can cause difficulty in differentiating recurrence from irradiated normal tissue (Coakley et al. 2001; Haider et al. 2008). Additional MR spectroscopy and dynamic contrast-enhanced MRI are advised and have shown promise in increasing overall imaging performance in the detection of local recurrence (De Visschere et al. 2010). The functional imaging techniques provide spatial information on the physiological tumour characteristics (Engelbrecht et al. 2010). Dynamic contrast-enhanced MRI (DCE-MRI) aims to analyze the extravasation to the interstitial space and the subsequent clearance of gadolinium-DTPA, a low-molecular weight paramagnetic contrast agent, after intravenous bolus injection. It is a non-invasive imaging tool providing both anatomical and physiological information on the tumour. The 'wash-in' is the speed of contrast uptake and suggested being the most accurate discriminator between malignant and benign prostatic tissue (Isebaert et al. 2011).

19.1.2.4 CT Scan, Bone Scan and PET Scan

Imaging studies including bone scan and CT scan have a limited role in the detection of local recurrence with early biochemical recurrence after brachytherapy. Metastasis may be detected by pelvic CT scan or bone scan, and a search for bone metastases or pelvic recurrence is only advised in men with a serum PSA of ≥10 ng/mL (≥5 ng/mL if prior ADT was administered), an increase of PSA >2 ng/mL per month or a PSA doubling time less than 6 months (Dotan et al. 2005). The role of PET imaging in prostate cancer is gradually evolving but still remains investigational. Different radiotracers (e.g. anti-3-^{18}F-FACBC, ^{111}In-capromab pendetide, (^{11}C)-choline and ^{18}F-choline) using PET/CT are being evaluated and may enhance the detection rate and the differentiation of prostatic from extraprostatic disease (Schuster et al. 2011).

In conclusion, evaluation in a patient with a PSA relapse after primary radiation therapy who might be candidate for secondary local salvage therapy should consist of a positive prostatic biopsy at least 18 months after the procedure and imaging studies preferably an endorectal MRI and exclusion of distant disease with isotope bone scan.

19.1.3 Timing of Salvage Treatment

There are significant challenges in counselling patients for treatment of local prostate carcinoma recurrence. The treatment of asymptomatic patients with recurrent disease in patients with associated comorbidity remains controversial. For patients with a biochemical recurrence following radical prostatectomy, it took 96 months to develop M+ disease without any further treatment, and median time to death was then 60 months. So a total of 13 years was observed between biochemical recurrence and death. The desire to intervene quickly after the formal definition of PSA failure can be tempered by some evidence that PSA rise, even with a positive local biopsy, does not inevitably lead to progressive clinical disease (Smathers et al. 2001). Patients that are nowadays selected for salvage therapy may differ from selected patients in the near future, since some older patients may have been treated with a treatment modality that was suboptimal possibly due to comorbidity. Therefore, the chance of local recurrence can be higher.

19.1.4 Patient Selection

Local salvage curative therapy may be considered in a patient with a good performance status if a local recurrence is documented by prostate biopsy, with no evidence of disseminated disease and a life expectancy >10 years. The morbidity for the therapeutic salvage options can be significant, and patients have to accept the possible side effects of second-line treatment. No consensus exists on optimal timing of salvage therapy for those recurrences thought confined to the prostate (Grossfeld et al. 1998). A suggestion has been made to initiate the treatment with a relatively low PSA level for which, however, further evaluation is needed.

As is true for most treatment modalities for prostate cancer with curative intent, no randomized trials comparing the different salvage treatments are available, and judgment on which treatment to start and at what time point is based on observational studies.

19.2 Treatment Options in Recurrence After Brachytherapy

19.2.1 Salvage Surgery

19.2.1.1 Salvage Open Radical Prostatectomy

Historically, salvage RP (SRP) for men with biopsy-proven local recurrence after RT has rarely been performed because of concerns regarding lack of efficacy and high morbidity. However, it remains a feasible option and improvements in surgical experience have led to improved functional outcomes with less side effects. Candidates for salvage radical prostatectomy must have an organ-confined recurrence with a good performance status, a life expectancy >10 years and a low post-implant PSA level (<10 ng/mL).

SRP is performed in a similar technique as the standard retropubic or perineal approach, with modifications in the surgical technique dependent on the extent of intrapelvic fibrosis, pelvic adhesions and scarring. These are the major challenges in the irradiated prostate and can make nerve sparing difficult in case the patient is still potent. However, nerve sparing is sometimes performed in very well-selected cases with preoperatively normal potency.

Salvage radical cystoprostatectomy may be necessary in case of bladder neck infiltration and can be considered in patients with severe urinary problems following radiation treatment, e.g. radiation-induced cystitis with severe haematuria and/or a contracted bladder.

Despite the willingness of most surgeons to consider post-radiation salvage surgery, the collective published experience is relatively small and relates more specifically to failures after EBRT than brachytherapy. The available data is limited to retrospective cohort studies, mostly single centre, and restricted by the small sample size and short follow-up. Brachytherapy is in most cases not described separately. The largest series was a recent retrospective, international, multi-institutional cohort analysis with a median follow-up of 4.4 years, 10 years after SRP was performed on 404 patients with radiation-recurrent PCa (Chade et al. 2011).

A summary of oncological and functional results after SRP in large series on SRP can be found in Tables 19.1 and 19.2.

Oncological Outcome

At a median follow-up of 35 months after surgery, biochemical disease-free survival is described in 75 % of patients ($n=32$) (Leonardo et al. 2009). At 5 years 47 % of patients were progression-free without androgen deprivation therapy. Among patients with pT2 disease, 100 % were progression-free at 5 years, compared with 35 % of patients with pT3N0 disease or higher and 0 % of patients with node-positive disease (Sanderson et al. 2006). At 6.9 years, the biochemical disease-free survival rate is reported to be 55 %, while overall and cancer-specific survival was 91 % (Darras et al. 2006). Ten-year BCR-free survival, metastasis-free survival and cancer-specific survival rates are 37, 77 and 83 %, respectively (Heidenreich et al. 2010).

Patients with lower pre-SRP PSA levels and lower post-radiation prostate biopsy Gleason score have the highest probability of cure from SRP (Heidenreich et al.

Table 19.1 Salvage prostatectomy after radiotherapy

	No. of pts	Mean age	Median FU (year)	Median pre-SRP PSA	PSA < 10 (%)	≤pT2b	BCR-free 5 year (%)	CSS 5 year (%)
Chade (2011)	404	65	4.4	4.5	NR	65	48	92
Leonardo (2009)	32	63	2.9	13	NR	78	75 % (2.9 year)	100 (2.9 year)
Sanderson (2006)	51	65	7.2	8	64	25	(47 % PFS)	NR
Darres (2006)	11	60.5	6.9	5.2	100	100	55 % (6.9 year)	91 % (6.9 year)
Heidenreich (2010)	55	65.3	5.6	7.8	81.8	80	74.5 (5.6 year)	NR
Gotto (2010)	98	NR	2.9	NR	NR	NR	NR	NR
van der Poel (2007)	27	64	3.6	8.6	NR	41	31 %	89 %
Bianco et al. (2005)	100	65	5	5.9	70	91	55 %	NR

Table 19.2 Salvage prostatectomy after radiotherapy

	Year	No. of pts	BNS (%)	Rectal injury (%)	Continency (%)	Potency (%)
Chade (2011)	1985–2009	404	NR	NR	NR	NR
Leonardo (2009)	2001–2004	32	12.5	0	21.9	9.3
Sanderson (2006)	1983–2002	51	41	2	73	22–25
Darres (2006)	1989–2004	11	18	0	81 (<2 pads)	0
Heidenreich (2010)	2004–2008	55	10.9	3.6	80 (<2 pads)	26.7
Gotto (2010)	1999–2007	98	26	9	30	25
van der Poel (2007)	1997–2005	27	3.7	3.7	37 (no pads)	7
Bianco (2005)	1984–2003	100	NR	NR	NR	NR

2010). Previous low-dose brachytherapy, <50 % positive biopsy cores and a PSA doubling time >12 months are predictors of organ-confined disease and to select patients who are most suitable for SRP (Heidenreich et al. 2010). A serum PSA of 5.0 ng/mL or less can best predict long-term recurrence-free survival preoperatively (van der Poel et al. 2007).

Surgical Outcome

Surgery is mostly difficult but no major intraoperative complications have been described recently, probably due to improved surgical techniques. Mean operative time in different series ranged between 119 and 122 min, and mean blood loss was between 360 and 550 mL (Gotto et al. 2010; Nuñez-Mora et al. 2009; Ahallal et al. 2011). There was no significant difference in median operative time, blood loss or transfusion rate when comparing open radical to salvage prostatectomy with 21 % of the patients having prior brachytherapy (Gotto et al. 2010). Hospital stay and catheterization time were 5 and 12 days, respectively (Gotto et al. 2010). There were no significant differences between the type of RT and surgical outcome. Scarring and fibrosis was less in the EBRT group compared with the EBRT/contemporary brachytherapy and permanent brachytherapy group (Ahallal et al. 2011).

Complications

There is a higher probability of medical and surgical complications, including urinary tract infection, bladder neck contracture, urinary retention, urinary fistula, abscesses and rectal injury. In 53 % minor complications and 7 % major complications are described (Jamal et al. 2008). Bladder neck stricture is the most common postoperative complication after SRP in radiorecurrent disease and described in 11–41 % patients (van der Poel et al. 2007; Nuñez-Mora et al. 2009; Ahallal et al. 2011). The brachytherapy group seems to have comparable complications to the EBRT group, but it is suggested that less postoperative bladder neck contracture and urine retention are seen (Jamal et al. 2008). Another series showed no correlation in the development of a bladder neck contracture and the type of RT (van der Poel et al. 2007).

Early complications following open salvage therapy are described in 27.3 %, e.g. UTI, epididymitis, superficial wound infection, prolonged transurethral catheter drainage, blood transfusion and rectal lesions. Late complications are described in 30.9 % of patients: urinary incontinence, bladder neck contracture and urethral stricture. The major late complication was rectourethral fistula in 1.8–2 % of patients. Anastomotic strictures could be treated with urethrotomy or with urethral dilation and temporary catheterization (Gotto et al. 2010; van der Poel et al. 2007; Ahallal et al. 2011).

Functional Results

Continence rates are worse than in primary radical prostatectomy. Complete continence and acceptable and stress incontinence rates are described in 22–80, 36 and 18 % of patients, respectively (Gotto et al. 2010; Nuñez-Mora et al. 2009; Ahallal et al. 2011). Improvement of continence strongly correlates with the type of previous RT: continence is restored in 9.5 % of patients following seed implantation. Incontinence remained in 21.1 % following EBRT and 33 % patients after EBRT plus brachytherapy. The median time to regain continence was 7.9 months (Ahallal et al. 2011). In patients requiring an artificial urinary sphincter, self-reported functional outcomes following prosthesis implantation were excellent. The impact of salvage surgery on QoL parameters can be greatly ameliorated in case of urinary incontinence with the use of prosthetic devices (van der Poel et al. 2007).

Full potency is described in only 0–9 % of patients; sparing bilateral bundles resulted in 25–26.7 % potency in men with unimpaired preoperative erections (Nuñez-Mora et al. 2009; Jamal et al. 2008). Ten to 40 % of patients achieved erections sufficient for sexual intercourse with the use of PDE5 inhibitors. In case of impotence, patients could be treated with either intracavernous self-injection of vasodilating agents or a penile implant (Gotto et al. 2010; Ahallal et al. 2011).

Data on the perineal surgical approach to the prostate as a salvage therapy after radiotherapy is limited. In 27 patients who underwent a perineal SRP after external beam or brachy-radiotherapy, 5-year biochemical recurrence-free survival was 31 %. Perioperative morbidity consisted of prolonged anastomotic leakage, urosepsis, prolonged haematuria and urinary and rectal perforation. Thirty-seven and 7 % of patients reported normal continence and erectile function, at least 1 year after perineal surgery, respectively (van der Poel et al. 2007).

19.2.2 Salvage Laparoscopic-Assisted Prostatectomy

The available data are limited to retrospective single-centre cohort studies and restricted by the small sample size and short follow-up. Brachytherapy is not described separately. In the patients treated no conversion to an open procedure was necessary.

In a series of nine patients (4 XRT, 5 BT), the positive surgical margin rate was 22.2 % (Nuñez-Mora et al. 2009). In a pilot study of 15 patients within an 8-month median follow-up, 11 patients were disease-free and 3 had persistent postoperative PSA elevation; the remaining patient experienced PSA recurrence after 21 months (Ahallal et al. 2011).

Mean operative time in different series ranged between 170 and 235 min, mean blood loss of 250 mL, and no ureteral or rectal or postoperative injuries were reported. The dissection in the salvage procedure was more difficult due to intense periprostatic fibrosis following radiotherapy. Postoperative complications were described in two patients (22.2 %): gross haematuria and a left pelvic lymphocele (Chauhan et al. 2011; Eandi et al. 2010).

At a minimum follow-up of 15 months in a series of nine patients, three (33.3 %) patients were pad-free. One of the five preoperatively potent patients achieved erections 16 months after the surgery (Chauhan et al. 2011). In a series of 15 patients, 1 presented with a rectal injury and 1 had an anastomotic leak. Seven patients achieved continence by 8.4 months, 1 patient had severe incontinence corrected by implanting an artificial sphincter, and 7 patients with a 12.6-month mean follow-up needed one or two pads per day. Erectile dysfunction was present in 5 patients before surgery and in 14 patients after surgery (Eandi et al. 2010).

19.2.3 Salvage Robot-Assisted Radical Prostatectomy (sRARP)

The feasibility of performing salvage radical prostatectomies has been extended into laparoscopy and more recently into robotic-assisted techniques. Only a few cases have been described in the literature until now. It is a safe and technically feasible salvage treatment modality and an emerging area of interest for prostate cancer for which primary radiotherapy has failed (Jamal et al. 2008; Boris et al. 2009). All sRARP cases were performed using a previously described six-port transperitoneal technique and no conversions were described. Most of the oncological and functional outcomes still need to be validated.

In a multi-institutional retrospective series of 15 patients with a median follow-up of 4.6 months, 4 patients (28.6 %) presented with biochemical recurrence after sRARP (Chauhan et al. 2011). In one series, 28 % had a positive surgical margin. The mean operative time ranged between 125 and 160 min, the mean blood loss was between 117 and 150 mL, and the mean hospital stay was 2–2.7 days (Stephenson and Eastham 2005; Eandi et al. 2010; Moman et al. 2010).

Perioperative complications occurred in 7 patients (39 %) of which the most common was urine leakage. Another series described 20 % of patients with postoperative complications consisting of deep vein thrombosis, a wound infection and an anastomotic leak developing into an anastomotic stricture. Continence rate was 71.4 % and none were potent after sRARP (Stephenson and Eastham 2005; Tharp et al. 2008). In a series of four patients, no major complications were described. Three patients were continent (Kaouk et al. 2008).

Salvage robot-assisted radical prostatectomy seems to be a safe procedure with no increase in perioperative morbidity and thus is an effective modality for salvaging patients with localized prostate cancer recurrence after radiation. However, morbidity remains high with this approach also. The benefits of the robot's improved three-dimensional vision and magnification should potentially further decrease the morbidity associated with SRP once more experience has been obtained.

The RALP in primary treatment of prostate carcinoma showed significantly lower blood loss and transfusion rates compared to the open approach, but the available data were not sufficient to prove the superiority of any surgical approach in terms of functional and oncologic outcomes (Ficarra et al. 2009). Further studies are warranted to validate the oncological and functional outcomes of SRP after radiation and/or brachytherapy.

19.3 Summary of the Role of Salvage Surgery

During the last decade, morbidity of SRP has strongly decreased with a percentage of rectal and ureteral injury at 2 %. It can be a curative option with moderate disease control rates. Salvage RP may favourably alter the natural history of biochemical recurrence after radiation therapy, but it must be instituted early in the course of recurrent disease to be effective (Stephenson and Eastham 2005). Nevertheless, the incidence of urinary incontinence and erectile dysfunction is an important factor in counselling the patient before deciding to start a salvage radical prostatectomy. Laparoscopic and robotic salvage prostatectomies seem to be interesting alternatives but need further investigation. Although few reports of salvage robotic prostatectomy have been published with limited long-term follow-up, initial oncologic results seem at least comparable to the salvage open prostatectomy series. sRALP should be performed by centres with a dedicated and well-experienced robotic urologic oncology programme as is also the case for the other approaches. As with all techniques, further long-term follow-up is needed.

19.4 Salvage Radiotherapy

19.4.1 Salvage External Beam Radiotherapy and Brachytherapy

Salvage radiotherapeutic treatments have not been widely used but are available and seem to have potential. Initial radiotherapeutic treatment may have been performed using an insufficient dose and leaves room for new radiation treatment (Table 19.3). No data exist on salvage external beam radiotherapy in prostate carcinoma. In brachytherapy, dependent on the activity of the source, a distinction is made between continuous low-dosage radiation (LDR) which is 0.4–2.0 Gy an hour and fractionated radiation with high dosage (HDR) which is >12 Gy an hour. HDR has been used for seed failure patients with doses varied at the discretion of the treating radiation oncologist, ranging from 600 to 900 cGy per HDR fraction in two to four total fractions (Tharp et al. 2008). The available data on brachytherapy is limited to retrospective cohort studies, mostly single centre and restricted by the small sample size and short follow-up (Tables 19.3 and 19.4).

One series reported a 58-month median follow-up of salvage HDR brachytherapy after EBRT or permanent seed implantation and a 71 % disease-free survival

Table 19.3 Salvage radiotherapy after radiotherapy

	No. of pts	Mean age (year)	Median FU (year)	Median PSA at SR Tx	Previous BT (%)	Type sBT	DFS (% at year)
Tharp (2008)	7	71	4.8	9.5	0	125I	71 (4.8)
Moman (2010)	31	69	9.0	11.4	36	125I	NR, BCR at 5 year: 51 %
Burri (2010)	37	70	7.2	5.6	14	103Pd or 125I	NR, BCR at 5 year: 96 %
Lee (2008)	21	72	3.0	3.8	0	103Pd	NR, BCR at 5 year: 38 %
Nguyen (2009)	25	65	3.9	5.5	0	125I	NR
Allen (2007)	12	NR	3.8	3.9	0	103Pd or 125I	67 (4.0)
Grado et al. (1999)	49	73	5.3	5.6	6	103Pd or 125I	34 (5.0)
Wallner et al. (1990)	13	NA	3.0	NA	NA	125I	51 (5.0)
Loening and Turner (1993)	31	NA	1.9	NA	0	198Au	40 (5)
Beyer (2004b)	30	NA	3.8	NA	0	103Pd or 125I	25–67 (3)
Koutrouvelis et al. (2003)	31	NA	2.5	NA	100	103Pd	NA, BCR at 2.5 year: 87 %
Wong et al. (2006)	17	68	3.6	4.7	0	125I and 103Pd	79 (4)

Table 19.4 Salvage radiotherapy after radiotherapy

	Year	No. of pts	Continency (%)	Rectal injury grade 3–4 (%)	Genitourinary grade 3–4 (%)	Institution
Tharp (2008)	2001–2006	7	71.5	0	28.5	Indiana Cancer Center
Moman (2010)	1994–2009	31	NR	7	19.0	UMC Utrecht
Burri (2010)	1994–2008	37	NR	3	0.1	Mount Sinai
Lee (2008)	1998–2005	21	100	0	14.0	Richmond University
Nguyen (2009)	2000–2005	25	100	24	16.0	DFCI/Brigham and Women's Hospital
Allen (2007)	1998–2003	12	77	0	0.0	Wisconsin
Grado et al. (1999)	1990–1996	49	94	4	14.0–24.0	Mayo Scottsdale
Wallner et al. (1990)	1978–1988	13	69	15	31.0	MSKCC
Loening and Turner (1993)	1984–1991	31	100	0	0.0	Iowa
Beyer (2004b)	1989–1994	30	76	NA	24.0	Arizona
Koutrouvelis et al. (2003)	1995–2002	31	100	5	<13.0	Uro-Radiology Prostate Institute
Wong et al. (2006)	1999–2004	17	94	6	47.0	Mayo Scottsdale

NA not available, *NR* not reported

rate (n=7) (Allen et al. 2007). After salvage I-125 implantation for prostate cancer recurrences, the 5-year freedom from biochemical failure rate was 23 %, and the DSS rate was 74 % after primary EBRT or I-125 implantation (n=31) (Moman et al. 2010). Thirty-seven men with local failure after initial prostate radiotherapy (32 EBRT and 5 BT) received a median dose to 90 % of the prostate volume of 122 Gy using either palladium-103 or I-125 seeds. The 10-year freedom from biochemical failure and CSS were 54 and 96 %, respectively (Burri et al. 2010). Twenty-one patients underwent salvage brachytherapy for local failure after EBRT. With a median follow-up of 36 months, the actuarial 3-year and 5-year overall survival rates were 81 and 81 %, and the biochemical failure-free survival rates were 94 and 38 %, respectively. There was no significant difference in biochemical failure-free survival and OS for patients who had androgen ablation (Lee et al. 2008). In 12 patients the median brachytherapy dose delivered was 97 Gy. At a median follow-up of 45-month post-salvage treatment in combination with 3-month ADT, the 4-year actuarial biochemical disease-free survival was 63 % and overall survival was 54 % (Allen et al. 2007).

With I-125 or palladium-103 salvage brachytherapy, up to 98 % of recurrences after EBRT may be locally controlled, and 5-year freedom from second relapse is approximately 50 %. With careful case selection, relapse-free rates up to 83 % may be achieved (Beyer 2004a).

19.4.1.1 Complications

After HDR brachytherapy in five patients (71 %), symptomatic urethral strictures developed. Salvage I-125 implantation after primary external beam radiotherapy had considerable genitourinary grade 3–4 rates of toxicity. Two of 7 patients with salvage HDR brachytherapy (previous seed failures) developed incontinence with urethral necrosis, followed by placement of an artificial urinary sphincter prosthesis. Severe toxicity of the gastrointestinal and genitourinary tract occurred in 29 % of the patients, most often in the late phase. These results seem to be in favour of HDR brachytherapy. Because of the high toxicity rates, patients for LDR should be selected with great care, based on individual risk factors (Allen et al. 2007; Beyer 2004a).

19.4.1.2 Functional Results

Results for lower urinary tract symptoms are usually described; however, incontinence and potency are reported rarely. One month after implantation, the median IPSS increased by 15 points and subsequently declined to the pretreatment level at last follow-up. Grade 2 urinary incontinence was reported in 25 % (Lam and Belldegrun 2004). After salvage I-125 implantation, 8 % grade 1 urinary symptoms and 46 % grade 2 toxicity are described. The latter complication required medications for symptom relief. Eight percent developed grade 3 toxicity such as obstructive uropathy requiring TURP or fulguration for gross haematuria. Grade 4 toxicity (prostatorectal fistula) was seen in 2.7 % (Nguyen et al. 2009). After palladium-103 salvage brachytherapy, 19 % of patients had grade 2 genitourinary adverse events, 9.5 % of patients had grade 1 genitourinary adverse events, and 4.7 % of patients had a grade 2 gastrointestinal adverse event (Onik et al. 1993).

Although bowel and urinary symptoms were described at 3 or 15 months, they recovered and there were no significant differences between baseline and 27-month urinary or bowel scores. An interval to re-irradiation of less than 4.5 years and prior brachytherapy were each associated significantly with the largest decrements in bowel function (Nguyen et al. 2009).

19.4.1.3 Conclusion

Initial radiotherapeutic treatment may have been performed using an insufficient dose and leaves room for new radiation treatment. No data exist on salvage external beam radiotherapy in recurrent prostate carcinoma. The disease control rates and complications of salvage brachytherapy treatment compare favourably with those reported using other modalities, and the salvage treatment is well tolerated. Effective salvage therapy can be achieved with a minimum of procedure-related morbidity and favourable long-term side effect profiles. It seems that prostate brachytherapy causes transient increases in urinary symptoms in the immediate posttreatment period that generally normalize on longer follow-up. Results seem to be in favour of HDR brachytherapy. Very limited data on salvage HDR or LDR brachytherapy are available and merit further investigation. Increased fractionation with lower doses of HDR could be considered as a method to reduce late toxicity. More research is needed to improve current patient selection procedures in the workup for salvage treatment, and functional results have to be prospectively evaluated.

19.5 Minimally Invasive Procedures

19.5.1 Salvage Cryotherapy

19.5.1.1 Technique

First-generation and second-generation cryosurgical systems utilized free-hand-placed liquid nitrogen to create an ice ball, which lacked precise control and monitoring, resulting in a high complication rate. Today's third-generation cryo units have transitioned to argon/helium-based systems using the Joule–Thomson principle to create precisely controlled isotherms through ultrathin needles mostly placed using a perineal grid as used for brachytherapy. Argon and liquid nitrogen achieve adequate freezing temperatures (resp. −187 and −196 °C); supercooled liquid nitrogen actually has a lower temperature (−209 °C). Considerable improvement in recent years has been made with the introduction of the transperineal ultrasound-guided implantation technique using thin needles, which has reduced toxicity and improved outcome for the treatment of primary diagnosed prostate cancer (Onik et al. 1993). Salvage cryosurgery can be performed in the patient with recurrent disease following EBRT as well as interstitial prostate brachytherapy. Previously placed radioactive seeds can be visualized quite well under TRUS and may cause some confusion as their sonographic appearance is similar to the tip of the cryoneedles, especially in the transverse view. Placing the needles in the sagittal plane can overcome this difficulty, since the length of the cryoneedles can be easily followed in this view (Lam and Belldegrun 2004; Finley and Belldegrun 2011).

Incorporating the routine use of multi-temperature sensing probes, double freeze–thaw cycles and urethral warming catheters has led to further improvements. Percutaneous lethal freezing of primary prostate carcinoma under image guidance has been described as minimally invasive, causing minimum pain and discomfort, with quick recovery time and no radiation, and is repeatable in case of failure. The technique has now also been introduced for salvage procedures because of the theoretical advantage that no radiation is being used and the formation of the ice ball can be monitored accurately. Few contraindications for cryotherapy have been described: these include involvement of seminal vesicles and anorectal absence or significant pathology. Although a history of brachytherapy may complicate needle placement, it is generally not considered to be a contraindication to salvage cryosurgical ablation. A major concern with the use of less aggressive tissue-preserving strategies is incomplete treatment of cancerous foci in remnant locations.

According to the 2008 American Urological Association (AUA) Best Practice Consensus Statement, ideal candidates for salvage cryoablation should have absence of seminal vesicle invasion, a PSA less than 10 ng/mL (preferably <4 ng/mL), a PSA doubling time of 16 months or more and at least a 10-year life expectancy.

The collective published experience is again relatively small and limited to retrospective cohort studies, single centre and restricted by the small sample size and short follow-up. It relates more specifically to failures of EBRT than to brachytherapy.

19.5.1.2 Oncologic Results

The Cryo On-Line Data Registry data from 279 patients treated with argon or nitrogen salvage cryoablation, of which only 47 had a minimum of 5-year follow-up, reported 5-year actuarial biochemical disease-free rates of 58.9 % (ASTRO) and 54.5 % (Phoenix), respectively (Pisters et al. 2008). There is a question as to whether these criteria can be used for salvage treatments, because they have not been tested in these patient cohorts. In 59 patients, a 59–69 % biochemical disease-free survival was reported with a median follow-up of 6.9 years (Bahn et al. 2003). In 176 patients undergoing salvage cryoablation with a mean follow-up of 7.46 years, overall DFS was 47, 39 and 39 % at 5, 8 and 10 years, respectively. In terms of prognostic factors, a PSA nadir above 1.0 ng/mL was significantly associated with poor prognosis (Williams et al. 2011). Another series of 56 patients showed 21 % biochemical disease-free survival at 5 years after primary radiation therapy, which is inferior to the salvage radical prostatectomy of 61 % at 5 years. The salvage radical prostatectomy resulted in superior biochemical disease-free survival. There was no significant difference in disease specific survival at 5 years for salvage cryotherapy (96 %) and salvage radical prostatectomy (98 %) (Pisters et al. 2009).

PSA at the time of salvage cryoablation seems to be a predictive factor for biochemical recurrence. Patients with pre-cryoablation PSA less than 4 ng/mL had a 5- and 8-year biochemical recurrence-free survival at a mean follow-up of 39 months of 56 and 37 %, respectively. In contrast, patients with pre-cryoablation PSA of 10 ng/mL or greater had a 5- and 8-year biochemical recurrence-free survival of only 1–7 %, respectively (Ng et al. 2007).

19.5.1.3 Complications

In the Cryo On-Line Data Registry, the rectal fistula rate was 1.2–3.2 % of patients that underwent transurethral prostate resection to remove sloughed tissue. Other complications included urethral stricture, urethral ulcer, urethrorectal fistulas in 2 %, acute urinary retention in 21 %, perineal pain in 14 % and haematuria in 11 % of patients (Eisenberg and Shinohara 2008; Ng et al. 2007).

19.5.1.4 Functional Results

Incontinence rate following salvage cryotherapy treatment was described in 4.4–40 % of patients (Yin et al. 2010; Anastasiadis et al. 2003; Murat et al. 2009) and erectile dysfunction rate of 90 % (Berge et al. 2010).

Focal Cryotherapy

To determine the efficacy of partial cryoablation, a retrospective analysis on salvage partial cryoablation was conducted. Nineteen patients had a biochemical recurrence-free survival rate of 89, 67 and 50 % at 1, 2 and 3 years, respectively (Uchida et al. 2011). Overall 5- and 8-year survival rates were described in 97 and 92 %, respectively (Murat et al. 2009).

19.5.1.5 Conclusion

Salvage cryoablation may potentially be curative and is a viable treatment option for patients with prostate cancer in whom radiation therapy has failed. However, it is associated with significant morbidity causing urinary incontinence and impotence.

19.5.2 Salvage HIFU

19.5.2.1 Technique

HIFU is a relatively new technique combining an imaging and treatment modality using ultrasound. It destroys tissue with rapid heat elevation, which essentially 'cooks' the tissue. Ultrasound energy is focused when the transrectal ultrasound is guided at a specific location, and at that focal point, the temperature rises to 90 °C in a matter of seconds. This technique is applied in some centres for the primary treatment of localized prostate cancer. Endorectal high-intensity focused ultrasound treatment is receiving increasing attention. Each shot consists of a burst of ultrasound waves, which entail 100 % acoustic power with a 5 s pulse of energy to create each discrete HIFU lesion, with a 5 s delay between the formation of each lesion. It has a very small focal volume which makes it possible to precisely define the lesion positioning and treatment delivery thus sparing surrounding tissue.

19.5.2.2 Available Data

The collective published experience is relatively small and relates more specifically to failures of EBRT. It is limited to retrospective cohort studies, mostly single centre and restricted by the small sample size and short follow-up. Brachytherapy is in most cases not described separately, because the interaction between heat formation and the seeds is unpredictable. HDR failures would therefore be more appropriate candidates for salvage HIFU.

Local cancer control in 194 salvage HIFU sessions for 167 patients was achieved with negative biopsy results in 73 %. The actuarial 5-year overall survival rate was 84 %. The actuarial 3-year progression-free survival rate was significantly lower in three circumstances: worsening of the pre-EBRT stage, increase in the pre-sHIFU PSA and use of AD during management (Murat et al. 2009). The biochemical disease-free survival rate in 22 patients at 5 years was 52 %. Rates of bDFS in low-, intermediate- and high-risk groups were 100, 86 and 14 %, respectively (Uchida et al. 2011). After EBRT specific survival in 72 patients was 94 % at 3 years and 90 % at 5 years, and progression-free survival was 50 % at 3 years and 44 % at 5 years (Poissonnier et al. 2008). In 46 patients treated with sHIFU, 39.1 % were classified as failures, while another study with 31 patients showed 29 % with evidence of recurrence after a mean follow-up of 7.4 months (Berge et al. 2010; Zacharakis et al. 2008).

Urethrorectal fistula was described in 2–4.5 %, urethral stricture or intervention for necrotic tissue in 18–36 %, urinary tract infection or dysuria syndrome in 26 %, an epididymitis in 4.5 %, and 4.3 % of patients developed urethrocutaneous fistulae (Ahmed et al. 2009; Trachtenberg et al. 2007; Nathan et al. 2002). Rectourethral fistula after salvage HIFU was higher after the failure of combined brachytherapy and EBRT. In another series, no rectal complications were observed (Zacharakis et al. 2008; Ahmed et al. 2009).

The urinary incontinence rate is described in 7–44 % divided in grade 1 (12–18 %) and grade 2/3 (17.3–32 %) (Ahmed et al. 2009; Shariat et al. 2005; Trachtenberg et al. 2007; Nathan et al. 2002). Urinary incontinence after HIFU has been treated by urinary sphincter implantations (Zacharakis et al. 2008). Erectile function sufficient for intercourse pre-HIFU was seen in 15.2 % of patients, and only two men (4.3 %) remained potent post-HIFU (Ahmed et al. 2009; Trachtenberg et al. 2007).

19.5.2.3 Conclusion

The first results on sHIFU are encouraging and indicate that the procedure is a reasonable treatment option, but better patient selection criteria are needed. Salvage HIFU appears most appropriate for those patients with histologically proven local recurrence only, with a life expectancy of at least 5 years and with some medical comorbidity contraindicating salvage prostatectomy. The side effects are not negligible but it is well tolerated by patients and seems to have lower morbidity than salvage RP and cryoablation. Mild to severe incontinence remains a major concern. This is most probably related to the treatment of the complete prostate, including the apex. Contraindications include anorectal absence or significant pathology, widespread prostate calcifications and high prostate volume. The role of salvage HIFU after LDR brachytherapy seems to be small since interaction between metal seeds and HIFU-induced heating is difficult to assess.

19.5.3 Salvage RITA

Radiofrequency interstitial tumour ablation has been described and is executed by ultrasound (TRUS)-guided transrectal needle delivery of thermal energy. It is a treatment technique that uses high frequency alternating electrical current to destroy tissue cells by heating them. Procedures can be performed in the cystoscopy suite of the outpatient clinic under intravenous sedation. The treatment plan consists of manually raising the power in a stepwise manner until the temperature at each individual hook reaches 100 °C (approximate duration, 2–3 min). The temperature is then maintained at 100 °C for 5 min. The urethra is cooled with normal saline through a 3-way Foley catheter placed just before the procedure. Cold saline is introduced within the rectum when the rectal thermocouple exceeds 41 °C.

The published experience is restricted to a pilot study in which 11 patients with biopsy-proven, hormone-naïve, clinically localized prostate cancer were enrolled in a prospective phase I/II trial. Eight patients had failed prior radiation therapy, and 3 were not candidates for curative primary therapy. Serum PSA levels decreased after RITA >50 % in 90 % of patients, >70 % in 72 % of patients, and >80 % in 46 % of patients. At 12 months after RITA, 50 % of patients with sufficient follow-up had no residual cancer on repeat systematic 12-core biopsy, and 67 % were cancer-free in biopsy cores sampled from the RITA-treated areas. In the radiation failure group, PSADT after RITA was 127.1 % longer than that before RITA. The placement of 1/4 lesions was aborted in two patients due to increasing rectal temperature. Complications included transient macrohaematuria (19 %), bladder spasms (9 %) and dysuria (9 %) (Shariat et al. 2005).

RITA is a minimally invasive, rapid, user-friendly, office-based procedure that is well tolerated. The efficacy of RITA is limited by the proper identification of cancerous prostatic lesions, and no patients with undetectable levels of PSA were described and only one patient had a sustained decrease in PSA of >80 %. This gives the impression that this technique is less effective than other minimally invasive salvage therapies, but only a few patients have been described with short-term follow-up.

19.5.4 Salvage Photodynamic Therapy

Photodynamic therapy is based on the administration of an energy source in the form of light of a specific wavelength, on a previously photosensitized tissue by a chemical compound, in the presence of oxygen, inducing the generation of free radicals and oxygen derivatives (hydroxyl compounds). Ultimately, these destructive reactions produce necrosis of the treated tissue and damage their blood supply. Improvements in technique during recent years have allowed its development as a therapeutic method for localized prostate cancer. It can be performed through transperineally placed catheters for light delivery and light dosimetry. Treatment response is assessed primarily by a hypovascular lesion formation on contrast-enhanced magnetic resonance imaging and transrectal ultrasound-guided biopsies targeting areas of lesion formation and secondarily by serum PSA changes.

The published experience is limited to a description of the technique and description of photosensitizers and the treatment is experimental.

Tookad, an intravascular photosensitizer, can be activated by 763 nm light wavelength. The light dose escalation demonstrated an increasing volume of effect with all six patients undergoing treatment responding at the highest light dose with lesions encompassing up to 70 % of the peripheral zone. There were no serious adverse events, and continence and potency were maintained (Trachtenberg et al. 2007). In a therapeutic trial in 14 men, treatment was well tolerated. PSA decreased in 9 patients and 5 had no viable tumour on posttreatment biopsies. Imaging showed necrosis involving up to 91 % of the prostate cross section. In 4 men stress incontinence developed which is slowly improving. Sexual potency was impaired in 4 of the 7 men able to have intercourse before photodynamic therapy. In one patient a urethrorectal fistula developed following a rectal biopsy shortly after therapy (Nathan et al. 2002). In concussion early histological and magnetic resonance imaging responses highlight the clinical potential of vascular-targeted photodynamic therapy to manage radiorecurrent prostate cancer. Precise light dosimetry has to be further evaluated and merits further investigation.

19.6 Conservative Management with or Without Hormonal Therapy

Expectant management, without initial hormonal therapy, may be a reasonable option for selected patients with prostate cancer who present with biochemical failure without metastatic disease after brachytherapy. It can be offered to patients with a less than 10-year life expectancy and patients desiring to avoid the complications associated with aggressive salvage therapy. Patients experiencing PSA failures after definitive radiotherapy who have a doubling time <3–6 months are at particularly high risk for developing distant metastases and experiencing prostate cancer-specific mortality, so a greater emphasis on androgen deprivation rather than local treatment is warranted (Pinover et al. 2003).

The start of early ADT results in improved OS but not improved prostate cancer-specific mortality and local failure in patients with prostate cancer who were not a candidate for curative treatment (Studer et al. 2006). A randomized trial to define the optimal timing for salvage hormonal therapy is warranted in this group of patients with PSA recurrence after RT (Souhami et al. 2010). In a prospective trial in 197 patients, the Swiss Group for Clinical Cancer Research (SAKK) was unable to show any major advantage of immediate compared with deferred hormonal treatment regarding quality of life or overall survival. Disabling complications were prevented in the deferred-treatment arm by careful follow-up; 42 % of these patients never required any tumour-specific treatment (Studer et al. 2004).

Conclusion

In patients with prostate carcinoma treated initially with brachytherapy who develop biochemical recurrence, evaluation should be performed to determine the extent of disease, and the patient should be informed about the different treatment options. These patients should be discussed in a multidisciplinary setting. The decision should be taken after balancing treatment outcomes with life expectancy.

Salvage radical prostatectomy may be the preferred treatment for many patients with recurrent prostate carcinoma after radiotherapy. It is reserved for motivated patients with organ-confined recurrence that accept the significant morbidity especially the chance on incontinence. Salvage brachytherapy seems to be a good alternative but merits further evaluation with focus on functional results. The minimally invasive treatment modalities can be used in patients with a wish for treatment, but these treatments should be considered investigational in this setting. Optimal timing of salvage hormone manipulation still has to be defined but may result in improved OS. Patient selection and timing of salvage treatment is an influencing factor in the choice of therapy. Longer follow-up and prospective trials are needed to evaluate the optimal patient selection and strategy.

References

Ahallal Y, Shariat SF, Chade DC, Mazzola C, Reuter VE, Sandhu JS, Laudone VP, Touijer KA, Guillonneau BD (2011) Pilot study of salvage laparoscopic prostatectomy for the treatment of recurrent prostate cancer. BJU Int 107

Ahmed HU, Ishaq A, Zacharakis E et al (2009) Rectal fistula after salvage high-intensity focused ultrasound for recurrent prostate cancer after combined brachytherapy and external beam radiotherapy. BJU Int 102:321–323

Allen GW, Howard AR, Jarrard DF, Ritter MA (2007) Management of prostate cancer recurrences after radiation therapy-brachytherapy as a salvage option. Cancer 110(7):1405–1416

Anastasiadis AG, Sachdev R, Salomon L et al (2003) Comparison of health-related quality of life and prostate-associated symptoms after primary and salvage cryotherapy for prostate cancer. J Cancer Res Clin Oncol 129:676–682

Bahn DK, Lee F, Silverman P et al (2003) Salvage cryosurgery for recurrent prostate cancer after radiation therapy: a seven-year follow-up. Clin Prostate Cancer 2:111–114

Baumert H (2010) Salvage treatments for prostatic radiation failure. Cancer Radiother 14(6–7): 442–445

Berge V, Baco E, Karlsen SJ (2010) A prospective study of salvage high-intensity focused ultrasound for locally radiorecurrent prostate cancer: early results. Scand J Urol Nephrol 44(4): 223–227

Beyer DC (2004a) Salvage brachytherapy after external-beam irradiation for prostate cancer. Oncology 18(2):151–158; discussion 158–160, 163–164

Beyer DC (2004b) Salvage brachytherapy after external-beam irradiation for prostate cancer. Oncology (Williston Park) 18(2):151–158

Bianco FJ Jr, Scardino PT, Stephenson AJ, Diblasio CJ, Fearn PA, Eastham JA (2005) Long-term oncologic results of salvage radical prostatectomy for locally recurrent prostate cancer after radiotherapy. Int J Radiat Oncol Biol Phys 62(2):448–453

Boccon-Gibod L, Djavan WB, Hammarer P et al (2004) Management of prostate-specific antigen relapse in prostate cancer: a European Consensus. Int J Clin Pract 58(4):382–390

Boris RS, Bhandari A, Krane LS, Eun D, Kaul S, Peabody JO (2009) Salvage robotic-assisted radical prostatectomy: initial results and early report of outcomes. BJU Int 103(7):952–956

Burri RJ, Stone NN, Unger P, Stock RG (2010) Long-term outcome and toxicity of salvage brachytherapy for local failure after initial radiotherapy for prostate cancer. Int J Radiat Oncol Biol Phys 77(5):1338–1344

Chade DC, Shariat SF, Cronin AM, Savage CJ, Karnes RJ, Blute ML, Briganti A, Montorsi F, van der Poel HG, Van Poppel H, Joniau S, Godoy G, Hurtado-Coll A, Gleave ME, Dall'oglio M, Srougi M, Scardino PT, Eastham JA (2011) Salvage radical prostatectomy for radiation-recurrent prostate cancer: a multi-institutional collaboration. Eur Urol 60:205–210

Chauhan S, Patel MB, Coelho R, Liss M, Rocco B, Sivaraman AK, Palmer KJ, Coughlin GD, Ferrigni RG, Castle EP, Ahlering TE, Parra-Davila E, Patel VR (2011) Preliminary analysis of the feasibility and safety of salvage robot-assisted radical prostatectomy after radiation failure: multi-institutional perioperative and short-term functional outcomes. J Endourol 25(6):1013–1019

Coakley FV, Hricak H, Wefer AE, Speight JL, Kurhanewicz J, Roach M (2001) Brachytherapy for prostate cancer: endorectal MR imaging of local treatment-related changes. Radiology 219:817–821

Cox JD, Gallagher MJ, Hammond EH, Kaplan RS, Schellhammer PF (1999) Consensus statements on radiation therapy of prostate cancer: guidelines for prostate re-biopsy after radiation and for radiation therapy with rising prostate-specific antigen levels after radical prostatectomy. American Society for Therapeutic Radiology and Oncology Consensus Panel. J Clin Oncol 17(4):1155

Darras J, Joniau S, Van Poppel H (2006) Salvage radical prostatectomy for radiorecurrent prostate cancer: indications and results. Eur J Surg Oncol 32:964–969

De Visschere PJ, De Meerleer GO, Fütterer JJ, Villeirs GM (2010) Role of MRI in follow-up after focal therapy for prostate carcinoma. AJR Am J Roentgenol 194(6):1427–1433, Review

Dotan ZA, Bianco FJ Jr, Rabbani F, Eastham JA, Fearn P, Scher HI, Kelly KW, Chen HN, Schöder H, Hricak H, Scardino PT, Kattan MW (2005) Pattern of prostate-specific antigen (PSA) failure dictates the probability of a positive bone scan in patients with an increasing PSA after radical prostatectomy. J Clin Oncol 23(9):1962

Eandi JA, Link BA, Nelson RA, Josephson DY, Lau C, Kawachi MH, Wilson TG (2010) Robotic assisted laparoscopic salvage prostatectomy for radiation resistant prostate cancer. J Urol 183(1):133–137

Eisenberg ML, Shinohara K (2008) Partial salvage cryoablation of the prostate for recurrent prostate cancer after radiotherapy failure. Urology 72(6):1315–1318

Engelbrecht MR, Puech P, Colin P, Akin O, Lemaitre L, Villers A (2010) Multimodality magnetic resonance imaging of prostate cancer. J Endourol 24(5):677–684

Ficarra V, Novara G, Artibani W, Cestari A, Galfano A, Graefen M, Guazzoni G, Guillonneau B, Menon M, Montorsi F, Patel V, Rassweiler J, Van Poppel H (2009) Retropubic, laparoscopic, and robot-assisted radical prostatectomy: a systematic review and cumulative analysis of comparative studies. Eur Urol 55(5):1037–1063

Finley DS, Belldegrun AS (2011) Salvage cryotherapy for radiation-recurrent prostate cancer: outcomes and complications. Curr Urol Rep 12(3):209–215

Gotto T, Yunis LH, Vora K, Eastham JA, Scardino PT, Rabbani F (2010) Impact of prior prostate radiation on complications after radical prostatectomy. J Urol 184:136–142

Grado GL, Collins JM, Kriegshauser JS et al (1999) Salvage brachytherapy for localized prostate cancer after radiotherapy failure. Urology 53:2–10

Grossfeld GD, Stier DM, Flanders SC, Henning JM, Schonfeld W, Warolin K, Carroll PR (1998) Use of second treatment following definitive local therapy for prostate cancer: data from the caPSURE database. J Urol 160(4):1398–1404

Haider MA, Chung P, Sweet J et al (2008) Dynamic contrast-enhanced magnetic resonance imaging for localization of recurrent prostate cancer after external beam radiotherapy. Int J Radiat Oncol Biol Phys 70:425–430

Hancock SL, Cox RS, Bagshaw MA (1995) Prostate specific antigen after radiotherapy for prostate cancer: a reevaluation of long-term biochemical control and the kinetics of recurrence in patients treated at Stanford University. J Urol 154(4):1412–1417

Heidenreich A, Richter S, Thüer D, Pfister D (2010) Prognostic parameters, complications, and oncologic and functional outcome of salvage radical prostatectomy for locally recurrent prostate cancer after 21st-century radiotherapy. Eur Urol 57:437–445

Isebaert S, De Keyzer F, Haustermans K, Lerut E, Roskams T, Roebben I, Van Poppel H, Joniau S, Oyen R (2011) Evaluation of semi-quantitative dynamic contrast-enhanced MRI parameters for prostate cancer in correlation to whole-mount histopathology. Eur J Radiol. doi:10.1016/j.ejrad.2011.01.107

Jamal K, Challacombe B, Elhage O et al (2008) Successful salvage robotic-assisted radical prostatectomy after external beam radiotherapy failure. Urology 72:1356–1358

Kaouk JH, Hafron J, Goel R, Haber GP, Jones JS (2008) Robotic salvage retropubic prostatectomy after radiation/brachytherapy: initial results. BJU Int 102(1):93–96

Kollmeier MA, Stock RG, Stone N (2003) Biochemical outcomes after prostate brachytherapy with 5-year minimal follow-up: importance of patient selection and implant quality. Int J Radiat Oncol Biol Phys 57:645–653

Koutrouvelis P, Hendricks F, Lailas N et al (2003) Salvage reimplantation in patient with local recurrent prostate carcinoma after brachytherapy with three dimensional computed tomography-guided permanent pararectal implant. Technol Cancer Res Treat 2:339–344

Lam JS, Belldegrun AS (2004) Salvage cryosurgery of the prostate after radiation failure. Rev Urol 6(Suppl 4):S27–S36

Lange PH, Ercole CJ, Lightner DJ et al (1989) The value of serum prostate specific antigen determinations before and after radical prostatectomy. J Urol 141(4):873–879

Lee HK, Adams MT, Motta J (2008) Salvage prostate brachytherapy for localized prostate cancer failure after external beam radiation therapy. Brachytherapy 7(1):17–21

Leonardo C, Simone G, Papalia R, Franco G, Guaglianone S, Gallucci M (2009) Salvage radical prostatectomy for recurrent prostate cancer after radiation therapy. Int J Urol 16(6):584–586

Loening SA, Turner JW (1993) Use of percutaneous transperineal 198Au seeds to treat recurrent prostate adenocarcinoma after failure of definitive radiotherapy. Prostate 23:283–290

Moman MR, van der Poel HG, Battermann JJ, Moerland MA, van Vulpen M (2010) Treatment outcome and toxicity after salvage 125-I implantation for prostate cancer recurrences after primary 125-I implantation and external beam radiotherapy. Brachytherapy 9(2):119–125, Epub 2009 Oct 21

Moul JW (2000) Prostate specific antigen only progression of prostate cancer. J Urol 163(6):1632–1642

Murat FJ, Poissonnier L, Rabilloud M, Belot A, Bouvier R, Rouviere O, Chapelon JY, Gelet A (2009) Mid-term results demonstrate salvage high-intensity focused ultrasound (HIFU) as an effective and acceptably morbid salvage treatment option for locally radiorecurrent prostate cancer. Eur Urol 55(3):640–647

Nathan TR, Whitelaw DE, Chang SC, Lees WR, Ripley PM, Payne H, Jones L, Parkinson MC, Emberton M, Gillams AR, Mundy AR, Bown SG (2002) Photodynamic therapy for prostate cancer recurrence after radiotherapy: a phase I study. J Urol 168(4 Pt 1):1427–1432

Ng CK, Moussa M, Downey DB, Chin JL (2007) Salvage cryoablation of the prostate: followup and analysis of predictive factors for outcome. J Urol 178(4 Pt 1):1253–1257

Nguyen PL, Chen RC, Clark JA, Cormack RA, Loffredo M, McMahon E, Nguyen AU, Suh WW, Tempany CM, D'Amico AV (2009) Patient-reported quality of life after salvage brachytherapy for radio-recurrent prostate cancer: a prospective phase II study. Brachytherapy 8(4): 345–352

Nuñez-Mora C, García-Mediero JM, Cabrera-Castillo PM (2009) Radical laparoscopic salvage prostatectomy: medium-term functional and oncological results. J Endourol 23(8): 1301–1305

Onik M, Cohen JK, Reyes GD, Rubinsky B, Chang Z, Baust J (1993) Transrectal ultrasound-guided percutaneous radical cryosurgical ablation of the prostate. Cancer 72:1291–1299

Patel C, Elshaikh MA, Angermeier K et al (2004) PSA bounce predicts early success in patients with permanent iodine-125 prostate implant. Urology 63:110–113

Pinover WH, Horwitz EM, Hanlon AL, Uzzo RG, Hanks GE (2003) Validation of a treatment policy for patients with prostate specific antigen failure after three-dimensional conformal prostate radiation therapy. Cancer 97(4):1127–1133

Pisters LL, Rewcastle JC, Donnelly BJ, Lugnani FM, Katz AE, Jones JS (2008) Salvage prostate cryoablation: initial results from the cryo on-line data registry. J Urol 180(2):559–563

Pisters LL, Leibovici D, Blute M, Zincke H, Sebo TJ, Slezak JM, Izawa J, Ward JF, Scott SM, Madsen L, Spiess PE, Leibovich BC (2009) Locally recurrent prostate cancer after initial radiation therapy: a comparison of salvage radical prostatectomy versus cryotherapy. J Urol 182(2):517–525

Poissonnier L, Murat FJ, Belot A, Bouvier R, Rabilloud M, Rouviere O, Chapelon JY, Gelet A (2008) Locally recurrent prostatic adenocarcinoma after exclusive radiotherapy: results of high intensity focused ultrasound. Prog Urol 18(4):223–229, Epub 2008 May 2

Potters L, Morgenstern C, Calugaru E, Fearn P, Jassal A, Presser J, Mullen E (2005) 12-Year outcomes following permanent prostate brachytherapy in patients with clinically localized prostate cancer. J Urol 173(5):1562–1566

Sanderson KM, Penson DF, Cai J, Groshen S, Stein JP, Lieskovsky G, Skinner DG (2006) Salvage radical prostatectomy: quality of life outcomes and long-term oncological control of radiorecurrent prostate cancer. J Urol 176(5):2025–2031; discussion 2031–2032

Schuster DM, Savir-Baruch B, Nieh PT, Master VA, Halkar RK, Rossi PJ, Lewis MM, Nye JA, Yu W, Bowman FD, Goodman MM (2011) Detection of recurrent prostate carcinoma with anti-1-amino-3-18F-fluorocyclobutane-1-carboxylic acid PET/CT and 111In-capromab pendetide SPECT/CT. Radiology 259(3):852–861

Seitz M, Gratzke C, Schlenker B, Buchner A, Karl A, Roosen A, Singer BB, Bastian PJ, Ergün S, Stief CG, Reich O, Tilki D (2011) Contrast-enhanced transrectal ultrasound (CE-TRUS) with cadence-contrast pulse sequence (CPS) technology for the identification of prostate cancer. Urol Oncol 29(3):295–301, Epub 2009 Jun 12

Shariat SF, Raptidis G, Masatoschi M, Bergamaschi F, Slawin KM (2005) Pilot study of radiofrequency interstitial tumor ablation (RITA) for the treatment of radio-recurrent prostate cancer. Prostate 65(3):260–267

Smathers S, Wallner K, Sprouse J et al (2001) Temporary PSA rises and repeat prostate biopsies after brachytherapy. Int J Radiat Oncol Biol Phys 50:1207–1211

Souhami L, Bae K, Pilepich M, Sandler H (2010) Timing of salvage hormonal therapy in prostate cancer patients with unfavorable prognosis treated with radiotherapy: a secondary analysis of Radiation Therapy Oncology Group 85–31. Int J Radiat Oncol Biol Phys 78(5):1301–1306

Stephenson AJ, Eastham JA (2005) Salvage RP may favorably alter the natural history of biochemical recurrence after radiation therapy, but it must be instituted early in the course of recurrent disease to be effective. J Clin Oncol 23(32):8198–8203

Studer UE, Hauri D, Hanselmann S, Chollet D, Leisinger HJ, Gasser T, Senn E, Trinkler FB, Tscholl RM, Thalmann GN, Dietrich D (2004) Immediate versus deferred hormonal treatment for patients with prostate cancer who are not suitable for curative local treatment: results of the randomized trial SAKK 08/88. J Clin Oncol 22(20):4109–4118

Studer UE, Whelan P, Albrecht W, Casselman J, de Reijke T, Hauri D, Loidl W, Isorna S, Sundaram SK, Debois M, Collette L (2006) Immediate or deferred androgen deprivation for patients with prostate cancer not suitable for local treatment with curative intent: European Organisation for Research and Treatment of Cancer (EORTC) Trial 30891. J Clin Oncol 24(12): 1868–1876

Taymoorian K, Thomas A, Slowinski T, Khiabanchian M, Stephan C, Lein M, Deger S, Lenk S, Loening SA, Fischer T (2007) Transrectal broadband-Doppler sonography with intravenous contrast medium administration for prostate imaging and biopsy in men with an elevated PSA value and previous negative biopsies. Anticancer Res 27(6C):4315–4320

Tharp M, Hardacre M, Bennett R, Jones WT, Stuhldreher D, Vaught J (2008) Prostate high-dose-rate brachytherapy as salvage treatment of local failure after previous external or permanent seed irradiation for prostate cancer. Brachytherapy 7(3):231–236

Trachtenberg J, Bogaards A, Weersink RA, Haider MA, Evans A, McCluskey SA, Scherz A, Gertner MR, Yue C, Appu S, Aprikian A, Savard J, Wilson BC, Elhilali M (2007) Vascular targeted photodynamic therapy with palladium-bacteriopheophorbide photosensitizer for recurrent prostate cancer following definitive radiation therapy: assessment of safety and treatment response. J Urol 178(5):1974–1979

Uchida T, Shoji S, Nakano M, Hongo S, Nitta M, Usui Y, Nagata Y (2011) High-intensity focused ultrasound as salvage therapy for patients with recurrent prostate cancer after external beam radiation, brachytherapy or proton therapy. BJU Int 107(3):378–382

van der Poel HG, Beetsma DB, van Boven H, Horenblas S (2007) Perineal salvage prostatectomy for radiation resistant prostate cancer Department of Urology, Netherlands Cancer Institute, Amsterdam, The Netherlands. Eur Urol 51:1565–1571

Verhagen PC, van Duijn PW, Hermans KG et al (2006) The PTEN gene in locally progressive prostate cancer is preferentially inactivated by bi-allelic gene deletion. J Pathol 208(5):699–707

Wallner KE, Nori D, Morse MJ et al (1990) 125Iodine reimplantation for locally progressive prostatic carcinoma. J Urol 144:704–706

Williams AK, Martínez CH, Lu C, Ng CK, Pautler SE, Chin JL (2011) Disease-free survival following salvage cryotherapy for biopsy-proven radio-recurrent prostate cancer. Eur Urol 60(3):405–410

Wong WW, Buskirk SJ, Schild SE et al (2006) Combined prostate brachytherapy and short-term androgen deprivation therapy as salvage therapy for locally recurrent prostate cancer after external beam irradiation. J Urol 176:2020–2024

Yin L, Choi WW, Gu X, et al (2010) Complications of primary vs salvage cryotherapy for prostate cancer. Presented at the American Urologic Association Meeting, San Francisco. 29 May–3 June 2010. Abstract #1066

Zacharakis E, Ahmed H-U, Ishaq A et al (2008) The feasibility and safety of high-intensity focused ultrasound as salvage therapy for recurrent prostate cancer following external beam radiotherapy. BJU Int 102:786–792

Prostate Cancer Brachytherapy: Radiation Protection Issues

20

Jean-Marc Cosset and Lawrence Dauer

20.1 Introduction

Brachytherapy for prostate cancer is not a new idea. As soon as 1913, Pasteau and Degrais (1913) started to treat prostatic cancers with radium tubes or needles, and almost in parallel, "radon seeds" were developed as treatment approaches using tiny glass capsules containing the "emanation" progeny of radium, the radon gas. In the 1970s, a few authors (Court and Chassagne 1977) proposed afterloading techniques based on the insertion of plastic tubes in or around the prostate, with a secondary loading of iridium wires, but apart from a few centers, the technique was abandoned. It were the problems of radiation protection associated with these techniques that played a significant role in their abandonment.

In current practice there are essentially two techniques used worldwide for prostate cancer brachytherapy. The first is the low-dose-rate (LDR) technique based on the permanent implantation of radioactive iodine (or palladium) seeds (performed as early as 1970 at Memorial Sloan-Kettering Cancer Center (Hilaris et al. 1987)). The second is the HDR technique, based on the temporary insertion through catheters, needles, or tubes, of a high-dose-rate miniaturized iridium (or more recently cobalt) radioactive source.

These two techniques have inherently different radiation protection issues and will be analyzed separately in specific sections below. However, in terms of overall radiation protection, there are a number of common issues, and for over 10 years, the International Commission on Radiological Protection (ICRP) has been developing guidance on the topic.

In 2000, ICRP released its Publication 86, on "Prevention of accidental exposures to patients undergoing radiation therapy" (*ICRP Publication 86.* Ann

J.-M. Cosset (✉) • L. Dauer
Department of Radiotherapy, Institut Curie, Paris, France

Memorial Sloan-Kettering Cancer Center (MSKCC), New-York, NY, USA
e-mail: jean-marc.cosset@curie.net

G. Kovács and P. Hoskin (eds.), *Interstitial Prostate Brachytherapy*,
DOI 10.1007/978-3-642-36499-0_20, © Springer-Verlag Berlin Heidelberg 2013

ICRP2000). The document addressed all types of accidents in radiotherapy and includes specific chapters devoted to brachytherapy. After a detailed analysis of the accidents which were reported worldwide prior to 2000, the task group charged with the writing of the document attempted to identify the causes and the factors contributing to accidental exposures in brachytherapy.

The document concentrated on the accidents occurring along the whole chain of procedures, with successive chapters on equipment problems, accidents linked to source ordering, delivery, calibration and acceptance, treatment planning, source preparation, treatment delivery, and source removal (with a number of accidents reported at this step associated with sources mistakenly remaining in the patient for various lengths of time); and also accidents involving public exposure and environmental contamination.

A large number of accidents/incidents/mistakes/errors were analyzed, allowing the writing group to extract a few generic lessons. Actually, in most of the accidents reported, it was a combination of contributing factors which allowed an initial mistake (sometimes a minor one) to escalate into an accidental exposure. Often, the lack of general radiation protection and poor safety concern of management was the underlying root cause.

Among the main contributing factors to those accidents ICRP Publication 86 listed were a lack of appropriate staff resources, insufficiently qualified or untrained staff, lack of an effective, systematic quality assurance program and procedures, and lack of effective communication procedures (as a number of accidents would have been avoided with improved communication between physician and physicists, physicists and technicians).

Having set the general principles for radioprotection in brachytherapy in ICRP Publication 86 in 2000, ICRP decided to more specifically address the problems of radioprotection in HDR and LDR brachytherapy. This was accomplished with the publication in 2005 of the two documents, ICRP 97 (*ICRP Publication 97.* Ann ICRP 2005a) and ICRP 98 (*ICRP Publication 98.* Ann ICRP 2005b). These two publications will serve as the core of the recommendations and suggestions in subsequent sections, with updates taking advantage of more recent studies.

20.2 HDR Prostate Brachytherapy: Radiation Protection Issues

20.2.1 Introduction

HDR is increasingly used in prostate cancer brachytherapy. Initially, it was used as a "boost" in conjunction with external radiotherapy. This enables escalation of the localized dose given to patients with intermediate- and high-risk localized prostate cancers. A number of centers are also proposing and evaluating HDR prostate brachytherapy as the sole treatment for selected low-risk, localized prostate cancers.

With an increasing number of patients expected to benefit from this technique, radioprotection problems are clearly to be kept in mind.

With adequate radiation protection design for the room where the HDR afterloader is operating, and with normal quality controlled operation, there should be minimal radiation protection problems for the patient (who should receive the correct dose in a precisely planned manner) or for the staff (who should not receive excess radiation doses, as they are located outside of the treatment room).

However, along the long chain of procedures for HDR treatment including source packaging and travel, the source preparation, the treatment planning, and the treatment delivery, a number of problems may (and actually unfortunately have) occurred. These events lead to more or less severe accidents depending on the specific situations.

20.2.2 Lessons Learned from a Severe HDR Accident

What is usually considered as the worst accident in HDR brachytherapy occurred in 1992 in the USA. It was not related to HDR brachytherapy for prostate cancer, but to treatment of an anorectal cancer. However, this accident is clearly a "model" (not to be reproduced!) for all types of HDR treatment.

What happened? (ICRP *Publication 97)* The source (HDR iridium-192) became detached from the drive mechanism at the moment of the planned retraction of the source (which therefore remained in the patient). Unfortunately, the physicians in charge had to deal with conflicting signals as the area radiation monitor actually detected the radiation, but the equipment (afterloader) indicated that the source had been shielded. In addition, radiation monitor malfunctions in the months leading up to the accident encouraged misinterpretation and induced the staff not to trust the indicators. Consequently, the wrong indication ("source shielded") of the equipment was accepted, and the patient, clothes, and room were not subsequently even checked with another radiation monitor or survey instrument.

The HDR source remained within the patient for 4 days, delivering a total dose of about 16,000 Gy (of note, the prescription was only for 18 Gy). The patient died on day 4. The catheter with the source went unrecognized, although it was removed from the patient along with necrotic tissues. This material was subsequently disposed of in a waste container, without identification of the source at that time. The waste container was picked up by a commercial medical waste disposal company 5 days later. It was then taken to an incinerator where the HDR radiation source was finally detected, recognized, and appropriately shielded. During the days the source remained in the patient or in the waste container, it irradiated 94 persons to various external dose levels.

20.2.3 ICRP Publication 97

This publication analyzed all the HDR brachytherapy accidents and incidents reported through 2005. Overall, they found that very few problems were associated with prostate treatment at that time. This was probably because the technique was still in its infancy in a number of countries or centers. However, the analysis of similar events to those published in ICRP Publication 97 can be useful as the lessons can clearly be extrapolated to prostate HDR brachytherapy. The "main points" from ICRP 97 are listed below:

- High-dose-rate (HDR) brachytherapy is a rapidly growing technique that has been replacing low-dose-rate (LDR) procedures over the last years in both industrialized and developing countries. It is estimated (in 2005) that about 500,000 procedures (treatment administrations) are performed by HDR units annually.
- LDR equipment (for temporary implantation) has been discontinued by many manufacturers over the last years, leaving HDR brachytherapy as the major alternative.
- HDR techniques deliver a very high dose rate, of the order of 1.6–5.0 Gy/min, so mistakes can lead to under- or overdosage with the potential for clinical adverse effects.
- More than 500 HDR accidents/incidents including the death described above have been reported along the entire chain of procedures from source packing to the delivery of dose. Human error has been the prime cause of radiation events.
- Many accidents could have been prevented if staff had functional monitoring equipment and paid attention to the results.
- Since iridium-192 has a relatively short half-life, the HDR sources need to be replaced approximately every 4 months. Over 10,000 HDR sources are transported annually (as estimated in 2005), with the resultant potential for accidents.
- A team of trained personnel following quality assurance (QA) procedures is necessary to prevent accidents. QA should include peer reviews of cases.
- Accidents and incidents should be reported, and the lessons learned should be shared with other users to prevent similar mistakes.

20.2.4 Recent Updates

Six additional years of experience have confirmed the validity of the "main points" of the ICRP 97 document.

Recently, several authors and societies have emphasized that there is only a very short time for avoiding catastrophic consequences in the case of a source sticking in the patient or in the tubes driving the mechanism, once such an event is recognized. It is estimated, due to the extremely high-dose rates, that in such a case, the staff must react to correct the problem within 1–2 min. This minimal opportunity for mitigation, by necessity, requires specific organization and emergency response training.

In a paper published in 1999, entitled "Emergency rescue in accidents with HDR afterloading units," it is noted that there are occasional reported troubles with afterloading units concerning the retraction of sources, which require immediate action to limit possible damage. In such a situation, they stress that the quickest possible rescue of a patient in an emergency demands clear definition of responsibilities. It is advised that (as the organizational structure of the clinic allows) the emergency physician should invariably be the physician who placed the applicator. The authors conclude that "A well-practiced emergency management can be of life-saving importance for the patient."

20.3 LDR Prostate Brachytherapy: Radiation Protection Issues

20.3.1 Introduction

Permanent LDR brachytherapy, with the implantation of iodine-125 (or palladium-103) seeds, preceded HDR techniques to treat selected localized prostate cancers. We now have more than 25 years of experience with seed implantation for prostate cancer, with very satisfactory long-term results, both in terms of relapse-free survival, overall survival, and toxicity.

Due to the characteristics of the seeds (e.g., each one emitting low-energy X-rays, 27 kV for iodine-125), even their loss does not represent a major external dose threat. That is the reason why the task group in charge of the writing of ICRP Publication 98 specified that, in 2005, "there was no report of accidents" with this brachytherapy technique [5]. However, there existed a number of radiation protection issues that the task group did address (Sect. 20.3.3), and since that time, a major series of events has been reported in a hospital in Philadelphia, USA, which is widely considered the worst case reported to date.

20.3.2 Lessons Learned from a Series of Severe LDR Events

The problem occurring in Philadelphia was not a specific "accident," but rather a succession of events leading to the eventual reporting of 97 medical errors out of 116 prostate cancer implants, performed over the course of 6 years, from 2002 to 2008. However, due to the number of patients involved, and to the consequences which can be expected for those patients, it is quite understandable that such a major problem has been treated as an "accident."

What happened? In February 2002, the Philadelphia Veterans Affairs Medical Center (PVAMC) initiated its prostate brachytherapy program. In February 2003, during a seed prostate implant, 40 out of 74 seeds were inappropriately implanted in the patient's bladder. Those seeds were subsequently expelled and recovered. In October 2005, 45 out of 90 seeds were again mistakenly implanted into the patient's bladder and recovered.

In May 2008, the Veterans Affairs National Health Physics Program (NHPP) notified the US Nuclear Regulatory Commission (USNRC) of a possible medical event involving a patient that received a dose less than 80 % of the prescribed dose.

This triggered a detailed internal and external on-site inspection. Based on the initial results of these investigations, the PVAMC prostate brachytherapy program was suspended in June 2008. In October 2008, the prostate cancer brachytherapy programs were suspended in three other VA hospitals in Cincinnati, Jackson, and Washington.

The initial survey identified 92 "medical events." Fifty-seven were due to a dose less than 80 % of the prescribed dose, and 35 were due to a dose to an organ or tissue outside the treatment site that exceeded 0.5 Sv (including overdoses of rectum, bladder wall, or tissues surrounding the prostate); of note, the overdose often considerably exceeded 0.5 Sv.

The official reports on this "accident" included detailed dose reconstructions, clearly showing that in some cases, almost all the seeds were implanted *below* the prostate, which resulted in overdosing the penile bulb and the prostate surrounding tissues. In one example, the "administered dose" to the prostate was as low as 24 Gy, for a "prescribed dose" of 160 Gy. In other cases, a number of seeds were implanted above the prostate, with about half of them expelled in the urine after being implanted directly into the bladder.

Investigative reports attempted to identify the causes for such mistakes. They listed incorrect placement of seeds, inadequate procedures, poor management oversight of contractors, inadequate training of licensee staff, poor management oversight of brachytherapy program, no peer review, observed poor placement of seeds and no corrective action taken, as well as a lack of a safety culture.

20.3.3 ICRP Publication 98

As emphasized previously, at the time of writing this document, no adverse effects to medical staff or the patient's family have been reported, as the events noted in Sect. 20.3.2 had not come to light. However, ICRP felt necessary to address a number of radiation protection issues associated with the LDR permanent seed implant brachytherapy procedure (ICRP 98 2005b). The document included several chapters that are summarized here.

20.3.3.1 Dose to People Approaching the Implanted Patients

When the task group began its work, there were surprisingly few precise data in the literature. The task group therefore initiated several complementary measurements in particular at Memorial Sloan-Kettering Cancer Center (USA), in Leeds (UK), and at the Institut Curie (France). Table 20.1 summarizes the direct measurements which were made in those centers.

The results of these direct measurements show that the doses to family and household members will remain very low, usually well below the 1 mSv limit for the public, and not even approach the constraint level of 5 mSv set for comforters and carers of such patients by the IAEA(1996).

For inclusion in the ICRP document, the Institut Curie group compared the calculated and measured doses at contact and at 20 cm for 47 patients (Fig. 20.1);

Table 20.1 Direct measurement from the patient

	Nb of patients	Anterior μSr/h						Lateral μSr/h					
		Surface	20 cm	25 cm	30 cm	50 cm	100 cm	Surface	20 cm	25 cm	30 cm	50 cm	100 cm
Smathers I 125	19	50 (22–89)					<0.3	0.06					<0.3
Leeds I 125	62	26.75 (2–67)				2.6 (0.2–5.1)	0.75 (0–1.6)	1.43 (0.1–17.4)				0.3 (0–1.9)	0.1 (0–0.5)
Curie I 125	47	115 (17–350)	22 (4–61)					0.8 (0.2–1.5)					
MSKCC I 125	545	37.3 (0.9–221)			6.0 (0.2–32.7)		<0.9	1.9 (0.9–16.8)					<0.9
Smathers PD 103	19	17 (5–49)					<0.3	0.19					<0.3
MSKCC PD 103	72	8.2 (0.9–63.6)			2.9 (0.2–15)		<0.3	1.4 (0.9–6.2)					<0.9

Reprinted from ICRP 98, with permission

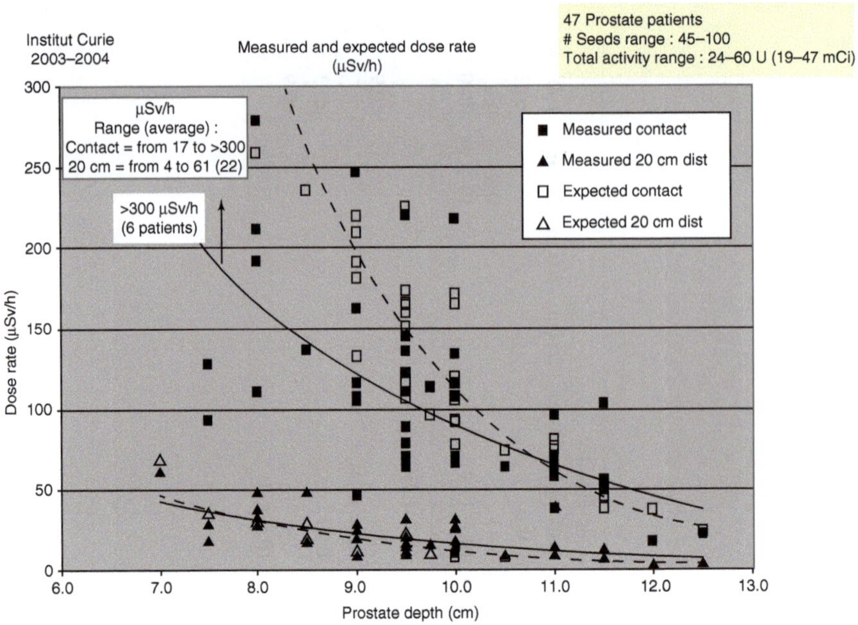

Fig. 20.1 Dose rate at abdomen surface (*squares*) and at 20 cm distance (*triangles*) for a series of 47 patients of Institut Curie for various patient thickness (the prostate depth was assumed to be half of the patient thickness) (Courtesy of Dr JC Rosenwald, reprinted from ICRP 98, with permission)

there was a very good agreement between measured and calculated doses at 20 cm, much less agreement at points of contact, perhaps because it is more difficult to measure a dose at a real skin "contact" location point.

Direct dose monitoring of family and household members has seldom been performed. In one of the few studies available, dosimeters were given to the patient, spouse, children, and pets of 44 patients, and 4 rooms frequently occupied by the patient were monitored (Michalski et al. 2003). He found very low levels of exposure, for example, the calculated mean lifetime dose to a spouse was 0.1 mSv for a ^{125}I implant.

Based on such data, ICRP Publication 98 set simple recommendations:

- No routine precautions are necessary, since doses to family or others will be below 1 mSv.
- Children should not to sit on lap of patient for 2 months.
- Avoid prolonged close contact with pregnant women.
- If a partner is pregnant consider individual risk assessment with dose rate measurements. (However, we will see below that these recommendations were subsequently slightly refined.)

20.3.3.2 Expelled Seeds

It is known that there is a slight risk that seeds may migrate after initial placement. In the rare case where they migrate to the lungs, they are "lost" from the prostate, but do not pose any radiation protection problem. Seeds may also be expelled from the patient's body in three ways: urine, semen, and gastrointestinal tract (this last case exceptional). Such migrations are usually considered to be more frequent with "free" individual seeds than with "stranded" seeds and appear to be correlated with lower levels of experience of the physician performing the implants.

In experienced teams, such migrations are now very rare (even with free seeds), but specific recommendations should be given to the patients:

- Sieve the urine while in hospital and for 3 days after implant.
- Wear condom for the first five ejaculations.
- If seed "found," do not touch. Put in protective container with spoon or tweezers and return to the department.
- If seed in lavatory bowl, flush away.

20.3.3.3 Cremation

Cremation may pose serious problems. While it is relatively uncommon in a number of countries, it is frequent in others (e.g., China, India), is the rule in Japan (Satoh et al. 2012), and is increasing in the USA (Dauer 2012).

The cremation of a patient previously implanted with iodine-125 or palladium-103 sources raises a number of issues related to the activity remaining in the patient's ashes, potentially responsible for irradiation of the crematorium staff and members of the family, and the possibility of airborne activity, potentially responsible for inhalation of radioactive particles by the crematorium staff and members of the public, and also for the triggering of some environmental monitors.

Only a few studies or reports focused on these issues (NCRP 1970; Yumoto et al. 2000; Que 2001). This dearth of information explained why the governmental recommendations were found to vary significantly between countries varying from 1 year or less (Japan, US, with precautions), to 2 years (Canada), and even 3 years (UK, France).

Finally, after evaluating and calculating the activity remaining in the patient's ashes and the potential airborne release, the ICRP considered that cremation can be allowed if 12 months have elapsed since an implantation performed with ^{125}I (3 months for ^{103}Pd). However, it must be kept in mind that some national authorities (UK, France), based on worse-case scenarios and using different types of calculations, are currently recommending much longer times (e.g., up to 3 years for ^{125}I).

In an excellent example of national verification of both the appropriateness and implementation of international radiological protection recommendations, deaths within 12 months after ^{125}I implantation for brachytherapy have been evaluated in an investigation of a unique radiation safety issue in Japan (Satoh et al. 2012). This impressive study of 15,427 patients from 2003 to 2010 offers a unique approach to such national verification finding that only 0.28 % of implanted patients died within the first 12 months and that the largest proportion of early deaths was due to cerebrovascular or cardiovascular disease, followed by malignant tumor and respiratory disease or infection. In addition, they find that in the overwhelming majority of early death cases. The prostate seeds were retrieved together with the prostate gland at autopsy (as suggested by the international recommendations).

20.3.3.4 Subsequent Pelvic or Abdominal Surgery

In rare cases, limited and careful transurethral resection (TURP) may be necessary after brachytherapy. This must be performed by an experienced surgeon, aware of the brachytherapy technique and necessary precautions. The removed seeds must be identified within the sieved prostate chips, put in a container using tweezers, and returned to the hospital in charge. Such an adapted TURP should not be performed sooner than 6 months after an ^{125}I implantation.

In case of subsequent abdominal or pelvic surgery, the surgeon has to be warned of the presence of the implant. To avoid potential problems, the patient should carry a card explaining that he had an implant and providing a hospital number for further clarification if necessary (of note, a specific appendix of ICRP 98 was devoted to the minimum information to be included on a "wallet card" to be given to the patient at hospital discharge).

20.3.3.5 Fathering of Children

Due to the reduction in volume and modified quality of the ejaculate, patients may think they are definitively infertile. Actually, the dose from the implant may not reach the threshold for castration, and a few cases of fatherhood have been reported after permanent implants. An extensive review of the literature estimated a dose of only 20 cGy for the dose to the testis (Mydlo and Lebed 2004). This strongly suggests that the effects of prostate brachytherapy on spermatogenesis in prostate cancer patients are minimal. After discussion within ICRP, it was postulated that current estimates of the genetic risks from radiation [14] suggest that a paternal testicular dose of 1 Gy to a patient would result in an excess of around 1 case in 300 live-born offspring. This is a small percentage increase (~4 %) over the natural incidence of these genetic effects, and these figures may serve to reassure patients on the relatively low risk of genetic effects in their children.

20.3.3.6 Triggering of Radiation Detection Monitors

Some radiation detection monitors are set at a very low alarm level (i.e., 1.5–2 times the natural background level in given locations). Such monitors can be found at the entry and exit of nuclear plants, nuclear research centers, waste areas, scrap metal factories and yards, and increasingly at airports and border crossings (e.g., as a security action associated with the detection of nuclear terrorism).

In case the patient triggers such a radiation detection monitor, he should be able to show the personal "wallet card" given to him at discharge from hospital (see above). It must be explained to the patient that current detection instruments are extremely sensitive and are able to detect radiation at levels well below those that are of concern to health.

20.3.3.7 Secondary Cancers

With almost no cases reported and very limited data available in 2005 about the risk of secondary cancers after prostate brachytherapy, ICRP Publication 98 concluded that "the risk of a second radio-induced cancer (in the lifetime of a patient) after prostate brachytherapy appears to be either nil, or extremely low; it seems that the benefits of the technique clearly outweigh by far the potential risk of second malignancies."

Prudently, the document noted that the follow-up was probably not long enough in some series to allow an unequivocal conclusion. However, more recent data confirm the 2005 position of the ICRP 98.

20.3.3.8 ICRP Publication 98 Appendices

A number of practical appendices appear in ICRP Publication 98:

- Appendix A: characteristics of the main permanently implanted radioactive sources used for prostate cancer
- Appendix B: dose measurement after an implantation of permanent sources for prostate cancer brachytherapy
- Appendix C: examples of minimum recommendations to be given to a patient undergoing prostate brachytherapy with permanently implanted sources
- Appendix D: personal identification card to be given to a patient undergoing permanent seed implantation

20.3.4 Recent Updates

20.3.4.1 Dose Received from the Patients

A key paper was published in 2010 by the Memorial Sloan-Kettering Cancer Center (MSKCC) group (Dauer et al. 2010) who studied a large cohort of patients (1,279 cases), for whom precise radiation exposure rate measurements have been obtained between 1995 and 2008. The first important message is that *no precaution* is necessary for a large panel of persons approaching the patients after a prostate implantation. Such is certainly the case for all implantations with palladium-103. After a typical implantation with iodine-125, no precaution at all is required for coworkers and nonpregnant adults (even those sleeping with the patient). Only a pregnant adult sleeping with the patient and children can in some situations reach the "limits" (of note, the limits chosen in the paper are still "conservative," since they were set at 50 % of the current international guidelines). The second message is that the authors propose an algorithm enabling the determination of the precaution time for a given patient, based upon the precise exposure rate measured at 30 cm from the patient. As mentioned above, those calculations are only useful for the case of a pregnant adult sleeping with the patient or in the case of children in the house who might be held by the patient. The result is that it is now possible to customize the duration of precautions for each patient. For example, at a median exposure rate of 5 μSv/h (0.5 mR/h) at 30 cm (for iodine-125), the authors report that the patient should avoid sleeping "in contact" with a pregnant adult for 84 days and avoid holding children on the lap for long periods of time (more than 1–3 h) for 42 days. Direct measurements on the patient and use of the algorithm now allow further refinements of the necessary precaution time. In the case of a very obese patient, with few seeds implanted, and with consequently a very low measured exposure rate, precaution times may be expected to be much shorter and even nil in some specific cases.

In contrast, for a skinny patient with a large number of implanted seeds, with a higher exposure rate (the authors report a maximum level of 36 μSv/h (3.6 mR/h)), the precaution time can be calculated to be significantly longer. These customized recommendations should serve to reassure both the patients and the authorities. In conclusion, in almost all cases, prostate brachytherapy patients should be considered as "normal people" (Cosset 2010).

Also in 2010, a Japanese group (Kono et al. 2011) reported on the dose received from implanted patients. From a series of measurements at a distance of 20, 50, and 100 cm, the authors conclude that the risk from the prostate brachytherapy patients to general public is quite low. Only in the case of close and prolonged contact with a pregnant woman or an infant should any risk be considered.

20.3.4.2 Secondary Cancers

Several papers addressing the risk of secondary cancers after brachytherapy have been published since the release of ICRP Publication 98.

In 2006, Moon et al. (2006) reported that patients who received external beam radiation therapy (EBRT) had significantly higher odds of developing second cancers both overall and in the areas that were exposed to radiation but that despite the higher doses of radiation delivered, patients who received radioactive implants had the lowest odds of developing second cancers.

Also in 2006, Liauw et al. (2006) found fewer second cancers after brachytherapy (1.6 %) than after EBRT (5.8 %), but the difference did not reach statistical significance ($p=0.06$). They concluded that "there may be an increased but small risk of developing a second malignancy after radiation therapy for prostate cancer."

Abdel-Wahab et al. (2008) found that the age-adjusted estimates of secondary primary cancers were greater with EBRT than with brachytherapy (2,178 vs. 1,901 SPCs/100,000; $p=0.025$) or with the no RT, no surgery group (1,971 SPCs/100,000; $p<0.0001$). However, this difference disappeared with time.

In 2009, Takam et al. (2009) reported that the average risk of developing secondary primary cancer was no greater than 0.6 % for all treatment techniques but was lower with either LDR or HDR brachytherapy alone compared with any EBRT technique.

In 2011, Huang et al. (2011) analyzed an RT cohort which consisted of 2120 patients matched on a 1:1 basis with surgical patients according to age and follow-up time. RT techniques consisted of conventional or two-dimensional RT (2DRT, 36 %), three-dimensional conformal RT and/or intensity modulated RT (3DCRT/IMRT, 29 %), brachytherapy (BT, 16 %), and a combination of 2DRT and BT (BT boost, 19 %). The authors found that of the different RT techniques, only 2DRT was associated with a significantly higher risk (HR 1.76, 95 % CI 1.32–2.35), but not BT boost (HR 0.83, 95 % CI 0.50–1.38), 3DCRT/IMRT (HR 0.81, 95 % CI 0.55–1.21), or BT (HR 0.53, 95 % CI 0.28–1.01).

Overall, the results obtained since the publication of ICRP Publication 98 support the conclusion reached in 2005 that the risk of a second malignancy linked to brachytherapy is at most very low and probably nil in most cases. This statement is in keeping with the large study of De Gonzalez et al. (2011) on the US SEER cancer registries, involving 647,672 cancer patients who were 5-year survivors and followed up for a mean of 12 years. The conclusion of this study, the largest ever published, is that a relatively small proportion of second cancers are related to radiotherapy in adults, suggesting that most are due to other factors, such as lifestyle or genetics.

Conclusions

The main concern in radiation protection in prostate cancer brachytherapy should be first to avoid accidents. Those accidents may involve the patients and also the medical staff and even members of the public in the case of loss of brachytherapy sources. This last risk cannot be neglected, when we keep in mind the number of sources involved (e.g., with about 10,000 HDR iridium-192 sources that have to be replaced and therefore transported for sometimes long distances) every year worldwide.

Analysis of past accidents and incidents has shown that the risks appear to be higher when working with HDR sources. Moreover, it has been stressed that in case of a problem with an HDR source, the most common being sticking of the source somewhere in the drive mechanism or disconnection of the source from the cable, corrective actions have to take place in the first few minutes following the event and ideally within 1 or 2 min. This will only be possible with well-trained and practiced staff, able to react rapidly, with a precise definition of responsibilities and a specific quality assurance program. The physician responsible for the corrective action should optimally be the same who placed the applicator.

LDR sources appear to pose less frequent problems. The seeds used for prostate cancer LDR brachytherapy are individually much less dangerous than their HDR counterparts, due to their lesser activity and to the low-energy X-rays emitted. If lost as in the case of expelled seeds in the urine, or seeds lost during the implantation procedure, the seeds raise limited and generally easily solved radiation protection problems; but prevention and corrective measures are necessary. The worst "accident" to date with implanted LDR seeds relates to a series of events in a US hospital, leading to a high percentage of patients receiving an inadequate dose from poor-quality implants.

Implanted seeds raise several other radiation protection issues that have been addressed by ICRP: the dose received from the patient is almost negligible and should not trigger any particular measure, except in the few cases where there may be contact with small children or pregnant women. Expelled seeds are now rarely seen by groups having a large experience with the technique. Simple recommendations should be given to the patient to adequately deal with the potential for seed expulsion. Cremation can raise problems of contamination or external irradiation of the incinerator staff and family if performed too soon following implantation. For iodine-125, ICRP set the minimum delay between brachytherapy and cremation at 1 year (while some countries set this delay at 3 years).

In all cases, the patient should receive oral and written information about these radioprotection issues and, for the months following the implantation, should carry a "wallet card" or a certificate including all details about his brachytherapy procedure.

It is extremely important that there be immediate local reporting and analysis of all accidents, incidents, errors, mistakes, or even simple "precursor events." This should be rapidly followed by the identification of causes, contributing factors, and extent of conditions, all of which should result in corrective measures. Responses to such situations should be followed by rapid and widespread circulation of the relevant information, to avoid similar problems being reproduced in another center. Adherence to clear regulations, as well as adequate quality assurance programs, should reduce the probability of most accidents. As such, an error prevention program should also minimize the risk that a small error, or mistake, escalates into a full-blown accident, with consequences to the patients, medical staff, and the public.

References

Abdel-Wahab M, Reis IM, Hamilton K (2008) Second primary cancer after radiotherapy for prostate cancer–a seer analysis of brachytherapy versus external beam radiotherapy. Int J Radiat Oncol Biol Phys 72(1):58–68

Cosset JM (2010) Prostate brachytherapy patients are (almost) normal people! Brachytherapy 9(2):112–113

Court B, Chassagne D (1977) Interstitial therapy of cancer of the prostate using iridium 192 wires. Cancer Treat Rep 61(2):329–330

Dauer LT (2012) Globalization, implantation, cremation…oh, my! Brachytherapy 11:197–198

Dauer LT et al (2010) Less-restrictive, patient-specific radiation safety precautions can be safely prescribed after permanent seed implantation. Brachytherapy 9(2):101–111

De Gonzalez AB et al (2011) Proportion of second cancers attributable to radiotherapy treatment in adults: a cohort study in the US SEER cancer registries. Lancet Oncol 12(4):353–360

Hilaris BS, Batata ME, Anderson LL (1987) Chapter 26. In: Pierquin B, Wilson JF, Chassagne D (eds) Prostate in modern brachytherapy. Masson, New York, pp 234–247

Huang J et al (2011) Analysis of second malignancies after modern radiotherapy versus prostatectomy for localized prostate cancer. Radiother Oncol 98(1):81–86

IAEA (1996) International basic safety standards for protection against ionizing radiation and for the safety of radiation sources. Safety series no. 115. International Atomic Energy Agency, Vienna

ICRP (2000) Prevention of accidental exposures to patients undergoing radiation therapy. ICRP Publication 86. Ann ICRP 30(3)

ICRP (2005a) Prevention of high-dose-rate brachytherapy accidents. ICRP Publication 97. Ann ICRP 35(2)

ICRP (2005b) Radiation safety aspects of brachytherapy for prostate cancer using permanently implanted sources. A report of ICRP Publication 98. Ann ICRP 35(3)

Kaulich TW et al (1999) Emergency rescue in accidents with HDR afterloading units. Strahlenther Onkol 175(10):524–529

Kono Y et al (2011) Radiation exposure to general public after permanent brachytherapy for prostate cancer. Radiat Prot Dosimetry 146:229–230

Liauw SL et al (2006) Second malignancies after prostate brachytherapy: incidence of bladder and colorectal cancers in patients with 15 years of potential follow-up. Int J Radiat Oncol Biol Phys 66(3):669–673

Michalski J et al (2003) Radiation exposure to family and household members after prostate brachytherapy. Int J Radiat Oncol Biol Phys 56(3):764–768
Moon K et al (2006) Cancer incidence after localized therapy for prostate cancer. Cancer 107(5):991–998
Mydlo JH, Lebed B (2004) Does brachytherapy of the prostate affect sperm quality and/or fertility in younger men? Scand J Urol Nephrol 38(3):221–224
NCRP (1970) Precautions in the management of patients who have received therapeutic amounts of radionuclides. NCRP report no. 37. National Council on Radiation Protection and Measurements, Bethesda
Pasteau O, Degrais P (1913) De l'emploi du radium dans les cancers de la prostate. J Urol (Paris) 4:341–345
Que W (2001) Radiation safety issues regarding the cremation of the body of an I-125 prostate implant patient. J Appl Clin Med Phys 2(3):174–177
Satoh T et al (2012) Deaths within 12 months after 125I implantation for brachytherapy of prostate cancer: an investigation of radiation safety issues in Japan (2003–2010). Brachytherapy 11:192–196
Takam R, Bezak E, Yeoh EE (2009) Risk of second primary cancer following prostate cancer radiotherapy: DVH analysis using the competitive risk model. Phys Med Biol 54(3):611–625
Yumoto Y, Hanafusa T, Nagamatsu T, Okada S (2000) Experimental incineration of low level radioactive samples. Health Phys 79(2 Suppl):S25–S32

Index

G. Kovács, P. Hoskin (eds.), *Interstitial Prostate Brachytherapy*,
DOI 10.1007/978-3-642-36499-0,

Q

R

S